IFMBE Proceedings

Volume 58

The International Federation for Medical and Biological Engineering, IFMBE, is a federation of national and transnational organizations representing internationally the interests of medical and biological engineering and sciences. The IFMBE is a non-profit organization fostering the creation, dissemination and application of medical and biological engineering knowledge and the management of technology for improved health and quality of life. Its activities include participation in the formulation of public policy and the dissemination of information through publications and forums. Within the field of medical, clinical, and biological engineering, IFMBE's aims are to encourage research and the application of knowledge, and to disseminate information and promote collaboration. The objectives of the IFMBE are scientific, technological, literary, and educational.

The IFMBE is a WHO accredited NGO covering the full range of biomedical and clinical engineering, healthcare, healthcare technology and management. It is representing through its 60 member societies some 120.000 professionals involved in the various issues of improved health and health care delivery.

More information about this series at http://www.springer.com/series/7403

Fatimah Ibrahim · Jadeera Phaik Geok Cheong
Juliana Usman · Mohd Yazed Ahmad
Rizal Razman · Victor S. Selvanayagam (Eds.)

3rd International Conference on Movement, Health and Exercise

Engineering Olympic Success: From Theory to Practice

Springer

Editors
Fatimah Ibrahim
Center for Innovation in Medical Engineering (CIME)
Department of Biomedical Engineering
Faculty of Engineering
University of Malaya
Kuala Lumpur
Malaysia

Jadeera Phaik Geok Cheong
Sports Centre
University of Malaya
Kuala Lumpur
Malaysia

Juliana Usman
Centre for Applied Biomechanics (CAB)
Department of biomedical Engineering
Faculty of Engineering
University of Malaya
Kuala Lumpur
Malaysia

Mohd Yazed Ahmad
Center for Innovation in Medical Engineering (CIME)
Department of Biomedical Engineering
Faculty of Engineering
University of Malaya
Kuala Lumpur
Malaysia

Rizal Razman
Sports Centre
University of Malaya
Kuala Lumpur
Malaysia

Victor S. Selvanayagam
Sports Centre
University of Malaya
Kuala Lumpur
Malaysia

ISSN 1680-0737 ISSN 1433-9277 (electronic)
IFMBE Proceedings
ISBN 978-981-10-3736-8 ISBN 978-981-10-3737-5 (eBook)
DOI 10.1007/978-981-10-3737-5

Library of Congress Control Number: 2017930934

The IFMBE Proceedings is an Official Publication of the International Federation for Medical and Biological Engineering (IFMBE)

Printed on acid-free paper

This Springer imprint is published by Springer Nature
The registered company is Springer Nature Singapore Pte Ltd.
The registered company address is: 152 Beach Road, #21-01/04 Gateway East, Singapore 189721, Singapore

About IFMBE

The International Federation for Medical and Biological Engineering (IFMBE) is primarily a federation of national and transnational societies. These professional organizations represent interests in medical and biological engineering. The IFMBE is also a Non-Governmental Organization (NGO) for the United Nations and the World Health Organization (WHO), where we are uniquely positioned to influence the delivery of health care to the world through Biomedical and Clinical Engineering.

The IFMBE's objectives are scientific and technological as well as educational and literary. Within the field of medical, biological and clinical engineering IFMBE's aims are to encourage research and application of knowledge, and to disseminate information and promote collaboration. The ways in which we disseminate information include: organizing World Congresses and Regional Conferences, publishing our flagship journal Medical & Biological Engineering & Computing (MBEC), our web-based newsletter – IFMBE News, our Congress and Conference Proceedings, and books. The ways in which we promote collaborations is through networking programs, workshops, and partnerships with other professional groups, e.g., Engineering World Health.

Mission

The mission of the IFMBE is to encourage, support, represent and unify the worldwide Medical and Biological Engineering community in order to promote health and quality of life through the advancement of research, development, application and management of technology.

Objectives

The objectives of the International Federation for Medical and Biological Engineering shall be scientific, technological, literary, and educational. Within the field of medical, clinical and biological engineering its aims shall be to encourage research and the application of knowledge, and to disseminate information and promote collaboration.

In pursuit of these aims the Federation may, in relation to its specific field of interest, engage in any of the following activities: sponsorship of national and international meetings, publication of official journals, co-operation with other societies and organizations, appointment of commissions on special problems, awarding of prizes and distinctions, establishment of professional standards and ethics within the field, or in any other activities which in the opinion of the General Assembly or the Administrative Council would further the cause of medical, clinical or biological engineering. It may promote the formation of regional, national, international or specialized societies, groups or boards, the coordination of bibliographic or informational services and the improvement of standards in terminology, equipment, methods and safety practices, and the delivery of health care.

In general, the Federation shall work to promote improved communication and understanding in the world community of engineering, medicine, and biology.

Foreword

This Conference Proceedings volume contains 36 accepted full papers that were presented at the 3rd International Conference on Movement, Health and Exercise 2016 (MoHE2016). It took place at the historic city of Malacca from 28th to 30th September 2016. The conference was jointly organized by Biomedical Engineering Department and the Sports Centre, University of Malaya. In total, there were 83 presenters and 140 participants that took part in this highly successful third edition of the conference. These papers provided practitioners, scientists, students, athletes and coaches with state-of-the-art information on sports, exercise and health.

The Conference provided a setting for discussing recent developments in a wide variety of topics under the ever-expanding area of sports and exercise science. It had been a good opportunity for participants coming from all over Malaysia and the Asia Pacific to present and discuss topics in their respective research areas.

The inclusion of seven highly distinguished plenary and invited speakers over the three-day period had served to be the highlight of MoHE2016. Participants had openly expressed their appreciation to the speakers for their efforts in presenting ideas and methods in a lively and accessible way.

The organizers would like to duly acknowledge the financial support from the Ministry of Higher Education, Malaysia as well as thank all the other sponsors of MoHE2016; United Akrab Sdn Bhd, DanMedik Sdn Bhd, Rahmat Dagangan Sdn Bhd, BioApps Sdn Bhd, and Melvita. Thank you to all participants for their contributions towards the Conference program and for their contributions to the Proceedings. Our appreciation also extends to all the speakers and reviewers for their extensive comments and insights that made this proceedings volume possible. Our special thanks are reserved for our colleagues from the University of Malaya for their devoted endeavours in the overall organization of the conference; and our sincere appreciation to Ms Soobia Saad Khan, Mr. Saad Jawaid Khan and Mr. Shah Mukim Uddin for their help in preparing this proceedings volume.

We are looking forward to meeting all the participants again in the fourth edition of this conference in Kuala Lumpur, 2017.

Dr. Juliana Usman and Dr. Rizal Mohd Razman
Chairpersons
MoHE2016

Conference Details

Name

3rd International Conference on Movement, Health and Exercise

Short name

MoHE2016

Venue

28th–30th September 2016
Swiss-Garden Hotel & Residences, Malacca, Malaysia

Proceedings Editors

Fatimah Ibrahim
Jadeera Phaik Geok Cheong
Juliana Usman
Mohd Yazed Bin Ahmad
Rizal Razman
Victor Selvarajah Selvanayagam

Organized by

Department of Biomedical Engineering,
Faculty of Engineering,
University of Malaya,
Malaysia

and

Sports Centre
University of Malaya,
Malaysia

Endorsed by

Malaysia's Society of Medical and Biological Engineering (MSMBE)
International Federation for Medical and Biological Engineering (IFMBE)

Supported by

Ministry of Higher Education, Malaysia
Movement, Health and Exercise Journal
Centre for Applied Biomechanics, University of Malaya

Malaysian Society of Biomechanics (MSB)
National Sports Institute, Malaysia
Innovative Manufacturing, Mechatronics and Sports Lab (iMAMS), Universiti Malaysia Pahang

Organizing Committee

Advisors

Hazally Bin Jali
Sport Department Secretary
Higher Education Ministry

Noor Azuan Abu Osman
Dean, Faculty of Engineering
University of Malaya

Zahari Taha
Chief Editor, Movement, Health and Exercise Journal
Innovative Manufacturing, Mechatronics and Sports Lab (iMAMS) Universiti Malaysia Pahang

Chairpersons

Juliana Usman
Rizal Mohd Razman

Secretary

Rumiza Ahmad

Treasurer

Mas Sahidayana Mohktar

Technical

Jadeera Phaik Geok Cheong
Mohd Yazed Ahmad
Victor Selvarajah Selvanayagam

Publication

Fatimah Ibrahim
Juliana Usman
Rizal Mohd Razman
Khairunnisa Binti Hasikin

Sponsorship and Exhibition

Wan Safwani Wan Kamarul Zaman

Publicity and Graphic Design

Khairunnisa Binti Hasikin

Website and Logistic

Nasrul Anuar Abd Razak

Secretariat Committee

Adhli Iskandar Putera Bin Hamzah
Ahmad Alif Abdullah
Cheah Pei Yi
Mohd Khairul Bin Amran
Nahdatul Aishah Mohd Sharif
Nor Eeza Zainal Abidin
Normazlianita Muhammad Alias
Nur Afiqah Binti Hashim
Saad Jawaid Khan
Sami Fathi Ali Khalil
Soobia Saad Khan
Yuslialif Bin Mohd Yusup

International Advisory Board

Abdul Halim Mokhtar	University of Malaya, Malaysia
Ahmad Munir Che Muhamed	Universiti Sains Malaysia, Malaysia
Ahmad Naim Ismail	Universiti Teknologi MARA, Malaysia
Aminuddin Yusof	Universiti Putra Malaysia, Malaysia
Andrew S. McIntosh	ACRISP, Australia
Ashril Yusof	University of Malaya, Malaysia
Fatimah Ibrahim	University of Malaya, Malaysia
Jadeera Phaik Geok Cheong	University of Malaya, Malaysia
Jay-Lee Nair	National University of Singapore, Singapore
Juliana Usman	University of Malaya, Malaysia
Khairunnisa Hasikin	University of Malaya, Malaysia
Kok Lian Yee	Universiti Putra Malaysia, Malaysia
Mas Sahidayana Mohktar	University of Malaya, Malaysia
Mohamad Razali Abdullah	Universiti Sultan Zainal Abidin, Malaysia
Mohamad Shariff Bin A. Hamid	University of Malaya, Malaysia
Mohd Khairi Zawi	Universiti Kebangsaan Malaysia, Malaysia
Mohd Yazed Ahmad	University of Malaya, Malaysia
Nasrul Anuar Abd Razak	University of Malaya, Malaysia
Nik Shanita Safii	Universiti Kebangsaan Malaysia, Malaysia
Nooranida Arifin	University of Malaya, Malaysia
Noor Azuan Abu Osman	University of Malaya, Malaysia
Nur Ikhwan Mohamad	Universiti Pendidikan Sultan Idris, Malaysia
Rizal Mohd Razman	University of Malaya, Malaysia
Roosfa Hashim	Universiti Islam Antarabangsa, Malaysia
Saidon Amri	Universiti Putra Malaysia, Malaysia
Saju Joseph	National Sports Institute, Malaysia
Shaharudin Abdul Aziz	Universiti Pendidikan Sultan Idris, Malaysia
Selina Khoo Phaik Lin	University of Malaya, Malaysia
Stephan Riek	University of Queensland, Australia
Teresa Wong	Singapore
Toby Mündel	Massey University, New Zealand
Victor Selvarajah Selvanayagam	University of Malaya, Malaysia
Wan Safwani Wan Kamarul Zaman	University of Malaya, Malaysia
Yeo Wee Kian	National Sports Institute, Malaysia
Zahari Taha	Universiti Malaysia Pahang, Malaysia
Zulqarnain Mohamed	University of Malaya, Malaysia

Contents

DietScore™: Sports Nutrition-based Mobile Application for Athletes and Active Individuals

S. Nik Shanita[1], A.H.A. Rahman[1], A. Azimah[2], M. Mohamad Izham[3], N.E. Othman[4], Y.T. Chan[1], B.S. Pushpa[1], B.K. Poh[1] and A. T. Ruzita[1]

[1] School of Healthcare Sciences, Faculty of Health Sciences, UKM, Kuala Lumpur, Malaysia
[2] Medical Faculty, National Defence University of Malaysia, Sungai Besi, Malaysia
[3] National Sports Institute, Bukit Jalil, Malaysia
[4] Faculty of Information Science and Technology, UKM, Bangi, Malaysia

Abstract— **Sports nutrition has been proven to contribute directly to athletes' performance. To help athletes improve their eating habits, sports nutritionists and dietitians need accurate information and feedback. In this paper, the framework of a sports nutrition app known as DietScore™ provides a reliable solution. Development of the tool was done by customizing energy requirement of athletes, types, duration and intensity of training and specific macro- and micro-nutrient needs based on local foods. Additionally, athletes would be able to plan their menu for the following day, which improves their knowledge on daily food intake. These daily activity data are stored in the local database in offline mode and later synchronized by mobile app, which also serves as a Knowledge Management System on Sports Nutrition. DietScore™ could expand into the public domain to increase its social and economic impact. It provides a useful and effective platform, which acts as virtual sports nutrition counselor for both athletes and sports nutritionists/dietitians. With the application of local foods content, DietScore™ will be unique as the first sports nutrition utility tool in the market.**

Keywords— **Sports nutrition, Athlete, Mobile App.**

I. Introduction

An athlete's optimum performance depends on several factors, including eating habits and nutrition. The roles of sports nutritionists and dietitians are to ensure that the athletes are provided with the correct information on sports nutrition and its impact on performance. The diets of many athletes are often deficient due to restrictive eating habits, obsession with weight and the types of food consumed. This is mainly due to lack of nutrition knowledge [1]. Any efforts on intervention strategies are futile unless the intervention strategy impacts on the nutritional status and physiological characteristics of the athletes. Thus, it is important to make available suitable tools that can enhance nutrition education among athletes. Unfortunately, sports nutrition education tools are not well-established in Malaysia.

In this digital era, most Malaysians use smartphones [2]; and mobile application (app) is preferred compared to web app due to its mobility and easy to use. Users tend not to use a system if it is inconvenient or when they find it cumbersome to activate, but mobile apps are convenient and can be applied as a personalized guide. Various mobile apps have been developed for specific purposes and target groups. Some focused on wellness diary [3], energy balance monitoring [4], weight loss [5] and other commercial products, such as DietPal. Others are designed for specific health issues such diabetics [6] and cardiac rehabilitation [7]. The use of electronic dietary assessment tools has been extended to primary healthcare. Kim [8] introduced the application of mobile devices in aiding dietary assessment and evaluation based on image analysis tools for identification and quantification of foods that are consumed during mealtimes. Furthermore, Lieffers and Hanning [9] conducted a review of dietary intake documentation in healthy populations and those trying to lose weight. It was concluded that applications versus conventional techniques such as paper records frequently resulted in better self-monitoring adherence. Bonilla et al. [10] highlighted that there was a strong interest among all disciplines in the use of electronic dietary assessment tools for the management of obesity, diabetes and heart disease. Although many mobile apps have been developed that are related to health and nutrition, none was specifically intended for athletes or active individuals in the Malaysian population.

There has been increasing awareness among Malaysians regarding the choice of food and its impacts on health status [11]. This awareness has led to increased participation in sports, gym, and walkathon or marathon in some segments of the population. Active individuals or athlete's energy requirement varies depending on the type of sport and intensity level of their sports activity [12]. For sports nutritionists and dietitians, it is useful to be able to set up their own groups or clusters in mobile apps. This limits the software potentiality, as a user management system allows tests on different groups, and thereby enabling the development of a more comprehensive body of knowledge for more meaningful and accurate analysis, both in the micro and macro aspects. From the athletes and active

F. Ibrahim et al. (eds.), *3rd International Conference on Movement, Health and Exercise*,
IFMBE Proceedings 58,
DOI: 10.1007/978-981-10-3737-5_1

individuals' point of view, the amount of daily energy requirement is important to balance dietary intake with physical activity routine [13].

In the next section, the aims, key features, main features and the snapshots of *DietScore*™ are described.

II. DIETSCORE™ - NUTRITION-BASED MOBILE APPLICATION

One of the added value of DietScore is the involvement of sports dietitian and nutritionist in the development process. As recommended by Lieffers and Hanning [8], this study included sports dietitians and nutritionists at the Malaysian National Sports Institute (ISN) as part of the development process to ensure that the information presented in this app are appropriate and effective.

The aim of DietScore™ is to formulate solutions to address issues that arose while undertaking our previous nutrition web-based tool. One of the key feature in DietScore™ is to introduce the first Malaysian sports nutrition mobile app developed using 100% local validated content, i.e. database containing local food choices, modules for menu planning by day and time, compendium of physical activities, household measurement units and basal metabolic rate (BMR) for Malaysian athletes and active individuals. App usage entered by the user are uploaded online and stored in a cloud-based database and feeding the development of a "Knowledge Management System on Sports Nutrition (KMSSN)". The KMSSN shall be the first to be developed in the country and will serve as an invaluable tool to access the country's fitness level of our elite athletes as well as active individuals in terms of food consumption and nutritional status.

As a mobile app, DietScore™ enables athletes and fitness enthusiasts to use it anytime, on the go, with or without an internet connection. Once it is online, data stored offline in the phone's database storage will be automatically transferred to the online database. The individual user, giving vital feedback according to user's sports profile, can view stored data as analytics. If the user is registered within a group or cluster, the data can also be used for analysis by nutritionists and dietitians. This invaluable source of data will be useful for informed intervention sessions and for helping the NSI to establish policies relating to food intake and improving the overall nutritional status of our country's top athletes.

An important feature in DietScore™ is the selection of a sports profile according to the sport selected during the user registration process. This enables more meaningful and accurate analysis, both in the micro and macro aspects. Sports profiling affect the energy requirement, recommended food intake goals, meal planning, and other features.

As the first and original sports nutrition app with validated local content, DietScore™ has no immediate competition. Most similar mobile app products developed elsewhere are mainly used to promote supplements, whereas DietScore™ emphasizes on local foods. Once completed, the app will be heavily promoted via word of mouth on social media (Facebook and Twitter). As the content is locally customized, it is hoped that the app will go viral via the NSI athletes. Promotional flyers will also be distributed through key fitness outlets, such as Fitness First, Clark Hatch Fitness, Fitness Gyms, as well as major sports associations or club.

III. DIETSCORE – MAIN FEATURES

The aim for the development of *DietScore*™ is not merely as a commercial app, but more importantly, to provide a new solution to the existing limitations of mobile apps currently available in the market. With this in mind, some new features are added to fulfill this expectation as follows:

A. Offline logging and data synchronization

This feature is important for users to be able to log in and log their food and training when the phone/device is offline (not connected to the internet). The sync facilities will be active for access across multiple gadgets such as smartphones, PC, and tablets. Synchronizing work will also be required for transferring data stored off-line (stored in SQL-Lite residing on the mobile gadgets) to the online database (MySQL, stored in an online server).

B. User Profiling

User profiles are incorporated and linked to analytics reports generated by the system. This facility will be used by sports nutritionists and dietitians during intervention sessions. Table 1 listed four different user profile categories which comprise of endurance, intermittent/power strength, skills and active individuals. *DietScore™ will take into account the sport's category based on* the type of sport selected by the user.

Table 1 Sport profile categories

Sport Categories	Description
Endurance	High-intensity sports which require high energy requirement.
Intermittent/Power Strength	Medium intensity for most of the sports.
Skills	Low-intensity sports which require low energy level.

Active Individuals	Same level as skills category (Low intensity)

The energy requirement (ER) for the user is depending on two parameters, basal metabolic rate (BMR) as mentioned by Wong et al. [14] and physical activity level (PAL). The ER is calculated based on the following equations,

$$BMR = 669 + 13 \times weight + 192 \times gender \quad (1)$$

$$ER = BMR \times PAL \quad (2)$$

where weight is measured in kilogram (kg) unit while for gender, numeric 1 is used for men and numeric 0 is used for women. The BMR unit is kcal/day. Next, the PAL is allocated based on the sports category [15]. For "Peak Season", the PAL is determined by the sports category while in "Off Peak" season, the same PAL factor is used for all categories. The snapshot of the User Profile page in *DietScore*™ is illustrated in Figure 1.

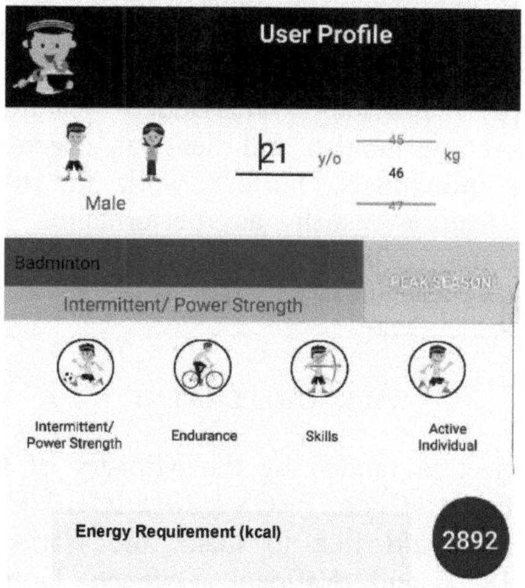

Fig. 1 User Profile page in Dietscore™

C. *Food, Training and Menu Planning Log*

Data integrity and improvement of accuracy is a feature that is given main priority in the development of this mobile app. Part of the process was to re-evaluate and re-validate foods data and measurement units to ensure that data generated are accurate and correct. *DietScore™ also* offers users the ability to plan their meals for the following day. It is developed as a dynamic feature so that users are allowed to mix and match the given menu sets from the database provided in the mobile app. Since different users have different energy requirements, the ER value is displayed as a benchmark for menu planning. To avoid misleading the users, *DietScore*™ only allows users to key-in menu planning for the following day and not other days.

D. *Daily Report*

All food and training logs can be viewed in two separate activity pages. One activity page is dedicated to showing the list of food and training activity for the day and navigation to any other date. The daily list is sorted by intake time to show the activity sequence. Another page referred to as pre-post training activity shows the food intake before and after each training session. Pre- and post-activity is created to educate and remind athletes and active individuals to consistently consume the respective foods before and after the training session to avoid lack of energy that is required to perform any intensive training. For effectiveness, the interval time for pre- and post-training is set as two hours.

E. *Food Intake Analysis*

Data analysis at micro levels is presented using web generated analytics whereas macro analysis analyses the trends in nutritional status and physiological characteristics of athletes. There are two other analysis features available in this app, which are Pre-post Training Analysis and Food Intake Analysis. Pre-post training analysis shows total food calories and total carbohydrate intake within 2-hour intervals before and after each training as shown in Fig. 2. This feature allows user and coaches to check and balance for knowledge and practice in daily routine. Advice on carbohydrate intake is important for coaches when monitoring athletes and active individuals' dietary intakes as highlighted by Jeukendrup [16].

Fig. 2 Pre –post training analysis pages

The second analysis displays the average caloric and macronutrient distribution and food intake analysis as illustrated in Fig. 3. In DietScore™, the analysis is performed by selecting the closest three days in the previous five days that had normal- or over-reporting status. The under-reporting days are indicated with red color while other days are indicated in blue for normal- or over-reporting. If the reporting status doesn't meet the criteria, the analysis will be not be performed. The recommended goal for food intake status, which depends on sports category, is displayed for comparison.

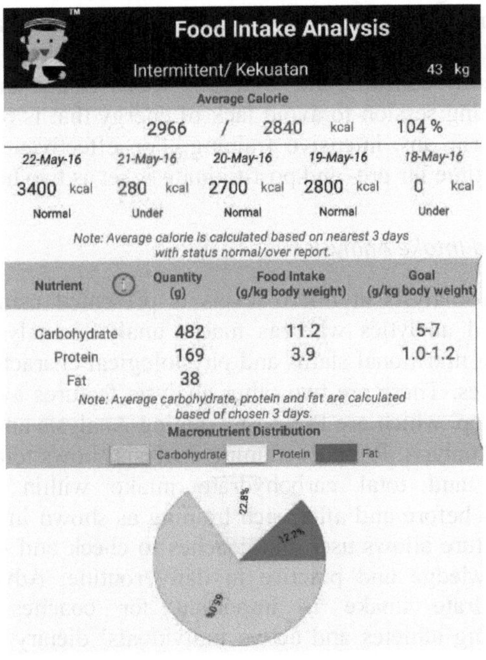

Fig. 3 Food Intake Analysis page

A. Nutrition Information

DietScore™ also provides nutrition facts to improve athletes' knowledge. A simple and effective way to present the facts is by using infographics. Infographics are easier to interpret and understand than wordy explanations. In order to do this, some key nutrition information is included in the *DietScore™* mobile app. Furthermore, the links to the information pages are provided in many pages to increase awareness. The information covered are as follows: (i) macronutrients (carbohydrate, protein, fat), (ii) micronutrients (minerals, vitamins) (iii) training (before, after, water intake) and (iv) menu planning (meal sets for 2400 kcal and 3200 kcal). The main menu for Nutrition Information page is shown in Fig. 4.

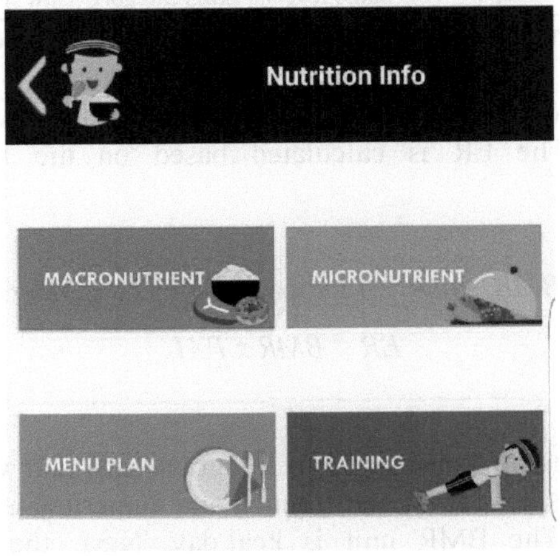

Fig. 4 Nutrition Infographics pages

I. Conclusions and Future Works

This paper highlights the development of DietScore™, a mobile app that suits the Malaysian context and that addresses the limitations of existing dietary and nutrition web application and other mobile apps.

DietScore™ provides value-added features, which benefits sports nutritionists and dietitians, coaches, athletes and active individuals. Further work to enhance the additional features, such as performance evaluation focusing on training analysis and extended databases, will be needed prior to mass distribution.

Acknowledgment

This project is funded by the Ministry of Education, Malaysia (Project Code: PRGS/1/2015/ICT01/UKM/02/6). The authors would like to thank the National Sports Institute (ISN) and National Defence University of Malaysia (UPNM) for their support and technical input. The authors also thank sports nutritionists and dietitians involved for the input and ideas.

Conflict of Interest

The authors declare that they have no conflict of interest.

References

1. Bryne S, McLean N. (2001) Eating disorders in athletes: A review of the literature. Journal of Science and Medicine in Sport, Vol.4(2):145-159
2. http://data.worldbank.org/indicator/IT.CEL.SETS.P2

3. Mattila E, Korhonen I, Salminen JH, Ahtinen A, Koskinen E, Särelä A et al. (2010) Empowering citizens for well- being and chronic disease management with wellness diary. IEEE Trans InfTechnol Biomed 2010 Mar;14(2):456-463
4. Hughes DC, Andrew A, Denning T, Hurvitz P, Lester J, Beresford S, et al. (2010) Balance (Bioengineering Approaches for Lifestyle Activity and Nutrition Continuous Engagement) : developing new technology for monitoring energy balance in real time. J Diabetes Sci Technol Mar;4(2):429-434
5. Lee W, Chae YM, Kim S, Ho SH, Choi I. (2010) Evaluation of a mobile phone-based diet game for weight control. J Telemed Telecare Jul;16(5):270-275.
6. Arsand E, Tatara N, Østengen G, Hartvigsen G. (2010) Mobile phone-based self-management tools for type 2 diabetes: the few touch application. J Diabetes Sci Technol . Mar;4(2):328-336
7. Varneeld M, Karunanithi MK, Särelä A, Garcia E, Fairfull A, Oldenburg BF, et al. (2011) Uptake of a technology- assisted home-care cardiac rehabilitation program. Med J Aust. Feb 21;194(4):S15-S19
8. Kim, S. (2010). The Use of Mobile Devices in Aiding Dietary Assessment and Evaluation. IEEE Journal of Selected Topics in Signal Processing, 4(4), 756–766.
9. Lieffers, J. R. L., & Hanning, R. M. (2012). Dietary assessment and self-monitoring: With nutrition applications for mobile devices. Canadian Journal of Dietetic Practice and Research, 73(3).
10. Bonilla, C., Brauer, P., Royall, D., Keller, H., Hanning, R. M., & DiCenso, A. (2015). Use of electronic dietary assessment tools in primary care: an interdisciplinary perspective. BMC Medical Informatics and Decision Making, 15(1), 14. http://doi.org/10.1186/s12911-015-0138-6
11. World Health Organisation. (2003). Health and Development Through Physical Activity and Sport. WHO Document Production Services, 3(2), 1–11.
12. Torun B (2015) Energy requirements of children and adolescents. Public Health 8(7A): 968-993
13. Haskell, W. L., Lee, I. M., Pate, R. R., Powell, K. E., Blair, S. N., Franklin, B. A., Bauman, A. (2007). Physical activity and public health: Updated recommendation for adults from the American College of Sports Medicine and the American Heart Association. Medicine and Science in Sports and Exercise, 39(8), 1423–1434.
14. Wong, J. E., Poh, B. K., Nik Shanita, S., Izham, M. M., Chan, K. Q., De Tai, M., Ismail, M. N. (2012). Predicting basal metabolic rates in Malaysian adult elite athletes. Singapore Medical Journal, 53(11), 744–749.
15. Institute of Medicine of the National Academies (2005) Dietary Reference Intakes For Energy, Carbohydrate, Fiber, Fat, Fatty Acids,Cholesterol, Protein, and Amino Acids THE NATIONAL ACADEMIES PRESS
16. Jeukendrup A. A Step Towards Personalized Sports Nutrition: Carbohydrate Intake During Exercise. Sports Medicine 2014 44(S1):25-33

Author:	Assoc. Prof. Dr. Nik Shanita Safii
Institute:	Dietetics Program,
	School of Healthcare Sciences,
	Faculty of Health Science, UKM
Street:	Jalan Raja Muda Abdul Aziz
City:	Kuala Lumpur
Country:	Malaysia
Email:	nikshanita@ukm.edu.my

Physical and Physiological Attributes Associated with Precision Sports Performance–A Novel Analysis Method

T.C. Loh[1], B.C. Cheah[2], M. Sanawi[2], L.H. Loo[2], S.H. Yaali[2] and L.V. Tang[3]

[1] Physiology Centre, National Sports Institute of Malaysia, Bukit Jalil, Malaysia
[2] Penang Satellite Centre, National Sports Institute of Malaysia, Penang, Malaysia
[3] Strength and Conditioning Centre, National Sports Institute of Malaysia, Bukit Jalil, Malaysia

Abstract— **Physical and physiological attributes have received limited attention in predicting precision sports performance. However, there were few studies showed a correlation between both attributes and performance in precision sports. Thus, greater physical and physiological characteristics might contribute the success in the precision sport. 26 athletes (10 males and 16 females) from archery, shooting, and tenpin bowling was recruited. The measured variables were body composition, muscular strength, flexibility, speed, agility, power, and cardiovascular endurance. The multivariate data analysis was performed using Unscrambler® X software. The Principal Component Analysis (PCA) showed bowlers were physically and physiologically different from archers and shooters. For male athletes, two factors were extracted involving lower body attributes and anthropometry. As for female athletes, anticorrelated variables were found in one of the two extracted factors. The prediction models were developed by the partial least squares (PLS) regression. The value obtained showed deviation for most references of performance score. In conclusion, each sport has its own explainable physical and physiological attributes as compared to other sports. The physical and physiological attributes could predict performance in precision sports athletes by using multivariate analysis. However, further refinement of the model is needed to predict it more precisely.**

Keywords— **Archery, Shooting, Tenpin Bowling, Sports Performance, Multivariate Data Analysis.**

I. Introduction

Physical and physiological assessment is to determine the readiness of the athletes for training and competition. It provides useful information to identify areas in need of improvement and assist the coach in team selection. However, how far do the physical and physiological attributes contribute to the success in precision sports?

Archery is a sport which requires strength and endurance of the upper body [10]. The repeatability of drawing the bow in short time may lead to specific sports injuries of the archers. A study reported one of the most common injury reason was inappropriate skills [8]. Thus, assigning suitable bows is important to achieve optimal performance. A study reported a positive relationship between maximal strength tests and bow weights, and it proposed consideration should be given to limb length in the assignment of equipment [6].

Prediction of archery performance was made on a high school archery team by collecting athletic performance test results. The Wherry-Doolittle (Test Selection) method was associated with four factors, including bow draw time, hand-eye coordination, 50-meter sprint, left-hand gripping ability. The multivariate regression equation showed 54% accuracy to predict archery performance [9].

Shooting is a sport involving fine motor skill, rather than gross motor skill. Despite that, significant correlations were found between shooting a score and dominant grip strength, combined grip strength, forearm girth and length of trigger finger. But a step wise multiple regression could not generate an equation to predict shooting score from the data [1].

As for bowling, this sport involves many repeated physical movements including running up to the lane, swinging a heavy ball, pushing off and lunging by lower limbs, and trunk and hip flexion. Aerobic capacity may contribute to bowling performance as aerobic power index correlated with bowling performance for female bowlers only [12]. Strength and flexibility may not correlate with performance in that study, but they may still be beneficial to the bowlers since they can decrease the risk of overuse injuries [5].

An attempt to distinguish tenpin bowlers' playing abilities through anthropometry and strength variables was made. The isometric strength tests showed a positive relationship with bowling performance. There was a 54% success rate for predicting playing level from anthropometric and strength discriminating variables, with forearm internal rotation strength being the best predictor [11].

Although the lack of physical fitness effect on cognitive performance during exercise, improvement in decisional performance has been observed immediately after the adrenaline threshold during incremental exercise. Consistent results indicated activities with greater energy demands require more attention to control movements [4].

Therefore, potentially greater physical and physiological attributes will yield greater performance for precision sports such as archery, shooting, and tenpin bowling. The aim of this study is to use multivariate data analysis to develop a prediction model for sports performance in precision sport.

F. Ibrahim et al. (eds.), *3rd International Conference on Movement, Health and Exercise*,
IFMBE Proceedings 58,
DOI: 10.1007/978-981-10-3737-5_2

II. Methods

A. Subjects

This was a retrospective study. Data taken was from monthly service assessment. Athletes had signed a consent form prior to assessment. 26 state athletes (10 males and 16 females) from archery, shooting, and tenpin bowling was recruited to participate in this study. All athletes competed in the 2014 Sukma Games.

B. Procedures

Body composition, muscular strength, flexibility, speed, agility, power, and cardiovascular endurance components were taken as health and skill related fitness. The data collection had been done weeks before the biannual event. All athletes have gone through baseline and retest prior to the actual test.

Measurement of body weight and height were obtained using a seca 767 electronic scale without shoes. A total of 6 fitness tests, including handgrip, sit and reach, 20m speed, Illinois agility test, standing long jump (SBJ), and yo-yo endurance level 1 (YE1) tests were performed. A Takei 5401 digital handgrip dynamometer was used to measure the maximum isometric strength of the hand and forearm muscles. A sit and reach Flex-Tester® box was used to measure the maximum flexibility of lower back and hamstrings muscle of an individual. The Brower Speed Trap II timing system was set up for both speed and agility tests. The 20m speed test was conducted to measure the ability to accelerate from a standing start and to cover 20m distance quickly. Meanwhile, Illinois agility test was used to assess a person's ability to perform an exercise that requires speed, coordination, change of direction, and balance all at the same time. A SBJ mat was used to identify an individual's level of explosive power of the lower limb musculature. Lastly, YE1 was conducted to evaluate the aerobic capacity or the ability to work continuously for a long period. Distance covered during the YE1 test was measured.

C. Statistical Analysis

Standard statistical analyses such as mean and standard deviation (SD) were calculated for each anthropology and fitness test according to genders.

The normality of the data was assessed with the ShapiroWilk test. Independent t-test was conducted to determine whether there was a gender difference in the variables using Statistical Package for the Social Sciences (SPSS) version 21.0 statistical program. A level of significance of $p < .05$ was used.

2014 Sukma Games results were collected. Due to the different performance scores in respective sports, athletes' performance Z-score was calculated. A Z-score is a numerical measurement of a value's relationship to the mean in a group of values, and it is expressed in terms of SD from the means. Thus mean and SD for completed events were generated to calculate their performance Z-score. The equation (1) was expressed as followed:

$$\text{Performance Z-score} = \frac{\text{Athlete's score - Mean score of event}}{SD\,\text{score of event}} \quad (1)$$

The experimental results were analyzed using Principal Component Analysis (PCA) with full cross-validation. All variables were standardized.

To assist in obtaining a better interpretation of components, assumption were checked by running KaiserMeyer-Olkin (KMO) Measure of Sampling Adequacy and Barlett's Sphericity Test, using SPSS version 21.0 statistical program.

The partial least squares (PLS) regression was built by physical and physiological factors as X-variable, and the Y-variable is associated to the performance Z-score.

The main purpose of all multivariate data analysis is to decompose the data to detect and model "hidden phenomena." All data were assessed using the Unscrambler® X 10.3 software version from CAMO Software AS (Oslo, Norway) unless stated otherwise.

Table 1: Mean and *SD* for profile and fitness test results of the sample

Variables	Male (n =10)		Female (n = 16)		
	Mean	*SD*	Mean	*SD*	*p*
Age (years)	17.9	1.4	16.9	2.1	.214
Weight (kg)	78.6	24.4	55.3	17.2	.009*
Height (cm)	171.5	5.6	156.6	4.4	< .001*
Sit and Reach (cm)	39.1	5.2	35.0	5.1	.620
Handgrip (kg)	84.8	12.6	55.8	8.7	< .001*
20m Speed (s)	3.52	0.19	3.97	0.27	< .001*
Illinois Agility (s)	17.74	1.54	18.79	1.25	.079
SBJ (cm)	188	44	164	29	.147
YE1 distance (m)	1044	382	790	288	.091
Performance Z-score	0.611	0.769	-0.498	0.844	.004*

III. Results and Discussion

The mean and SD for profile and fitness test results of the sample were presented in Table 1. No statistical difference in age was shown between gender as the minimum age for athletes who can participate in the Sukma Games is 21 years old. There were statistically significant differences in physical weight and height by gender, as the male athletes were heavier and taller than the female athletes. Although the result did not reveal statistically significant difference by gender in flexibility, agility, power, and aerobic, male athletes had the significantly greater strength and speed compared to female athletes. In general, the mean values in all variables showed the male athletes were superior to the female athletes in terms of profile and fitness level. Thus, separate PCA analysis between genders was suggested.

It is interesting to note that despite female group contributed more medals (6 medals) than male group (1 medal), only 2 female athletes delivered the medals, whereas 3 male athletes clinched the gold medal in a team event. Overall, male athletes statistically achieved better results in the competition as compared to female athletes.

PCA is a data reduction method used to identify a small set of variables that account for a large portion of the total variance in the original variables. It computes linear combinations of variables. The first linear combination of variables accounts for the largest amount of variation in the sample, the second for the next largest amount of variables in a dimension independent of the first; and so on.

Generally, a KMO index greater than 0.5 is considered acceptable to proceed with factor analysis [7]. The Bartlett test should be significant, which means that the variables are correlated highly enough to provide a reasonable basis for factor analysis as in this case [3].

After removal of strength variable in male samples (KMO = .244), the KMO measure confirmed the sufficient sampling number for the analysis, KMO = .568. Also the Bartlett"s test $\chi 2$ (21) = 46.704, $p < .001$, indicated that correlations between variables were sufficiently large for a PCA analysis. As for female samples, the KMO measure confirmed the sampling adequacy for the analysis, KMO =

.725. The Bartlett"s test $\chi 2$ (28) = 62.812, $p < .001$, indicated that correlations between variables were sufficiently large for a PCA analysis.

The PCA models were formed and interpreted by evaluating the scores and loadings plots simultaneously. For a group of men's athletes, the items that cluster on the same components suggest that PC-1 represented lower body limbs and PC-2 represented anthropometry. It explained 90% the data variation, and 72% for the validation of results.

From Figure 1, it could be observed that PC-1 is given by variation of flexibility, aerobic, power, agility, and speed. It involved legs movement and explained 61% of data variation. The state bowlers possessed agility and speed attributes, while archers and shooters had flexibility, aerobic and power attributes. Bowlers require to take off and accelerate while approaching the lanes before throwing the ball. Archers and shooters perform static standing position for a long duration during training and competition.

PC-2, which explained 31% of data variation, was given by the weight and height. PC-2 was named anthropometry because these were the most influential variables. The results showed most skill sports athletes had lighter and shorter.

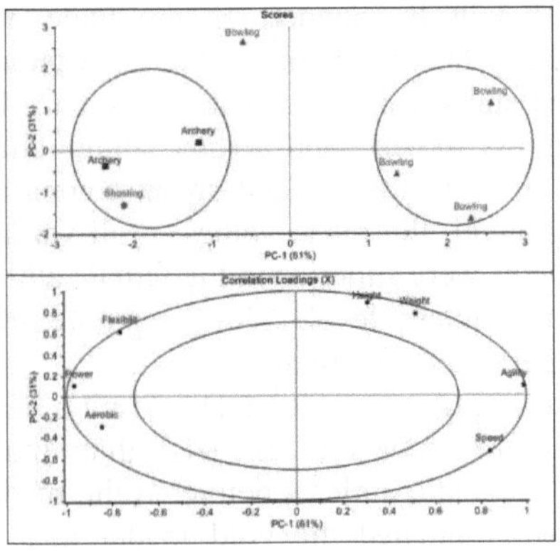

Fig 1: Male scores and loadings plots for PC-1 and PC-2

Meanwhile, as for female athletes, 2 main components explained over 77% of the data variation, with a result validation of 43%, as shown in Figure 2.

There were two group variables contribute to PC-1. The first group was aerobic and power, while the second group consists of strength, speed, agility, weight, and height. Those variables are anti-correlated. This means that if the female skills athletes who are aerobically fit and have powerful legs, it was most likely that they are light in weight, short stature, have weaker handgrips, and slow in running and also a change of direction.

PC-2, which explained 14% of data variation, was given by flexibility. The result showed female shooters had better flexibility as compared to other skill sports athletes. Poor flexibility can result in unnecessary and unwanted activation of the trunk or lower extremity muscles to maintain a balanced standing position for shooters [2].

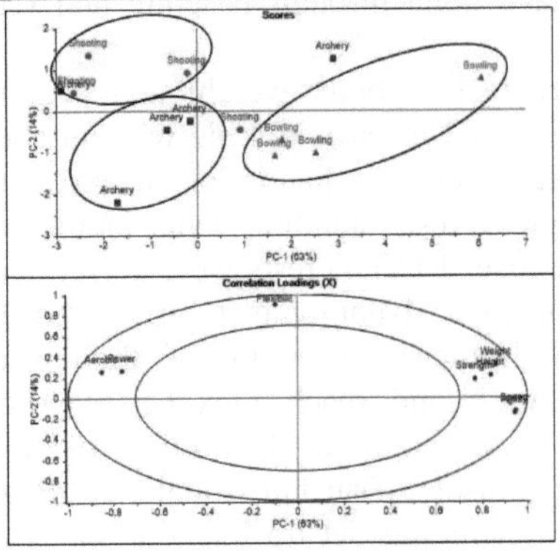

Fig 2: Female scores and loadings plots for PC-1 and PC-2

PLS regression generalizes and combines features from PCA and multiple linear regressions. Its goal is to analyze or predict a set of dependent variables from a set of independent variables or predictors. This prediction is achieved by extracting from the predictors a set of orthogonal factors called latent variables, which have the best predictive power.

The resulting model seemed to be able to determine performance Z-score. In Figure 3, it indicated the Z-score error of 0.309 and 0.472 when predicting a new subject. R2 of 0.719 and 0.675 for regression of predicting male and female athletes' performance model. To validate the model, 3 random samples for each gender that kept apart from calibration step. Not all predicted samples with deviation were close to reference z-score performance.

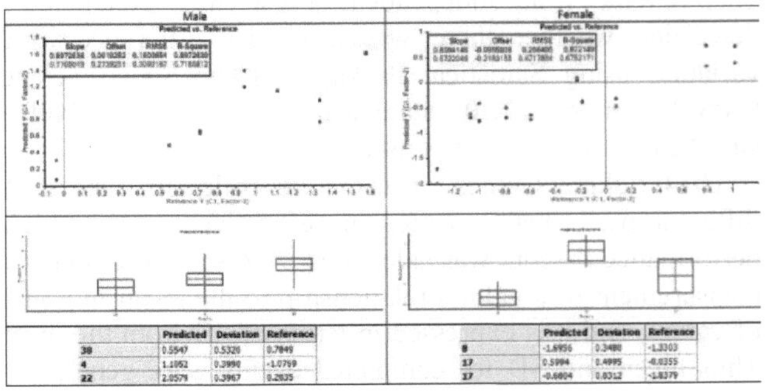

Fig 3: Validating model of performance

This analysis method can be used by sports scientists and coaches to detect major attributes that influence the performance outcome. To run the analysis, first, identify and collect X-variables and Y-variable. Next, check the assumption before running the PCA to build a model. Then, compute PLS regression with validation, and lastly, refine the model whenever there is additional new data. With this, coaches can then continuously identify areas that need further enhancement in order to increase chances of better performance during competition.

IV. CONCLUSION

Multivariate data analysis is a new approach to explain the relationship and identify main factors among measured variables. Application of PCA was attempted to build a model of fitness for skill sports athletes to predict performance. It showed physical and physiological attributes could be differential among different precision sports according to genders. Additional new data and further refinement of the model is needed to predict performance more precise.

ACKNOWLEDGMENT

The authors would like to extend their thanks to the National Sports Institute of Malaysia, Penang State Sports Council, coaches, and the athletes for the given support.

CONFLICT OF INTEREST

The authors declare that they have no conflict of interest.

REFERENCES

1. Anderson GS & Plecas DB (2000) Predicting shooting scores from physical performance data. Policing, 23(4): 525 – 537. DOI 10.1108/13639510010355611
2. Arnot C (2010) Postural Considerations for Rifle and Pistol Shooters.http://www.usashooting.org/library/Instructional/Physical_Train ing/jan_feb_2010_postural_considerations.pdf
3. Bartlett MS (1950) Tests of Significance in Factor Analysis. British Journal of Statistical Psychology, 3: 77 – 85. DOI 10.1111/j.2044-8317.1950.tb00285.x
4. Brisswalter J, Collardeau M, & René A (2002) Effects of acute physical exercise characteristics on cognitive performance. Sports Med. 32(9): 555 – 566. DOI 10.2165/00007256-200232090-00002
5. Duda M (1988) Pinning Down Bowling Injuries. The Physician and Sports Medicine. 16(4): 173 – 180. DOI 10.1080/00913847.1988.11709490
6. Eckert H & Wendt D (1965) Various Measures related to Draw in Archery. Research Quarterly. American Association for Health, Physical Education and Recreation, 37(1): 145 – 147. DOI 10.1080/10671188.1966.10614748
7. Kaiser, H.F., 1974. An index of factorial simplicity. Psychometricka. 39: 31 – 36. DOI 10.1007/BF02291575
8. Liu MH, Wu TY, Li YC, Huang MH, & Wu SK (劉明煌, 吳聰義, 黃明祥, 李曜全&吳昇光)(2008) Youxiu Shejian Xuanshou zhi Yundong Shanghai Diacha (優秀射箭選手之運動傷害調查) [An Investigation of Sports Injuries in Elite Archers]. JIANKANG CUJIN KEXUE (健康促進科學). 3(2): 67.
9. Gao WL (高文良) (2006) Shejian Yundong Nengli Ceyan Xiangmu Bianzhi zhi Yanjiu (射箭運動能力測驗項目編製之研究) [A study on selection of archery ability test items]. Unpublished manuscript. National Hsinchu University of Education (國立新竹教育大學).
10. Mann DL & Littke N (1989) Shoulder injuries in archery. Canadian Journal of Sport Sciences. 14(2): 85 – 92.
11. Razman R, Cheong JPG, Wan Abas WAB, & Abu Osman NA (2012) Anthrpometric and strength characteristics of tenpin bowlers with different playing abilities. Biol. Sport, 29: 33 – 38.
12. Tan B, Aziz AR, & Teh KC (2000) Correlations between physiological parameters and performance in elite ten-pin bowlers. J.
13. Sci. Med. Sport. 3(2): 176 – 185.

Author: Loh Tze Chien
Institute: National Sports Institute of Malaysia
Street: Kompleks Sukan Negara
City: Bukit Jalil
Country: Malaysia
Email: chien@isn.gov.my / tze.chien_loh@hotmail.com

The Effects of Rest Interval on Electromyographic Signal on Upper Limb Muscle during Contraction

N.U. Ahamed[1], M.F. Rabbi[2], Z. Taha[1], K. Sundaraj[3] and T. Sikandar[2]

[1] Imams Lab, Faculty of Manufacturing Engineering, Universiti Malaysia Pahang, Pekan-26600, Pahang, Malaysia
[2] VisIS Lab, Faculty of Electrical and Electronics Engineering, Universiti Malaysia Pahang, Pekan-26600, Pahang, Malaysia
[3] Department of Electronics & Computer Engineering Technology, Universiti Teknikal Malaysia Melaka (UTeM), Melaka, Malaysia

Abstract— **In this paper, the Electromyographic (EMG) signal was investigated on the Biceps Brachii muscle during dynamic contraction with two different rest intervals between trials. The EMG signal was recorded from 10 healthy right arm-dominant young subjects during load lifting task with a standard 3-kg dumbbell for 10 seconds. Root mean square (RMS) has been used to identify the muscle function. The resting period was 2- and 5-minutes between each trial. The statistical analysis techniques included in the study were *i)* linear regression to examine the relationship between the EMG amplitude and the endurance time, *ii)* repeated measures ANOVA to assess differences among the different trials and *iii)* the coefficient of variation (CoV) to investigate the steadiness of the EMG activation. Results show that EMG signal is more active after 5 minutes rest period compare to 2-minutes gap. On the other hand, EMG signals were steady during 2-minutes rest (7.59%) compare to 5-minutes resting interval (16.14%). Results suggest that moderate interval between each trial is better to identify the muscle activity compare to a very short interval. The findings of this study can be used to improve the current understanding of the mechanics and muscle functions of the upper limb muscle of individuals during a contraction which may prevent from muscle fatigue.**

Keywords— EMG, RMS, Rest interval, Muscle, Contraction.

I. Introduction

Analysis of EMG signal has been given a lot of attention in the last few decades since the investigation and processing of the signal have a huge influence in developing adaptive control of prosthetic devices in the rehabilitation program as well as in diagnosis of neuromuscular diseases. Generally, EMG signal is used to identify the electrical activity from the skeletal muscles during contraction and body movement and, it provides the detail information about the structure and function of these muscles [1]. One of the major issues reported in recent years is the signal effect on muscle during inter-trial rest intervals [2, 3] while recording the EMG signal. The reason is, lack of proper knowledge of the approximate time duration, between each trial, may result in muscle fatigue, soreness, stiffness and even muscle cramp.

Usually, researchers prefer different time duration as rest interval between each trial during EMG recording. For example, 5 seconds [4], 1 minute [4-6] and 5 minutes [7]. However, few studies have shown the significance of same time duration between each experimental trial. For example, Maia et al., investigated EMG signals from lower limb muscle with a rest interval of 30 seconds, 1 minute, 3 minutes and 5 minutes [8]. Authors found that no rest or relatively shorter rest intervals (30 seconds and 1 minute) might be more effective to stimulate greater agonist repetition enhancement and muscle activation. Furthermore, Pincivero et al. examined the effects of rest interval (5, 40 and 160 sec) on quadriceps femoris muscle activation [17]. The effects of rest interval length on bench press performance with an interval of 1, 2 and 3 minutes were investigated in [19]. In this consequence, the effect of rest interval for muscle characteristics identification during different contractions (isometric, eccentric, concentric and isokinetic) and from different muscles (upper- and lower-limb) were investigated in [9-11].

Moreover, having the significant effect of rest interval on EMG activity, a number of feature extraction methods have been used to investigate the muscle functions. For example, root mean square (RMS), zero crossing, mean frequency, median frequency, average-rectified value (ARV), integrated EMG (IEMG), mean absolute value (MAV), normalized spectral moments, wavelet transforms, increase in synchronization (IIS) index and fractal dimension are more frequently used time and frequency domain techniques. However, no study has been found investigating muscle activity as well as the signal variation on Biceps Brachii (BB) muscle during two rest interval period of 2 and 5 minutes. Also, RMS feature extraction technique has not been given much attention in such kind of investigation. In this study, we investigated the muscle activity during specific rest interval period using RMS feature extraction method. We investigated the effects of two different rest intervals on EMG activity in upper limb muscle during dynamic contraction. Specifically, 2 min and 5 min rest

F. Ibrahim et al. (eds.), *3rd International Conference on Movement, Health and Exercise*,
IFMBE Proceedings 58,
DOI: 10.1007/978-981-10-3737-5_3

interval were considered as a short and a long resting period respectively. Results found in this study may be useful for further investigation on muscle fatigue during contraction.

II. Subjects and Methods

A. Subjects

Ten right-hand dominated subjects (8 males, mean age: 23.25 yrs, weight: 64.88 kg, height: 16.75 cm; and 2 females mean age: 23.12 yrs, weight: 54.5 kg, height: 151.2 cm) voluntarily participated in the study and gave their written informed consent. Subjects did not have any history of disorder or pain in biceps muscles. Note that, the subjects were treated in accordance with the ethical standards of the Declaration of Helsinki.

B. Testing Procedures

At first, each subject was asked to stand straight holding the weight handle bar while it was resting. Then dynamic contraction was performed by lifting a standard 3-kg weight dumbbell. During the arm movement the elbow was swung (flexion and extension) almost at the same speed and generates pendulum arm motion within 0° to 90° angles. The angle was measured using a goniometer at shoulder-to elbow and elbow-to-palm respectively. Each subject underwent two sessions for 10 sec and each session consisted of three trials. There was 2 min rest interval between each trial of the first session and then 30 min break before starting the second session. In the second session, there was 5 min rest between each trial. Note that, EMG signal was recorded after each trial. Fig. 1 depicts the experimental protocol from the subject while lifting the load to generate the dynamic contraction.

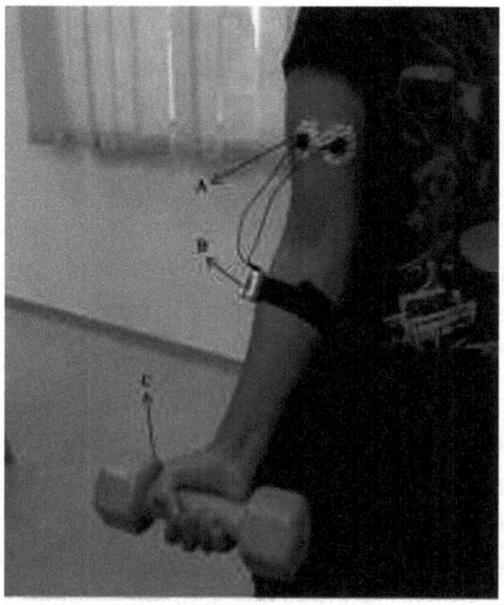

Fig 1 Experimental protocol setup. A) electrode placed on BB muscle, B) wireless EMG sensor, C) a 3-kg dumbbell.

C. EMG Data Recording

In this experiment, a wireless three-channel EMG signal storage device, called SHIMMER™ (Model SHSHIM-KIT-004) was used to record the EMG signal from the muscles. The device is also touch-proof and Bluetooth enabled as well. Two channels were used for the EMG recording and another one was used as the reference channel. The built-in frequency range of the device is 5-482 Hz including an EMG amplifier gain of 682 dB. A sampling frequency of 1 kHz was used to record the raw EMG signal which was preamplified with a band-pass filter with a frequencyy range of 10–500 Hz. There was a distance of 5 feet between the EMG daughter board device and a laptop with Bluetooth facility.

Pre-gelled Ag/AgCl non-invasive electrodes with biopotential sensors were used in this study. The electrodes can identify the flow of ions through a nerve fiber in the human body. In addition, these electrodes have few more advantages such as *i*) non-polarized, *ii*) allow free current flow across the electrode junction, *iii*) quiet, generating noise levels lower than 10μV, *iv*) disposable, *v*) large (30 mm) and *iv*) have low adhesiveness. To obtain better EMG signals with reduced artifacts and bioimpedance, the skin of the muscle was prepared using a skin cleaning gel (sigma gel) and an alcohol swab. The skin preparation and the electrode placement procedures were similar to [12-14].

D. EMG Feature

The root mean square (RMS) is one of the most popular features used to interpret the amplitude of an EMG signal. Consequently, in this study, the RMS was used to calculate and analyse the surface EMG signals. RMS is used to statistically investigate the magnitude of a time-varying signal and can be defined according to [15, 16]:

$$\mathrm{RMS} = \sqrt{\frac{1}{N}\sum_{i=1}^{N} x_i^2} \tag{1}$$

E. Statistical Analysis

After recording, statistical analysis was performed using Minitab® software (version 13.32). Significant differences in the resting interval and EMG amplitudes were detected through repeated-measures analysis of variance (ANOVA) using a significance level of α=0.05 and 95% (P<0.05) confidence intervals for all of the variables. The variation in the muscle activity which can be identified by the steadiness of the EMG signal was characterized by the coefficient of variation (CoV). CoV can be defined as the ratio of the standard deviations divided by the means, CoV = σ/μ.

III. Results and Discussion

Table 1 and Fig. 2 present the effectiveness of EMG signal on the BB muscle after two (2- and 5-min) rest intervals. Most of the individual results show that EMG amplitude is increased after a 5-min interval in comparison with a 2-min interval. Accordingly, combined result also demonstrates the same scenario where a 5-min rest period generated higher signal activity (0.23±0.04mV) than a 2-min rest period (0.18±0.02 mV). This implies that muscles involved in higher activity after 5-min interval than the 2-min interval. Almost similar result was reported in [18] where it was demonstrated that adequate recovery of muscle force within trials is a prerequisite for the generation of tension in subsequent contractions. Thus, setting a relatively longer rest interval will allow sufficient time for the muscles to refill the intramuscular stores. However, different results found in terms of CoV calculation. Here, EMG signals were more variable (16.14%) during 5 min interval than 2 min interval period (7.61%). This indicates that the muscle activity is steadier with the short rest interval than with the longer resting period. Finally, significant differences ($p<0.05$) observed from both the interval periods between time and signal amplitude.

Table 1: Summarized RMS results from entire subjects from 2- and 5-minute experiments

Subject	2-minutes Interval				5-minutes Interval			
	Mean±SD		CoV (%)		Mean±SD		CoV (%)	
Sub 1	0.22±0.03		5.52		0.23±0.03		5.64	
Sub 2	0.33±0.04		11.13		0.41±0.03		6.26	
Sub 3	0.11±0.01		4.83		0.12±0.01		6.26	
Sub 4	0.12±0.02	0.18±0.02	18.86	7.61	0.13±0.03	0.23±0.04	11.64	16.14
Sub 5	0.11±0.01		4.76		0.29±0.06		19.94	
Sub 6	0.24±0.02		6.45		0.25±0.01		0.71	
Sub 7	0.16±0.01		1.69		0.15±0.02		9.07	
Sub 8	0.18±0.02		10.02		0.23±0.03		11.12	
Sub 9	0.01±0.01		3.58		0.15±0.12		80.49	
Sub10	0.25±0.03		9.12		0.29±0.04		10.27	

Assessment of the muscle activity and signal variations based on resting interval is a major unresolved challenge in human ergonomics, biomechanics, biomedicine, and rehabilitation. Keeping this in mind, the aim of the study was to identify the impact of the EMG signal on upper limb muscle during two different rest intervals. It has been found that the EMG activity was higher during 5-min rest interval than 2min rest period. But, according to the signal

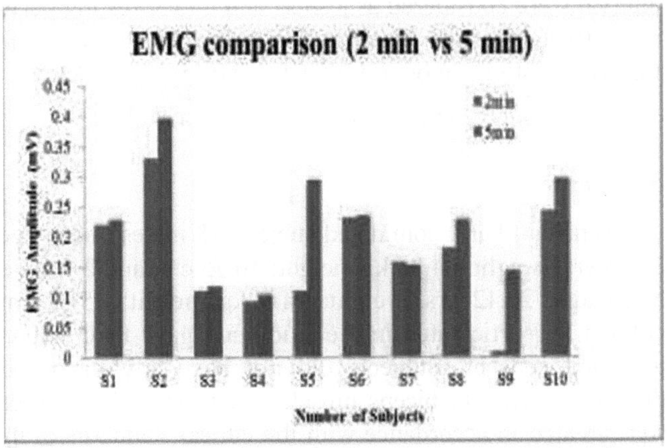

Figure 2: Comparison graph between two interval periods during each trial

consistency, the result was vice versa, *i.e.,* EMG signal was more consistent with a 2-min interval in comparison with that with a 5-min interval. So, the result suggests that the EMG data will be acquired more accurately if the interval between each trial is neither too short nor too long.

The findings can be helpful for developing an appropriate procedure of EMG data recording. Also, the results can be applied in diverse areas of biomedical applications such as the neuromuscular system analysis, ergonomics, biomechanics and rehabilitation engineering. Also, the result of the study may expand the current understanding of the mechanics of the upper limbs of subjects involved in the physical exercise with short and long rest interval. However, there were few limitations of the current study as well. For example, only two rest interval duration, one muscle (BB) and dynamic contraction have selected in this study. Thus, there is a need for further investigation examining different rest intervals in other muscles of the human body with different experimental protocols.

IV. Conclusions

In the EMG-based assessment, it is crucial to set an appropriate interval period between the trials. Otherwise, it may disrupt the signal recording from the subjects and thus, the recorded muscle activity may be erroneous. The results of the current study suggest that a very short interval may cause the muscle fatigue or cross talk among the adjacent muscle. On the other hand, a very long interval may cause the subject uninterested in data recording, which may affect the signal impedance or even signal-to-noise ratio. Thus, moderate rest interval (not very short or very long period) is the best way to record the EMG signal. However, our future research will be directed to study a large number of interval periods for further investigation on finding optimum value of the rest interval.

ACKNOWLEDGMENT

This research was supported by the University Malaysia Pahang (UMP) Research Grant Scheme (No. RDU1403157).

CONFLICT OF INTEREST

The authors declare that they have no conflict of interest.

REFERENCES

[1] R. M. Rangayyan and N. P. Reddy, "Biomedical signal analysis: a case-study approach," *Annals of Biomedical Engineering*, vol. 30, pp. 983-983, 2002.

[2] M. Henselmans and B. J. Schoenfeld, "The effect of inter-set rest intervals on resistance exercise-induced muscle hypertrophy," *Sports Medicine*, vol. 44, pp. 1635-1643, 2014.

[3] J. C. Jambassi Filho, L. T. Gobbi, A. L. Gurjão, R. Gonçalves, A. K. Prado, and S. Gobbi, "Effect of different rest intervals, between sets, on muscle performance during leg press exercise, in trained older women," *Journal of sports science & medicine*, vol. 12, p. 138, 2013.

[4] I. G. Amiridis, D. Mani, A. Almuklass, B. Matkowski, J. R. Gould, and R. M. Enoka, "Modulation of motor unit activity in biceps brachii by neuromuscular electrical stimulation applied to the contralateral arm," *Journal of Applied Physiology*, vol. 118, pp. 1544-1552, 2015.

[5] S. I. Khan, J. L. Taylor, and S. C. Gandevia, "Unexpected factors affecting the excitability of human motoneurones in voluntary and stimulated contractions," *The Journal of physiology*, 2016.

[6] M. N. Houston, V. E. Hodson, K. K. Adams, and J. M. Hoch, "The effectiveness of whole-body-vibration training in improving hamstring flexibility in physically active adults," *Journal of sport rehabilitation*, vol. 24, 2015.

[7] F. J. Vera-Garcia, I. Ruiz-Pérez, D. Barbado, C. Juan-Recio, and S. M. McGill, "Trunk and shoulder emg and lumbar kinematics of medicine-ball side throw and side catch and throw," *European Journal of Human Movement*, vol. 33, pp. 93-109, 2014.

[8] M. F. Maia, J. M. Willardson, G. A. Paz, and H. Miranda, "Effects of different rest intervals between antagonist paired sets on repetition performance and muscle activation," *The Journal of Strength & Conditioning Research*, vol. 28, pp. 2529-2535, 2014.

[9] C. A. Vieira, J. B. Ferreira-Junior, A. Vieira, S. J. Martins, M. F. Vogt, R. Freitas Junior, *et al.*, "Effect of Rest Interval on Muscle Performance in Breast Cancer Survivors," in *MEDICINE AND SCIENCE IN SPORTS AND EXERCISE*, 2014, pp. 128-129.

[10] P. W. Stratford, A. Bruulsema, B. Maxwell, T. Black, and B. Harding, "The effect of inter-trial rest interval on the assessment of isokinetic thigh muscle torque," *Journal of Orthopaedic & Sports Physical Therapy*, vol. 11, pp. 362-366, 1990.

[11] J. Edge, N. Eynon, M. J. McKenna, C. A. Goodman, R. C. Harris, and D. J. Bishop, "Altering the rest interval during high intensity interval training does not affect muscle or performance adaptations," *Experimental physiology*, vol. 98, pp. 481-490, 2013.

[12] H. J. Hermens, B. Freriks, R. Merletti, D. Stegerman, J. Block, and G. R. e. al., "SENIAM: European recommendations for surface electromyography Roessingh Research and Development, Enschede," *http://www.seniam.org/*, 1999.

[13] P. Zipp, "Recommendations for the standardization of lead positions in surface electromyography," *European Journal of Applied Physiology and Occupational Physiology*, vol. 50, pp. 41-54, 1982.

[14] E. Delagi and A. Perotto, "For the electromyographer: the limbs," *Physical Medicine and Rehabilitation*, 1980.

[15] R. Boostani and M. H. Moradi, "Evaluation of the forearm EMG signal features for the control of a prosthetic hand," *Physiological measurement*, vol. 24, p. 309, 2003.

[16] A. Phinyomark, P. Phukpattaranont, and C. Limsakul, "Feature reduction and selection for EMG signal classification," *Expert Systems with Applications*, vol. 39, pp. 7420-7431, 2012.

[17] Pincivero, D. M., Campy, R. M., & Karunakara, R. G., "The effects of rest interval and resistance training on quadriceps femoris muscle. Part II: EMG and perceived exertion," *Journal of sports medicine and physical fitness*, vol. 44, no. 3, pp. 224, 2004

[18] Pincivero, D. M., Lephart, S. M., & Karunakara, R. G. "Effects of rest interval on isokinetic strength and functional performance after short-term high intensity training," *British journal of sports medicine*, vol. 31, no. 3, pp. 229-234, 1997.

[19] Willardson, J. M., & Burkett, L. N.,"The effect of rest interval length on bench press performance with heavy vs. light loads," *The Journal of Strength & Conditioning Research,*,vol. 20, no. 2, pp. 396-399, 2006.

Author:	Nizam Uddin Ahamed
Institute:	Manufacturing Engineering, University Malaysia Pahang
Street:	Pekan
City:	Pahang
Country:	Malaysia
Email:	ahamed1557@hotmail.com

Comparison between Stair Climbing and 1 Mile Walking in Relation to Cardiorespiratory Fitness among Sedentary Adults

I.A. Mir, V.L.K. Yu, F.S. Jia and A.H. Bhat

Department of Physiotherapy, Faculty of Medicine & Health Sciences, Universiti Tunku Abdul Rahman, Kajang, Malaysia

Abstract— **Background: The worldwide epidemic of physical inactivity is rising significantly and the main cause is a sedentary lifestyle. The purpose of this study was to evaluate the effectiveness of stair climbing and 1 mile walking in relation to cardiorespiratory response and to estimate the VO2max and Physical Fitness Index (PFI). Methods: PAR-Q and IPAQ were used to assess the level of physical activity and sedentary status of participants. 37 participants were initially assessed for their maximal oxygen uptake (VO2 max) and physical fitness index (PFI) using modified Harvard step test after randomly assigning them into 3 groups. Only 34 participants completed the study; 12 Full stair climbing, 11 half stair climbing, and 11 one mile walking. Full stair climbing group ascended 7 stories of 154 steps, each step 17.5cm in height (vertical distance =27m), whereas half stair climbing group ascended 3.5 stories of 77 steps (vertical distance =13.5m) and 1-mile walk group walked for 1 mile on the treadmill without any inclination. VO2max, PFI, heart rate recovery and rate of perceived exertion were measured at baseline and at the end of week 4. Results: Repeated measures 2 way ANOVA was used to find the differences between 3 groups. Full stair climbing group showed an increase in VO2 max and 1-mile walking group exhibited improvement in HRR. Conclusion: 1-mile walking group and full stair climbing groups exhibited improvement in cardiorespiratory fitness. Full stair climbing group took less than 2 minutes and 1-mile walking group took 11 to 14 minutes to complete the exercise, hence one can opt to perform either stair climbing or walking depending on their preference to improve cardiorespiratory fitness.**

Keywords— **VO2 Max, Physical Fitness Index, Heart Rate Recovery, Stair Climbing, Walking**

I. INTRODUCTION

Most people are leading a sedentary lifestyle as they indulge themselves in their never-ending work. The incidence of cardiovascular disease is escalating rapidly and fingers are being pointed to physical inactivity and poor cardiorespiratory fitness as one of the few main predictors of premature mortality[1]. Individuals who are insufficiently physically active have 20%-30% increased the risk of all causes of mortality factors compared to those who perform a minimum of 30 minutes of moderate-intensity physical activity most of the days[2]. A study was conducted to determine the physical activity index among Malaysian youth. 1801 participants were recruited in this study and researchers found that 23% (men) and 31.3% (women) live a sedentary lifestyle[3].

A. Cardiorespiratory Fitness

Cardiorespiratory fitness is a component of physical fitness and is defined as the ability of the respiratory, circulatory and muscular systems to supply oxygen during sustained physical activity[4]. Cardiorespiratory fitness is usually expressed in maximal oxygen uptake (VO_2 max) and physical fitness index (PFI) and both are an acceptable indexes of cardiorespiratory fitness. Higher cardiorespiratory fitness level permits continuous physical activity without a decline in performance and reduces the risk of cardiovascular and non-communicable diseases. In a healthy individual, 30 minutes moderate intensity physical activity 5 days per week, or 20 minutes of vigorous physical activity 3 days per week is required to develop and maintain cardiorespiratory fitness[5].

B. Stair climbing & walking

Stair climbing is a moderate physical activity that is sufficient to improve cardiorespiratory fitness in sedentary individuals[6]. Regular stair climbing has been shown to be effective in increasing VO_2 max, reduce low-density lipoprotein cholesterol and increase the strength of lower limbs[6,7,8,9,10]. Individuals who perform stair climbing at work decreases 25% of the risk of mortality compared to those who are less physically active[10]. Besides stair climbing, walking is one of the most common moderate intensity physical activity which requires no special skills or facilities. American College of Sports Medicine (ACSM) recommends walking as an exercise to increase cardiorespiratory fitness and have suggested 50-85% of VO_2 max as the appropriate intensity, 20–60 minutes/day and 5 days a week[11]. There are certain barriers that make people reluctant to walk or use stairs and these include the sedentary style of work and passive mode of transportation. People are opting for lifts and elevators rather than the staircase and driving rather than going for public transport or walking. Not only that, lack of infrastructure became a reason as most people have the perception that physical activity must be done with proper equipment's[12].

To the author's knowledge, no study has been conducted to compare the stair climbing and walking to find its effectiveness on cardiorespiratory fitness, although literature supports benefits of these activates separately. In addition, to overcome the barriers in young sedentary adult's researchers choose to conduct this study so that they get motivated to take up these activities.

F. Ibrahim et al. (eds.), *3rd International Conference on Movement, Health and Exercise*,
IFMBE Proceedings 58,
DOI: 10.1007/978-981-10-3737-5_4

II. MATERIALS AND METHODS

A. Participants

37 participants were recruited initially but only 34 completed the study. Inclusion criteria were: (a) age group of 18-25 years old; (b) height of 160cm-170cm; (c) healthy with no apparent illness; (d) living a sedentary lifestyle. All the participants were recruited using physical activity readiness questionnaire (PAR-Q) and international physical activity questionnaire (IPAQ). The purpose of physical activity readiness questionnaire was to determine the health status of participants while international physical activity questionnaire was used to determine whether participants are sedentary or active[13,14]. Participants were randomly divided into three groups using lottery method; 12 participants in full stair climbing group, 11 in half stair climbing group and 11 participants in 1 mile walking group. Participants signed the informed consent after being informed thoroughly of the research study. The study was approved by Universiti Tunku Abdul Rehman scientific and ethical review committee (Malaysia).

B. Familiarization of stair climbing and standardized warm-up protocol

The main purpose of familiarization was to help the participants to establish their own pace when climbing the stairs and to prevent delayed onset muscle soreness[15]. Participants practiced stair climbing with a brisk and self-selected pace that was quick and sustainable throughout the entire climb. Running or double step was not allowed. Familiarization trial lasted for 5 days. Participants were given a standardized warm-up protocol for 5 minutes before starting the exercise. This protocol included stretching of gluteus, psoas, adductors, quadriceps, hamstring, gastrocnemius and peroneus muscles[16]. Same warm up protocol was used before starting the following exercises every time.

C. Exercise Protocol

Participants in full stair climbing group climbed up 7 floors (14 flights of stairs). Each flight was separated by a horizontal connecting platform with a minimum area of $4m^2$. The total numbers of steps were 154, each step 17.5cm in height with a total vertical distance of 27 meters. Participants of half stair climbing group were required to climb up 3.5 floors (7 flights of stairs). Each flight was separated by a horizontal connecting platform with a minimum area of $4m^2$. The total numbers of steps were 77, each step 17.5cm in height with a total vertical distance of 13.5 meters. In 1-mile walking group, participants walked as fast as possible on a treadmill for 1 mile with no inclination and no running was allowed. Each group carried out respective exercises twice a day, for 5 days a week for 4 weeks.

D. Measures

Baseline measurements (PFI, VO_2 max and HRR) for each group were taken 1 day before starting the exercise protocol and at the end of 4 weeks of exercise training. PFI was measured by using the Modified Harvard step test. Participants stood barefoot close to the stepping bench which was 40 cm in height and was instructed to do few cadences, "up-up-down-down" on the bench to familiarize with the beat of a metronome at 90 steps/minute[19]. Then the actual test was carried out which lasted for 5 minutes. Participants were allowed to change the lead leg during the test, but they were instructed not to break the rhythm of the test. After completion of the test, heart rate recovery (HRR) between 1-1.5 minutes, 2-2.5 minutes, and 3-3.5 minutes was recorded with participants in sitting position. PFI score was determined using following equation[17,18].

$$\text{PFI} = \frac{(100 \times \text{test duration in seconds})}{(2 \times \text{sum of heart beats in the recovery periods})}$$

However, participants who could not complete 5 minutes of the test, PFI were determined using the following equation[19].

$$\text{PFI} = \frac{(100 \times \text{test duration in seconds})}{(5.5 \times \text{HRR in period during 1-1.5mins})}$$

Table 1. Physical Fitness Score

Physical Fitness Index (PFI)	Fitness category (Physical condition)
>90	Excellent
80-90	Good
55-79	Average
<55	Poor

HRR measured was used to estimate VO_2 max using Astrand-Rhyming nomogram. On the nomogram, HRR (0-1 min at any time for 15s) and weight of participant (kg) was accurately marked on the designated scales. A line was drawn between the two marks and the point where it intersected indicates VO_2 max (L/min)[18,20]. VO_2 max adjusted for body mass ($ml.kg^{-1}min^{-1}$) was computed as follows[21, 22].

$$VO2max\,(ml.kg-1min-1) = \frac{VO2max\,(L.min-1)\,x\,1{,}000}{Body\,mass\,(kg)}$$

E. Statistical Analysis

The data was analyzed using descriptive statistics of mean and standard deviation. The differences among subjects, between subjects and between groups for VO_2 max, PFI and HRR were determined with repeated measures 2 way ANOVA. For all tests, the level of significance was set at 0.05 and confidence interval for the difference at 95%.

III. Results

34 (91.89%) of 37 participants (12 full stair climbing group, 11 half stair climbing group and 11 one mile walking group) completed the study and only their data were analyzed. 23 (67.65%) of the participants were female and 11 were male (32.35%). Table 2 summarizes the physical characteristics of participants and table 3 and 4 show the changes within and between subjects for full stair climbing and 1-mile walking groups. There was a significant difference ($p<0.001$) within and between subjects in each group for all measures but pairwise comparisons (table 5) showed statistically significant difference in VO2 max ($p=0.19$) only in full stair climbing although there was an improvement in the VO2 max in other 2 groups but it was not significant.

Table 2. Physical characteristics of the participants

	FSC (n=12)	HSC (n=11)	1MW (n=11)	P
Gender	6M, 6F	3M, 7F	3M, 7F	
Age Range	21 ±1.04 19 - 22	20.55±0.82 19 - 22	20.09 ±1.14 19 - 22	0.115
Height (cm)	164.44 ± 3.88	164.82 ±3.66	164.55 ±3.15	0.968

Table 3. Difference in cardiorespiratory fitness between subjects

Source	Measure	df	F	Sig.	Partial Eta Squared	Obserd Power[a]
(FSC) Intercept	VO2 max	1	1508.357	.000	.979	1.000
	PFI	1	1004.259	.000	.968	1.000
	HRR	1	3088.777	.000	.989	1.000
Error	VO2 max	33				
	PFI	33				
	HRR	33				
1MW Intercept	Vo2 max	1	2312.001	.000	.987	1.000
	PFI	1	1050.668	.000	.971	1.000
	HRR	1	3662.693	.000	.992	1.000
Error	Vo2 max	31				
	PFI	31				
	HRR	31				

Table 4. Difference in cardiorespiratory fitness within subjects

Source	Measure	df	F	Sig.	Partial Eta Squared	Obserd Power
FSC	VO2 max	1	26.347	.000	.444	.999
	PFI	1	31.777	.000	.491	1.000
	HRR	1	21.686	.000	.397	.995
Error (FSC)	VO2 max	33				
	PFI	33				
	HRR	33				
1MW	VO2 max	1	25.741	.000	.454	.998
	PFI	1	30.259	.000	.494	1.000
	HRR	1	21.245	.000	.407	.994
Error (1MW)	VO 2max	31				
	PFI	31				
	HRR	31				

Table 5. Pairwise Comparisons of VO2 max between groups

(I) GROUP	(J) GROUP	Mean Difference (I-J)	Std. Error	Sig.[b]
FSC	FSC			
	HSC	11.366*	3.858	.019
	1MW	7.092	3.762	.207
HSC	FSC	-11.366*	3.858	.019
	HSC			
	1MW	-4.274	3.537	.709
1MW	FSC	-7.092	3.762	.207
	HSC	4.274	3.537	.709
	1MW			

*. The mean difference is significant at the .05 level.
b. Adjustment for multiple comparisons: Bonferroni.

Table 6. Pairwise Comparisons of HRR between groups

(I) GROUP	(J) GROUP	Mean Difference (I-J)	Std. Error	Sig.[b]
FSC	FSC			
	HSC	8.809	4.076	.116
	1MW	12.750*	4.008	.010
HSC	FSC	-8.809	4.076	.116
	HSC			
	1MW	3.941	3.996	.996
1MW	FSC	-12.750*	4.008	.010
	HSC	-3.941	3.996	.996
	1MW			

*. The mean difference is significant at the .05 level.
b. Adjustment for multiple comparisons: Bonferroni.

No significant difference ($p>0.5$) was found in the PFI scores between the groups even though full stair climbing group showed an increase of 7.1 and 6.3 compared to the half stair climbing and 1-mile walking group respectively. Pairwise comparisons of HRR (table 6) showed a significant difference ($p=0.10$) in 1-mile walking group, with a mean difference of -12.8 and -3.9 in comparison to full stair climbing and half stair climbing groups.

There was a significant reduction ($p<0.005$) in time taken to complete the intervention in the all three groups. The time taken to complete 14 flights of stair climbing decreased significantly (107.42s to 98s), 7 flights of stair climbing (55.09 to 52.64s) and one mile walking (869.45s to 785.64s).

IV. Discussion

The main finding of our study was that at the end of 4 weeks of exercise training full stair climbing group exhibited improvement in terms of VO_2 max and 1-mile walking group showed improvement in HRR. Our findings were consistent with the previous literature where authors found an increase in VO_2 max among sedentary adults following 8 weeks of stair climbing[7,8]. In the present study 2 minutes of stair climbing produced almost same effects as 6 minutes of stair climbing and 45 minutes of walking[8, 24]. Thus it can be concluded that stair climbing of approximately 6 minutes per day or 45 minutes of walking per day improves VO_2 max same way as 2 minutes of stair climbing per day. Improvement in VO_2 max may be due to metabolic changes, such as an increase in blood supply and oxidative capacity within the muscles, and the changes of the cardiovascular system that enhances blood and oxygen delivery to the active working muscles during the exercise[25].

PFI determination is one of the important criteria to assess the cardiorespiratory efficiency of an individual. It is estimated using Harvard step test and time is taken to complete the test is 5 minutes. In the present study 33 (97.1%) of 34 participants completed the test after 4 weeks of the intervention compared to pre-test where only 25 (73.5%) of 34 participants completed the test indicating that participants PFI score improved after undergoing exercise training. Clinically all 3 groups showed an increase in the PFI scores but it was not statistically significant. Full stair climbing group gained most benefit in the PFI and it could be due to the high-intensity impact of the given exercise. With training, there is an increase in the number of mitochondria. The enzyme system that is involved in oxidative metabolism becomes more efficient. The number of capillaries to muscle cells also increases with better distribution of blood to the muscle fibers. The net effect is a complete extraction of O_2 and consequently, for a given workload, less increase in lactate production. This effect leads to less hypoxia that in turn leads to less chemoreceptor trigger and less cardio-acceleratory effect making individual to continue the exercise for longer duration[26, 27]. The result of the current study is similar to a previous study in which stair climbing group showed higher PFI when compared to the control group[17].

Heart rate recovery indicates the well-being of body's autonomic system and is an implication of health of heart muscles[28]. Heart rate would not immediately return to its baseline after exercise, instead, it remains elevated and slowly returns to its resting rate. In the present study, there was a significant reduction in HRR seen in I mile walking group. Improvement of heart rate recovery following exercise is due to increase in vagal activity or decreased in sympathetic activity[29, 30]. After sub-maximal training, heart rate returns to its resting rate quickly after an exercise compared to before training. Hence, HRR has been proposed as an indirect index of CRF and is used to estimate CRF[23].

This study has provided an idea on how walking and stair climbing will improve the cardiorespiratory fitness of sedentary adults and indirectly reduce the risk of mortality. People spend an estimated two-thirds of their lives at school/college/university or place of employment; walking and stair climbing are the cheapest ever and most convenient physical exercises that can be incorporated at these places. Through this research work, we hope that more people will come to the realization that physical activity is important and in order to commit themselves, they can push aside the barriers of physical activity and start applying walking and stair climbing to their daily routines.

Conflict of Interest

Authors have no conflict of interest to declare.

References

1. Lakka, T., Laaksonen, D., Lakka, H., Niskanen, L., Rauramaa, R., & Salonen, J. (2003). Sedentary Lifestyle, Poor Cardiorespiratory Fitness, and the Metabolic Syndrome. Medicine & Science In Sports & Exercise, 35(8), 1279-1286.
2. Mendis, S., P. Puska and B. Norrving, (2011). Global atlas on cardiovascular disease prevention and control. World Health Organization/World Heart Federation and the World Stroke Organization, Geneva.
3. Salamudin, N., & Harun, M. (2013). Physical Activity Index among Malaysian Youth. Asian Social Science, 9(12).
4. Lee, D., Artero, E., Xuemei Sui, & Blair, S. (2010). Review: Mortality trends in the general population: the importance of cardiorespiratory fitness. Journal of Psychopharmacology, 24(4Suppl), 27-35.
5. Garber CE, Blissmer B, Deschenes MR, Franklin BA, Lamonte MJ, Lee IM, Nieman DC and Swain DP. (2011). American College of Sports Medicine position stand. Quantity and quality of exercise for developing and maintaining cardiorespiratory, musculoskeletal, and neuromotor fitness in apparently healthy adults: guidance for prescribing exercise. Med Sci Sports Exerc 43: 1334-1359.
6. Teh KC, Aziz AR. (2002) Heart rate, oxygen uptake, and energy cost of ascending and descending the stairs. Med Sci Sports Exerc; 34:695–699
7. Boreham, C., Kennedy RA, Murphy MH, Tully M, Young IS, Wallace WFM. (2005). Training effects of short bouts of stair climbing on cardiorespiratory fitness, blood lipids, and homocysteine in sedentary young women. British Journal of Sports Medicine, 39(9), 590-593.
8. Kennedy RA, Colin A.G. Boreham, Marie H. Murphy, Ian S. Young, Nanette Mutrie, (2007) Evaluating the Effects of A Low Volume Stairclimbing Programme on Measures of Health-Related Fitness in Sedentary Office Workers. Journal of Sports Science and Medicine (06), 448 – 454
9. Meyer P, Kayser B, Kossovsky MP, Sigaud P, Carballo D, Keller PF, Martin XE, Farpour-Lambert N, Pichard C and Mach F. (2010). Stair instead of elevators at workplace: cardioprotective effects of a pragmatic intervention. Eur J Cardio Prev Rehab 17: 569-575
10. Cohen, SM. (2013). Examining the Effects of a Health Promotion Intervention on the Use of Stairs. Journal of Articles In Support Of the Null Hypothesis, 10(1), 17-24.
11. American College of Sports Medicine (2010) ACSM's Guidelines for Exercise Testing and Prescription. Philadelphia, PA: Lippincott Williams & Wilkins
12. WHO (2015). WHO | Physical activity. Retrieved 14 December 2015, from http://www.who.int/mediacentre/factsheets/fs385/en/

13. Craig, C., Marshall, A., Sjostrom , M., Bauman, A., Booth, M., & Ainsworth, B. et Al. (2003). International Physical Activity Questionnaire: 12-Country Reliability and Validity. Medicine & Science In Sports & Exercise, 35(8), 1381-1395. 14. Dinger, M., Behrens, T., & Han, J. (2006). Validity and Reliability of the International Physical Activity Questionnaire in College Students. American Journal Of Health Education, 37(6), 337-343.
15. Tan HYF, Aziz AR, Teh KC, Chia YHM. (2004) Reliability of the stair-climb test (SCT) of cardiorespiratory fitness. Adv Exerc Sports Physiol 10:77-83.
16. Ayala, F., De Ste Croix, M., Sainz de Baranda, P., & Santonja, F. (2015) Acute effects of two different stretching techniques on isokinetic strength and power. Revista Andaluza De Medicina Del Deporte, 8(3), 93-102.
17. Mankar, S. R., Shenvi, D. N., Hegde, G. R. & Kaurani, V. G. (2015) Promotion of Stair Climbing Exercise in Medical Students to Achieve Physical Fitness with Student's Normal Routine.. International Journal of Health Sciences and Research (IJHSR), 5 (9), 254-259
18. Gupta, R., Chaudhuri, A., Bansod, N., & Gupta, R. (2015). Effects of exercise training on correlation of VO2 max and anthropometric parameters, physical fitness index in young adult Indian males. Saudi J Sports Med, 15(2), 153.
19. Bajaj,A, S. Appadoo, S, Bector, CR, Chandra,S. (2008) Measuring physical fitness and cardiovascular efficiency using harvard step test approach under fuzzy environment
20. Astrand, P. (2003). Textbook of work physiology. Physiological bases of exercise. 4.ed.. Champaign: Human Kinetics.
21. Brown, Stanley P & Miller, Wayne C & Eason, Jane M (2006). Exercise physiology : basis of human movement in health and disease. Lippincott Williams & Wilkins, Philadelphia
22. Werner W. K. Hoeger & Hoeger, Sharon A (2011). Lifetime physical fitness and wellness a personalized program (11th ed). Australia Wadsworth, Cengage Learning
23. Kenney, W. Larry & Wilmore, Jack H., 1938- & Costill, David L & Wilmore, Jack H., 1938-. Physiology of sport and exercise (2012). Physiology of sport and exercise (5th ed). Champaign, Ill. Human Kinetics
24. Duncan J.J., Gordon N.F., Scott C.B. (1991) Women walking for health and fitness - how much is enough? JAMA 266, 3295-3299
25. Abernethy, B. (1997). The biophysical foundations of human movement. Champaign, IL: Human Kinetics.
26. Das B, Ghosh T, Gangopadhyay A. Comparative study of Physical Fitness Index (PFI) and predicted maximum aerobic capacity (VO2 max) among the different groups of female students in West Bengal, India. Int J Appl Sports Sci 2010;22:13-23.
27. Das KK, Dhundasi SA. Physical fitness: A longitudinal study among Muslim children of Bijapur (Karnataka). Indian J Physiol Pharmacol 2001;45:457-62.
28. Cho, L (2011). Heart Rate Recovery Can Be Improved with Exercise - Heart Advisor Article - The Cleveland Clinic Heart Advisor. http://www.heartadvisor.com/issues/14_12/features/Heart-Rate Recovery_9901.html#.Vm7TQtJ97Mx
29. Voulgari, C., Pagoni, S., Vinik, A., & Poirier, P. (2012). Exercise improves cardiac autonomic function in obesity and diabetes. Metabolism clinical and experimental, 62, 609 – 621.
30. Prado, D., M., Silva, A., G., Trombetta, I., C., Ribeiro, M., M., Guazzelli, I., C., ..., Villares, S., M. (2010). Exercise training associated with diet improves heart rate recovery and cardiac autonomic nervous system activity in obese children. International Journal of Sports Medicine, 21(12), 860-865.

Author:	Imtiyaz Ali Mir
Institute:	Department of Physiotherapy, Faculty of Medicine & Health Sciences, Universiti Tunku Abdul Rahman.
Street:	Sungai Long
City:	Kajang
Country:	Malaysia
Email:	imtiyaz@utar.edu.my

Moderation Role of Attitude on the Relationship between Participation in Competitive Sports and Academic Performance of Student-Athletes in Saudi Arabia

M.I. Alahmed[1], A. Yusof[2], A. Saidon[2], A. Borhannudin[2] and K. Prihadi[3]

[1] King Faisal University/Department of Physical Education, Al-Ahsa, Kingdom of Saudi Arabia
[2] University Putra Malaysia / Department of Sport Studies, Selangor, Malaysia
[3] Universiti Sains Malaysia / School of Educational Studies, Penang, Malaysia

Abstract— **Background: This study aims to investigate how the attitude of undergraduate student-athletes moderates the relationship between competitive sports participation (CSP) and academic performance (AP). In this study, a type of attitudes was taken into consideration namely academic performance attitude (APT). Methods: A total of 102 undergraduate student athletes from 3 academic institutions in the Eastern Region of Kingdom of Saudi Arabia (KSA) were selected as subjects. Instruments used were Survey of Study Habits & Attitudes (SSHA) by Brown and Holtzman, to obtain the data on APT. The instrument was back-translated to the Arabic language in order to improve its comprehensiveness. Data on sports participation and academic performance were obtained by directly enquire the participants about their grade point average (GPA) and the frequency of their CSP in a self-developed questionnaire. Results: The results showed that APT significantly alters the effect of CSP on AP, because the β value of the effect of CSP on AP was greater than the β value of the effect of CSP and APT on AP. Because β value shows the strength of the influence, it can be concluded that the actual strength of the CSP is not as strong as when it is combined with APT. In other words, APT moderates the influence of CSP on AP. Discussion: The moderation effect occurs due to the regulation of Saudi Universities Sports Federation (SUSF) that only student-athletes with required academic scores can participate in competitive sports. Without such a regulation, APT might not be correlated to the AP because some students might pay more focus on CSP and less on AP. It can be concluded that the regulation helped student-athletes in KSA to improve their AP without sacrificing their CSP and vice versa.**

Keywords— **Sports Participation, academic performance, Competitive Sports, Moderation, Attitude.**

I. Introduction

Recently, university sports are becoming the common part of students' life [3]. In the context of the Kingdom of Saudi Arabia (KSA), this matter is handled by Saudi University Sports Federation (SUSF), who aims to improve the number and level of students' competitive sports participation (CSP), without jeopardizing the academic performance (AP) of the student-athletes. However, university students in (KSA) showed the inconsequential percentage of participation in competitive sports events [2].

With aims to keep the balance between AP and CSP among student-athletes, SUSF produced a regulation that a student athlete has to report good AP in order to participate in competitive sport events [2]. Studies behind the relationship between AP and CSP had been done for long, nonetheless, they reported inconsistence results; for instance, some studies reported negative influence of CSP on the students' AP [3; 4]. On the other hand, other studies also reported that CSP positively influences the students' AP [5; 6; 7]. It was even reported that there is no significant correlation between CSP and AP among student-athletes [8]. It is sensible to note that none of those studies have been done in KSA. Thus, in order to determine whether the SUSF's regulation can help them to achieve the balance between CSP and AP, it is significant to study any common elements between the variables above that might play some moderating roles on the relationship between CSP and AP among student-athletes in KSA. In order to do so, it is important to pay attention to the factors that affects CSP and AP.

In the context of university student-athletes, some factors or elements have been reported to predict CSP, such as perceived supports from the parents [9], availability of sports facility in the university [10], motivation [11], socio-economic status [12], perceived injury risk [13], psychosocial health [14], attitude [15] promotions from management of sports institutions [16], and many more.On the other hand, AP among student-athletes have also been studied and the knowledge yielded from the studies were that it has its factors such as health behavior [17], high-arts participation and reading habits [18], parental socialization [19], attitude [15] and many more.

Some of the literatures above advocated that academic performance attitude (APT) is a factor that might alter the effect of student-athletes on their AP (for example: [17; 18; 19]. Theoretically, attitude can be affected by three attitudinal bases; affective, cognitive, and behavior [15; 20]. In the context of this study, APT can be explained by all the three bases, (1) a student-athlete might feel good when he

F. Ibrahim et al. (eds.), *3rd International Conference on Movement, Health and Exercise*,
IFMBE Proceedings 58,
DOI: 10.1007/978-981-10-3737-5_5

score high in an exam, therefore he develops positive APT, (2) a student-athlete might think that scoring high in an exam will be good for his future, and he can also participate in sport competition. Therefore he develops positive APT, (3) a student-athlete might develop positive APT because he used to score high in the exam. In other words, it is safe to hypothesize that APT significantly predicts AP.

As a conclusion of the introduction, it can be said that this current study will rely on the theory of attitude and some previous studies which support an argument that CSP might affect the AP among student-athletes in KSA, and the effect is moderated by APT. Thereby, the possibility of having a balanced CSP and AP might be achieved by altering the APT; hence the significance of this current study is to discover a moderating role of APT on the effect of CSP on AP. The obtained knowledge is expected to serve as a foundation for further research in improving APT in order to keep CSP and AP in balance. The overarching aim of this present study is to investigate the moderating effect of APT on the influence of CSP on AP. In order to develop the hypotheses, some literature have been reviewed and discussed in the next subsections.

II. Interrelationship between CSP and AP

It is universal for every university, including in KSA to put AP as the main goal, and hence pay more consideration to any strategies, actions, and activities related to the enhancement of AP [18]. Therefore, if CSP is seen as a positive predictor of AP, it is more likely for the universities to endow with better infrastructure to improve CSP among students, such as promoting parental socialization related to their children's CSP and AP [19], or promoting healthy lifestyle and behavior [17]. The relationship between CSP and AP has been reported some studies; [21] found that students' CSP is associated with 5 percent increase in Bachelor's degree attainment expectations, while [22] reported that their the data support the link between physical activity, cognitive function, and AP. At the same tone, research by [5] yielded a result that physical activity, including competitive sports numbers, has positive prediction on academic results.

Furthermore, [23] acclaimed that certain kinds of CSP mediate the relationship between the availability of sports activity and academic achievements, which means that CSP predicts AP. Accordingly, CSP can also be viewed as an indicator of several factors such as self-confidence, teamwork ability, and the ability to succeed despite high pressure in competitive situations [21], which can be seen as an important determiner for AP. CSP has also been reported as positive predictors of overall well-being of student-athletes due to its positive influence on factors such as general academic achievements [21], homework completion [24], socialization levels [25], social health [14], as well as the reduce the illegal behaviour frequency [26].

On the other hand, some studies had also reported no positive relationship between CSP and AP. [3] reported the limited influence of CSP on AP. Similarly, another study [8], reported no significant relationship between AP and CSP. At the extreme end, another finding indicated that CSP has a negative prediction on AP for certain ethnic groups [4]. Accordingly, some society members developed the idea that CSP might contribute to negative developments of student-athletes' life, such as violence [27], and drug abuse [28]. Despite the inconsistency of the finding, it was confirmed that CSP positively predicts the constructs related to AP. The confirmation was derived from the results of a meta-analysis study by [5] from Department of Exercise Science, University of South Carolina.

Inconsistency stated in the previous two paragraphs represent the problems around the thoughts of keeping the balance between positive AP and positive CSP, as aimed by SUSF in the context of KSA [2]; hence, deeper understanding of the interrelationship between AP and CSP, along with its elements. Moreover, consistent with the theory of attitudinal base [20], and studies which reported that APT (or can also be called attitude towards academic performance) is a common element between AP and CSP, thus, further research is highly required to see how altering APT in order to maintain the positive balance between CSP and AP among student-athletes in KSA.

III. Attitude

Attitude refers to a positive or negative evaluation of people, objects, event, activities, ideas, or just about anything in an individual's environment, and it can be formed from a person's past and present [29]. The definition was renewed by [30] as a psychological tendency that is expressed by evaluating a particular entity with some degree of favour or disfavour. Attitude can also be defined as an evaluation of individuals on other individuals, events, or phenomenon around them based on three attitudinal bases: affective, cognitive, and behavioral [20]. Cognitive attitudinal base refers to logical reason in evaluating an object, for example to like a subject due to its financial benefits; affective attitudinal base refers to the involvement of emotions in evaluating an object, for instance to like some individuals due to their good looks; at last, behavioural attitudinal base refers to the evaluation of objects due to habituation process, such as preferring certain food during certain time due to the cultural or habitual reason [15]. In this current study's context, APT or attitude towards academic performance can be formed cognitively, such as how significant it is to have high academic performance; affectively, how 'cool' it feels when the individuals are active in sports, yet having good academic performance; or behavioural, where at the beginning, scoring high AP was seen as a way to get involved in competitive sports, but when scoring high AP had become a habit, the individuals will get used to studying hard and attempt to score high AP every year.

The distinctive feature of this study is that it is conducted in the KSA, where the regulation of SUSF allows only student-athletes with required AP to competitively participate in sports events [2]. The regulation itself plays an important role in linking CSP to AP; students who want to be an athlete (developing high CSP) must have a positive APT, and at the end, their positive APT predicts their AP. Therefore, the regulation might contribute to the cognitive attitudinal base towards the AP. Theory of operant conditioning [31] supports the statements in the previous paragraph. The theory mentioned that a behavior that paired with positive reinforcement (rewards) would likely to be repeated and getting better in quality. In the context of this current study, high AP (behavior) is paired with CSP (rewards), hence the more the students want to participate in competitive sports, the more positive their APT because the AP will reward them with what they want (more involvement in competitive sports).

IV. Interrelationship among Variables

The previous subsections had discussed that there is a link involving CSP and AP, including the one where it involves APT. The attitudinal base theory [20] had suggested that achieving high AP frequently can be a behaviour attitudinal base that contributes to the next achievement of high AP, which rewards the students with the allowance of CSP and more positive APT.

At the same tone, Skinner's operant conditioning theory [31] supports the fact that CSP is a positive reward that is paired with high AP (due to the SUSF regulation). Therefore, student-athletes will be more determined to get high AP in order to be able to participate in competitive sports. Additionally, APT or general academic attitude is formed by the individuals without any influence of CSP, or it might be contributed by other external factors, such as the regulation of SUSF.

The connection among variables in this study led to a hypothetical statement that APT moderate the relationship between CSP and AP. Theoretically, in the context of this study, the criterion (AP) is affected by both the predictor (CSP) and moderator (APT), as illustrated in Figure 1.

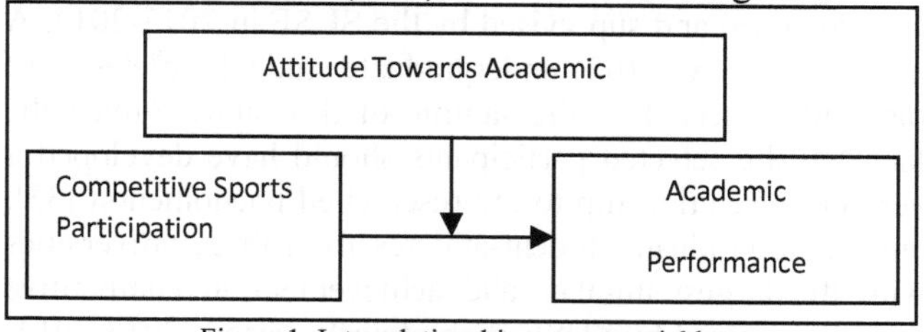

Figure 1: Interrelationship among variables.

Interrelationship among variables in this current study is captured in Figure 1. It was hypothesized that CSP predicts the AP, while APT plays a moderating role that alters the prediction above. Additionally, supported by the operant conditioning theory [31] and attitudinal theory [20], as well as many studies, the influence of CSP on AP can be conceptualized. Moreover, the regulation of SUSF led to the situation where APT can be added to the picture.

Interconnection among variables in this study, where the influence of CSP on AP might be altered by APT, met the conditions of moderation effect (Baron & Kenny, 1986); APT is the moderator between the CSP and AP. While the concept was drawn theoretically, it is required to obtain some data in order to understand the significance of the influence and the beta value (power) of the influence of the variables above. Therefore, data was collected and analyzedin order to test the hypotheses "There is no moderation effect of APT on the causal relationship between CSP and AP".

V. Gap of Literature

There are two gaps in literature that will be filled by this paper. First, the attitude towards the academic performance of the student-athletes in this study might be affected by the SUSF regulations, which might be the moderation of APT more significant than in other settings where high AP is not required for student-athletes to have active CSP. The second gap to cover is that while the causal relationship between CSP and AP has been commonly studied, there has not been any research investigating the moderating effect of APT on their relationship.

VI. Methods

A. Population and Sampling

Samples of the student-athletes were taken from the eastern region of KSA. The area consists of 3 government universities, two private universities and one private college participated in the fourth season of competitive sports which are organized and supervised by the SUSF in 2013-2014. A hundred and two (102) undergraduate student-athletes were randomly selected as the sample of this study. Given the fact that the selected participants should have developed a very close relationship to the researched phenomenon [33], only undergraduate student-athletes from three universities with the largest number and achievement in competitive sports organized by SUSF in the fourth season (2013-2014) were recruited as samples. According to the statistical data from SUSF, three universities; King Faisal University, Prince Mohammad bin Fahd University and Dammam University have the highest number of participant and highest achievement in the sports event 2013-2014 [34].

B. Instruments

A questionnaire that consisted of four sections was administered for data collection of the current study. In the context of this study, CSP is defined as frequency of the students' participation in sports activities in a week [10] and whether the students are active members of sport university teams [11] or sport clubs [12]; therefore, the three studies have been utilized in order to develop the instrument to measure CSP for this current study. Academic performance (AP) is measured by the formal academic report from the student-athletes' respective university. Finally, Survey of Study Habits & Attitudes (SSHA) by [35] was adapted to measure APT in this current study. Translations to Arabic have been done by [36]. The instrument was back-translated to Arabic due to the objectives of this current study. The following subsections discuss the instrument in detail.

The first section targeted at demographic data that was prepared to get information on academic performance (GPA value), academic institution, faculty, department, education level, name and ID. Most of the universities and colleges in Saudi Arabia are very similar to the United States except the way the grades are said. Participants were asked about their AP scores in order to collect the data. Therefore, the scores can be analysed by using statistical software such as SPSS.

The second section targeted at data of competitive sports participation (CSP). In the context of this study, CSP is defined as frequency of the students' participation in sports activities in a week [10] and whether the students are active members of sport university teams [11] or sport clubs [9].

The third section is the "Survey of Study Habits & Attitudes (SSHA)" by [35], which was adapted for this study. Translations to Arabic has done by [36]. This scale is used to measure some of the studying habits and attitudes toward study among the students of colleges and institutes of higher education, and high school students. In addition, the scale consists of six dimensions; "delay avoidance, work methods, teacher approval, education acceptance, study habits and study attitudes" and consisted of 100 phrases. Validity obtained ranged from .27 to .66 for men and from .26 to .65 for women using one-semester grades as a criterion. In the questionnaire texts, it is written that the purpose of this survey is to furnish an inventory of study habits and attitudes to serve as a foundation for self-improvement. Furthermore, the reliability of the original SSHA was reported to be .93, .91, .88, and .90, respectively for the four week interval of test-retest coefficient. The scale has been adapted into Likert scale in order to fit the requirements of this current study.

VII. Results

Table 1 indicates the result of the linear regression analysis with AP as the criterion and CSP as the predictor

Table 1 Influence of CSP on AP

Model		Unstandardized Coefficients		Standardized Coefficients		
		B	Std. Error	Beta	t	Sig.
1	(Constant)	.860	.273		3.153	.002
	TOTAL CSP	.255	.025	.717	10.273	.000

a. Dependent Variable: AP

Table 1 indicates the significance of CSP in predicting AP in the form of GPA. The influence is positive and significant Beta=.717(.025,.273);p<.00 which means that when the CSP can explain about 72% of the variance of the AP among the sample.

Table 2 indicates the result of the linear regression analysis with AP as the criterion and APT as the predictor

Table 2 Influence of APT on AP

Model		Unstandardized Coefficients		Standardized Coefficients		
		B	Std. Error	Beta	t	Sig.
1	(Constant)	2.077	.276		7.530	.000
	TOTAL APT	.010	.002	.498	5.745	.000

a. Dependent Variable: AP

Table 2 indicates the significance of APT in predicting AP in the form of GPA. The influence is positive and significant Beta=.498(.002,.276);p<.00 which means that when the APT can explain almost50% of the variance of the AP among the sample.

Finally, to determine the moderation effect of a moderator, a causal relationship between the moderator variable and the dependent variable should be proven significant. Table 3 indicates the influence of CSP on AP through APT.

Table 3 Influence of CSP on AP through APT

Model		Unstandardized Coefficients B	Std. Error	Standardized Coefficients Beta	t	Sig.
1	(Constant)	.860	.273		3.153	.002
	CSP	.255	.025	.717	10.273	.000
2	(Constant)	.549	.284		1.937	.056
	CSP	.220	.027	.617	8.201	.000
	APT	.005	.002	.221	2.931	.004

a. Dependent Variable: AP

In Table 3, it is shown that the Beta value of the CSP with APT (.617) is smaller than the one without (.717). In other words, without APT, the CSP explained 71.7% of the AP variance; with APT included, the proportion of explained variance decreased to 61.7%. Thus, because the Beta value in the equation 3 showed smaller value than in the equation 1, it can be considered that moderation does occur. Therefore, the null hypothesis is rejected. In other words, when the student athletes develop a positive attitude towards academic performance, the level of their participation in the sport competition will affect their academic scores.

VIII. Discussions and Recommendations

Finding of this current study indicates the moderating role of APT on the influence of CSP on AP. In other words, the more the subjects participate in competitive sports, the higher their academic performance is; especially when they have a positive attitude towards the academic performance. Theoretically, this phenomenon might also be supported by the regulation from SUSF, where only students with required AP can participate in the competitive sports.

Findings of this paper confirmed the reports of [21] and the Meta-analysis study of [5], that CSP has a positive influence on AP. Findings of this paper do not support the findings of [3], who stated that CSP has limited influence on AP, [8], who reported that there is no significant relationship between AP and CSP, as well as the report of [4], that CSP has negative prediction of AP.

In the context of this current study, the student-athletes' willingness to study and to achieve higher AP can be considered as positive APT. Thus, it can be assumed that because they had already paired AP and CSP in the minds, hence they are ready to 'pay the price' in order to participate in competitive sports. When the SUSF released their regulations, this willingness to participate in competitive sports was connected to their effort in achieving high AP (positive APT). Nonetheless, because the moderation seemed to occur due to the SUSF regulation, it is worthwhile to question whether without the regulation, findings of this current study will support the finding of [8] that there is no correlation between CSP and AP.

Accordingly, it is imperative to bear in mind that the results have yielded from the data that was collected in KSA, where the SUSF regulation is applied. Therefore, same results might not be yielded from different data taken from different locations and demographics. It is also significant to remember that the data was only taken from several universities in KSA, hence different results might be obtained from the larger sample within the population of student-athletes in KSA. It is recommended to the future researchers to look into the variables which are not included in this current research, for instance, the involvement of female athletes, the participation in recreational sports, or even the attitude towards the AP itself. However, it is expected that the finding of this current study can be taken as a foundation for further researchers to improve the sport performances or AP among student-athletes in KSA.

References

1. Mozafari, S.A.A. (2002). A study of attitudes of people toward sport and physical activities in Iran. Unpublished research project, TarbiatMoallem University.
2. Saudi University Sports Federation (2010) Profile and Regulation. Ministry of higher education: Riyadh.
3. Rees, D. I., &Sabia, J. J. (2010). Sports participation and academic performance: Evidence from the National Longitudinal Study of Adolescent Health. Economics of Education Review, 29(5), 751-759.
4. Zeiser, K. L. (2011). Examining racial differences in the effect of popular sports participation on academic achievement. Social Science Research,40(4), 1142-1169.
5. Howie, E. K., & Pate, R. R. (2012). Physical activity and academic achievement in children: A historical perspective. Journal of Sport and Health Science.
6. Lambourne, K., & Donnelly, J. E. (2011). The role of physical activity in pediatric obesity. Pediatric Clinics of North America, 58(6), 1481-1491.
7. Rasberry, C. N., Lee, S. M., Robin, L., Laris, B. A., Russell, L. A., Coyle, K. K., &Nihiser, A. J. (2011). The association between school-based physical activity, including physical education, and academic performance: a systematic review of the literature. Preventive Medicine, 52, S10-S20.
8. Eide, E. R., & Ronan, N. (2001). Is participation in high school athletics an investment or a consumption good?: Evidence from high school and beyond.Economics of Education Review, 20(5), 431-442.
9. Dollman, J., & Lewis, N. R. (2009). Interactions of socioeconomic position with psychosocial and environmental correlates of children's physical activity: an observational study of South Australian families. International Journal of Behavioral Nutrition and Physical Activity, 6(1), 56.
10. Wicker, P., Hallmann, K., & Breuer, C. (2013). Analyzing the impact of sport infrastructure on sport participation using geo-

coded data: Evidence from multi-level models. Sport Management Review, 16(1), 54-67.
11. Lindner, K. J., & Kerr, J. H. (2000). Metamotivational orientations in sport participants and non-participants. Psychology of Sport and Exercise, 1(1), 7-25.
12. Dollman, J., & Lewis, N. R. (2010). The impact of socioeconomic position on sport participation among South Australian youth. Journal of Science and Medicine in Sport, 13(3), 318-322.
13. Siesmaa, E. J., Blitvich, J. D., White, P. E., & Finch, C. F. (2011). Measuring children's self-reported sport participation, risk perception and injury history: Development and validation of a survey instrument. Journal of Science and Medicine in Sport, 14(1), 22-26.
14. Belanger, L. J., Plotnikoff, R. C., Clark, A. M., &Courneya, K. S. (2012, May). Sports Participation in Young Adult Cancer Survivors. In medicine and science in sports and exercise (Vol. 44, pp. 510-510). 530 Walnut St, Philadelphia, PA 19106-3621 USA: Lippincott Williams & Wilkins.
15. Aronson, E., Wilson, T., & Akert, R. (2005). Social Psychology, Fifth Edition. New Jersey, NY: Pearson Education, Inc.
16. Rowe, K., Shilbury, D., Ferkins, L., &Hinckson, E. (2013). Sport development and physical activity promotion: An integrated model to enhance collaboration and understanding. Sport Management Review, 16(3), 364-377.
17. Bradley, B. J., & Greene, A. C. (2013). Do Health and Education Agencies in the United States Share Responsibility for Academic Achievement and Health? A Review of 25 Years of Evidence About the Relationship of Adolescents' Academic Achievement and Health Behaviors. Journal of Adolescent Health,52(5), 523-532.
18. Gaddis, S. M. (2013). The influence of habitus in the relationship between cultural capital and academic achievement. Social science research, 42(1), 1-13.
19. Strambler, M. J., Linke, L. H., & Ward, N. L. (2013). Academic identification as a mediator of the relationship between parental socialization and academic achievement. Contemporary Educational Psychology, 38(1), 99-106.
20. McGuire, W. J. (1969). The nature of attitudes and attitude change. In G. Lindzey, & E. Aronson, The Handbook of Social Psychology (pp. 136-314). Reading, MA: Addison-Wesley Theory of attitude.
21. Lipscomb, S. (2007). Secondary school extracurricular involvement and academic achievement: A fixed effects approach. Economics of Education Review, 26(4), 463-472.
22. Donnelly, J. E., &Lambourne, K. (2011). Classroom-based physical activity, cognition, and academic achievement. Preventive Medicine, 52, S36-S42.
23. Stearns, E., &Glennie, E. J. (2010). Opportunities to participate: Extracurricular activities' distribution across and academic correlates in high schools. Social Science Research, 39(2), 296-309.
24. Bembenutty, H., & White, M. C. (2013). Academic performance and satisfaction with homework completion among college students. Learning and Individual Differences, 24, 83-88.
25. Devecioglu, S., Sahan, H., Yildiz, M., Tekin, M., &Sim, H. (2012). Examination of socialization levels of university students engaging in individual and team sports. Procedia-Social and Behavioral Sciences, 46, 326-330.
26. Davis, B. S., & Menard, S. (2013). Long term impact of youth sports participation on illegal behavior. The Social Science Journal, 50(1), 34-44.
27. Sønderlund, A. L., O'Brien, K., Kremer, P., Rowland, B., De Groot, F., Staiger, P., ... & Miller, P. G. (2014). The association between sports participation, alcohol use and aggression and violence: A systematic review.Journal of science and medicine in sport, 17(1), 2-7.
28. Smith, A. C., Stewart, B., Oliver-Bennetts, S., McDonald, S., Ingerson, L., Anderson, A., ... &Graetz, F. (2010). Contextual influences and athlete attitudes to drugs in sport. Sport management review, 13(3), 181-197.
29. Allport, G. (1935). Attitudes. In C. Murchison, A Handbook of Social Psychology (pp. 789–844). Worcester, MA: Clark University Press.
30. Eagly, A. H., & Chaiken, S. (1998). Attitude Structure and Function. In D. Gilbert, S. T. Fisk, & G. Lindsey, In Handbook of Social Psychology (pp. 269–322). New York: McGowan-Hill.
31. Skinner, B. F. (1953). Science and Human Behavior. New York: MacMillan.
32. Baron, R. M., & Kenny, D. A. (1986). The Moderator-Mediator Variable Distinction in Social Psychological Research: Conceptual, Strategic, and Statistical Considerations. Journal of Personality and Social Psychology, 51 (6), 1173-1182.
33. Woods, T. (1999). Beginning postmodernism. Manchester University Press.
34. Saudi University Sports Federation (2014) Profile and Regulation. Ministry of higher education: Riyadh.
35. Brown, W. F., & Holtzman, W. H. (1967). Survey of Study Habits and Attitudes: SSHA; Forms C and H; Manual. Psychological Corporation.
36. Jaber, A. J.,&Al Sheikh, S. A. (1979). Instructions brochure for scale of study habits and attitudes toward study. Cairo: House of Arab Renaissance.

Author: Mustafa Al-Ahmad
Institute: Department of Physical Education, King Faisal University
City: Al-Ahsa
Country: Kingdom of Saudi Arabia
Email: alahmed.m@hotmail.com

The Instantaneous Effects of Generic Foot Arch Support Insoles on Centre of Foot Pressure in Various Modes of Static Balance

C. L. Chok and S. Joseph

Centre of Sports Biomechanics, National Sports Institute of Malaysia, Kuala Lumpur, Malaysia

Abstract— **The purpose of this study was to investigate the effects of generic foot arch support insoles on the center of foot pressure in Double-Legs-Stance, Single-Right-Leg-Stance, Single-Left-Leg-Stance, Right-Tandem-Stance and Left-Tandem-Stance. Ten healthy adults (male=5, female=5) were recruited in this study. Subjects were tested with no insole (Insole-A) and three different design of insoles (Insole-B, C and D) in each type of standing mode. A pressure mapping sensor (25Hz) with 25 sensel per square inch spatial resolution and pressure measurement system software were used to collect the center of pressure movements. A non-parametric Friedman test with Wilcoxon singed-ranks post-hoc test were used to analyze the significant effects of the insole. There was significant differences in anteroposterior travelled distance across insole support types, $\chi^2(3, N=10) = 9.120$, $p < 0.05$ for Double-Leg-Stance. However post-hoc test indicates that no significant difference in between any paired of insoles. Significant differences were found in anteroposterior peak velocity across insole support types for Single-Right-Leg-Stance, $\chi^2(3, N=10) = 8.567$, $p < 0.05$. Post-hoc test indicates that there is a significant difference in anteroposterior peak velocity for the Insole-C and Insole-B, $z = -2.550$, $p < 0.05$, with lower peak velocity of center of pressure shift in Insole-C. There were no significant differences found in Single-Left-Leg-Stance and Right-Tandem-Stance across all types of insole support. The results of the Friedman test indicate that no significant differences exist in Right-Tandem-Stance across all types of insole sup-port. There were significant differences in mediolateral range, mediolateral travelled distance and mediolateral position standard deviation across insole support types with chi square, $\chi^2(3, N=10) = 9.0$, $\chi^2(3, N=10) = 9.12$, $\chi^2(3, N=10) = 8.04$, $p < 0.05$ respectively for Left-Tandem-Stance. This study concludes that the generic foot arch support insoles have no significant positive instantaneous effects on center of pressure measurement.**

Keywords— **Balance, Body Sway, Centre of Pressure, Insole, Stability.**

I. Introduction

Balance is to maintain the center of gravity of the body over its base of support. It required a continuous sensory feedback from the proprioceptive, vestibular and visual system to evaluate the body's balance state. Neuromuscular system generates appropriate responses to the feedback inputs in order to maintain the body balance [1]. In past decades, studies show the ability of dynamic and static balance was significantly associated to the risk of injury especially in elderly population. In additional, studies also show that the balance ability has a significant correlation to the sports related injury [2]–[6].

Static balance is the ability to maintain the center of gravity over the base support with minimal body movement. The most typical assessment for static balance in laboratory setup is analyzing the center of pressure (CoP) movement within a specific duration on a force platform. The common CoP variables reported in past studies were the range of CoP shifted, total distance travelled by the CoP and velocity of CoP in a certain time frame. It is logically accepted that with a minimal CoP movement is revealing good balance ability. Accelerometry method was also used in past studies to measure the body sway patterns as an indication for the balance ability [7], [8]. In field static balance ability is assessed through the timed single-leg-stance test. However, there are several variation of testing methods/protocols exist in field. The most predominant methods could be the Stork balance test and Flamingo balance test [9]. For all the existing tests, the most common testing duration used for balance tests is from 10 to 30 seconds [10]–[16]. Regardless of laboratory or field test, balance test was widely used in skill precision sports to assess athletes' level which has significant correlation associated to sports performance, especially in archery and shooting [17]–[19]. Study also has shown that balance ability was associated to the maximum skating speed performance in ice hockey players [20].

Gymnastics sport requires balance ability to the utmost in comparison to other sports. Davlin (2004) conducted a study to compare the dynamics balance performance in between elite gymnasts, elite swimmers, elite footballers and non-athlete group. Result show that gymnast group outperformed all the other groups. Besides, Davlin also reported that moderate to high negative correlations were found between dynamic balance performance and height and weight [15]. However, there was no difference in dynamics balance between collegiate gymnasts and collegiate footballers [21]. Bressel et al. (2007) also reported that collegiate gymnast was outperformed by collegiate basketballer's in static balance. In additional, collegiate footballer had superior dynamics balance than collegiate basketball player [21]. Matsuda, Demura, & Uchiyama (2008) reported a similar finding where the footballer had exhibited lower sway velocity, lower

F. Ibrahim et al. (eds.), *3rd International Conference on Movement, Health and Exercise*,
IFMBE Proceedings 58,
DOI: 10.1007/978-981-10-3737-5_6

anterior-posterior sway and lower horizontal sway than the basketball players [22].

Comparison of balance ability in different levels of competition athletes within the same sports have been widely studied. Paillard et al. (2006) compared the postural performance between soccer players at different levels of competition. Their result shows that the CoP surface area and CoP velocity were significantly smaller in national player compare to regional player [23]. Sell, Tsai, Smoliga, Myers, & Lephart (2007) reported a high proficiency golfer group (handicap index < 0) had significantly greater single leg balance than low proficiency golfer group (handicap index = 10-20) [24]. In air rifle shooting, the balance ability test shows experienced rifle shooters were significantly better than less experienced rifle shooters [25]–[27]. Conversely, Chapman, Needham, Allison, Lay, & Edwards (2008) reported there was no significant difference between groups of elite and recreational surface in area of 95th centile ellipse and sway path length. In addition, Paillard, Costes-Salon, Lafont, & Dupui (2002) also reported that there was no significant difference between international Judo player and regional Judo players in postural balance performance [28]. These implied that balance ability of elite athletes is being likely to outperform than the non-elite athletes could be based on sports specificity.

In the recent years, footwear technology and intervention in elderly population has received a high attention. Perhaps the footwear aids can reduce the risk of fall injury, improve movement and balance mechanics in elderly population. Gross, Mercer, & Lin (2012) investigated the effects of two weeks custom-moulded orthotics intervention on 13 elderly participants. They reported that there were significant greater stance times in single-leg stance and tandem stance [29]. Palluel, Olivier, & Nougier (2009) studied the effects of after five minute standing/walking with spike insole on 19 elderly people. Results show significant improvement in CoP surface area and root mean square of anterior-posterior axes in eyes closed double-leg stance condition [30]. Wang & Yang (2012) reported there was a significant reduction of de-trended fluctuation analysis scaling exponent in anterior-posterior axes after 10 minutes of walk with vibrating insole [31]. Qiu et al. (2012) assessed the immediate sway effects of hard and soft textured surface insoles in elderly adults. They concluded that both textured surface insoles were able to reduce postural sway of elderly people on a foam surface with eyes closed condition [14]. Conversely, Hatton, Dixon, Rome, Newton, & Martin (2012) reported that use of textured insoles have no significant effects on anterior-posterior or medial-lateral standard deviation and CoP velocity with eyes opened and closed conditions in elderly female population [13].

Despite of positive or no effects found in various types of footwear insole, limited studies have been conducted to discover the instantaneous effects of generic foot arch support insole. Therefore, present study was to investigate the instantaneous effects of generic foot arch support insole on center of foot pressure in various types of static balance stance. The null hypothesis was there is no significant difference in the center of pressure movement between without use of foot arch support insole and with use of generic foot support insole. The alternatives hypothesis was defined as there is significant difference in the center of pressure movement between without used of foot arch support insole and with used of generic foot support insole.

II. Methodology

A. Subjects

Ten healthy volunteer adults (age: 23.4 ± 1.9 years, height: 1.63 ± 0.09 m, weight: 63.1 ± 15.9 kg) without any musculoskeletal injuries (for the past one year) were recruited for this study. Dominant leg of all recruited subjects was right leg. A written informed consent was obtained from the subjects prior to the experiments.

B. Experimental Setup

Experiments were carried out on a pressure mapping sensor (Model: 7101, Tekscan, USA) with sensel spatial resolution of 25.0 sensel per square inch. Pressure mapping sensor was connected to a wireless datalogger (VersaTek, Tekscan, USA) via two connectors (VersaTek Cuff, Tekscan, USA) and two CAT5e cables. A laptop integrated with pressure measurement system (HR MAT Research version 7.00-65, Tekscan, USA) was connected to the wireless datalogger via IEEE 802.11 g protocol to receive the scanning data. The sampling frequency of the pressure measurement system was set at 25 Hz. Noise threshold was set at 3 raw units. Pressure mapping sensor was calibrated (step calibration) according to the standard procedure provided by manufacture.

C. Experimental Protocol

The subjects were given 15 minutes to choose their best comfort insole support size and familiarizing with the different type of static stances. Pressure mat sensor was calibrated prior the experiment starts. In each type of insole support, subject was required to complete the Double-Legs-Stance, Single-Right-Leg-Stance, Single-Left-Leg-Stance, Right-Tandem-Stance and Left-Tandem-Stance. Single leg stances and tandem stances were conducted in alternating leg attempt. Three valid attempts were required in each type of stance. An attempt was considered valid if the subject performed the test without a fall, and if any leg step and hands were released from the waist. The duration of each attempt was 12 seconds with a rest interval of 10 seconds in between attempts. The testing order of no insole support (Insole-A) and three insole support types (Insole-B, Insole-C and Insole-D) was randomized in each participant.

D. Data Analyses

First and last 25 frames data were excluded from analyses. Averaged data of three attempts were used for statistical analyses.

E. Statistical Analyses

All statistical tests were carried out in IBM SPSS (Statistical Package for Social Sciences) Statistics version 21.0 (IBM Corp., Armonk, NY). The association of four types of insole support used and CoP variables were tested using non-parametric Friedman test. Wilcoxon signed-ranks post-hoc test was applied to determine the insole pair difference whenever significant was detected in Friedman test. Statistical alpha level was set to $\alpha = 0.05$.

Table 1 Terminology

Term	Acronym	Defined
Double-Legs-Stance	DLS	Standing erect, hands on side waist, feet parallel pointing forwards and approximately shoulder width apart.
Single-Right-Leg-Stance	SRLS	Standing erect, hands on side waist, right leg lifted up at malleolus' height.
Single-Left-Leg-Stance	SLLS	Standing erect, hands on side waist, left leg lifted up at malleolus' height.
Right-Tandem-Stance	RTS	Standing erect, hands on side waist, left foot placed in front of the right foot with heel of the left foot touching the toe of the right foot.
Left-Tandem-Stance	LTS	Standing erect, hands on side waist, right foot placed in front of the left foot with heel of the right foot touching the toe of the left foot.
Centre of pressure	CoP	Centre of foot pressure.
Range	Rng	Maximum value minus minimum value of COP coordinates.
Travel Distance	TvlDist	Total distance travelled by the COP.
Peak Velocity	PkVel	The highest shift velocity of COP.
Position Standard Deviation	PoSD	The standard deviation of COP's coordinate.
Anteroposterior	AP	Direction in forwards-backwards.
Mediolateral	ML	Direction in left-right.
Insole-A	Insole-A	No additional insole support.
Insole-B	Insole-B	With insole designed B.
Insole-C	Insole-C	With insole designed C.
Insole-D	Insole-D	With insole designed D.

III. RESULTS

A. Comparison of CoP within Double-Leg-Stance

The results of the Friedman test indicate that significant differences do exist in TvlDist AP across insole support types, $\chi^2(3, N=10) = 9.120$, $p < 0.05$. However, Wilcoxon test indicates that no significant difference in TvlDist AP for between paired of insoles. The result of descriptive statistic (Table 2) indicates that Insole-A has a lowest mean value of TvlDist Total in anteroposterior direction. Insole-B has lowest value of TvlDist ML and PkVel ML.

No significant differences were found across insole support types regarding a) Rng AP ($\chi^2(3, N=10) = 5.4$, $p > 0.05$), b) Rng ML ($\chi^2(3, N=10) = 2.758$, $p > 0.05$), c) TvlDist ML ($\chi^2(3, N=10) = 2.76$, $p > 0.05$), d) TvlDist Total ($\chi^2(3, N=10) = 4.92$, $p > 0.05$), e) PkVel AP ($\chi^2(3, N=10) = 5.061$, $p > 0.05$), f) PkVel ML ($\chi^2(3, N=10) = 2.212$, $p > 0.05$), g) PkVel Resultant ($\chi^2(3, N=10) = 4.08$, $p > 0.05$), h) PoSD AP ($\chi^2(3, N=10) = 2.28$, $p > 0.05$), i) PoSD ML ($\chi^2(3, N=10) = 0.12$, $p > 0.05$) and j) PoSD Resultant ($\chi^2(3, N=10) = 1.92$, $p > 0.05$)

Table 2 Center of Pressure Variables in Double-Leg-Stance Position

Variable	Mean (SD)			
	Insole-A	Insole-B	Insole-C	Insole-D
Rng AP, cm	1.39 (0.60)	1.58 (0.71)	1.64 (0.71)	1.64 (0.50)
Rng ML, cm	0.98 (0.56)	1.06 (0.63)	1.11 (0.69)	1.03 (0.48)
TvlDist AP $^{\alpha}$, cm	6.20 (2.12)	6.71 (3.03)	6.91 (2.20)	7.10 (1.90)
TvlDist ML, cm	9.79 (5.61)	9.54 (3.96)	11.12 (6.60)	11.40 (5.69)
TvlDist Total, cm	12.84 (5.98)	12.93 (4.91)	14.47 (6.72)	14.88 (5.53)
PkVel AP, cm/s	3.27 (1.26)	3.89 (2.58)	3.87 (1.65)	4.11 (1.57)
PkVel ML, cm/s	4.76 (3.05)	4.42 (1.90)	5.67 (3.80)	5.75 (2.93)
PkVel Resultant, cm/s	5.33 (3.06)	5.40 (2.79)	6.38 (3.75)	6.72 (2.72)
PoSD AP, cm	0.35 (0.14)	0.38 (0.16)	0.40 (0.17)	0.41 (0.14)
PoSD ML, cm	0.21 (0.13)	0.24 (0.16)	0.24 (0.17)	0.22 (0.10)
PoSD Resultant, cm	0.43 (0.17)	0.46 (0.20)	0.48 (0.21)	0.48 (0.15)

$^{\alpha}$ indicates significant difference, $p < 0.05$ (Friedman Test)

B. Comparison of CoP within Single-Right-Leg-Stance

The results of the Friedman test indicate that significant differences do exist in PkVel AP across insole support types, $\chi^2(3, N=10) = 8.567$, $p < 0.05$. The post-hoc Wilcoxon test results indicate that there is a significant difference in PkVel AP for the Insole-C and Insole-B, z = -2.550, p < 0.05, with lower peak velocity of center of pressure shift in Insole-C. Descriptive statistic (Table 3) indicates that Insole-B has lowest mean value of Rng ML. On the other hand; Insole-C has lowest mean value of Rng, TvlDist, PkVel and PoSD in anteroposterior direction. Furthermore, result also indicates that Insole-C has lowest resultant mean value in PkVel and PoSD.

No significant differences were found across insole support types regarding a) Rng AP ($\chi^2(3, N=10) = 5.545$, p > 0.05), b) Rng ML ($\chi^2(3, N=10) = 7.8$, p > 0.05), c) TvlDist AP ($\chi^2(3, N=10) = 4.92$, p > 0.05), d) TvlDist ML ($\chi^2(3, N=10) = 3.24$, p > 0.05), e) TvlDist Total ($\chi^2(3, N=10) = 3.48$, p > 0.05), f) PkVel ML ($\chi^2(3, N=10) = 1.56$, p > 0.05), g) PkVel Resultant ($\chi^2(3, N=10) = 3.36$, p > 0.05), h) PoSD AP ($\chi^2(3, N=10) = 6.6$, p > 0.05), i) PoSD ML ($\chi^2(3, N=10) = 5.16$, p > 0.05) and j) PoSD Resultant ($\chi^2(3, N=10) = 2.76$, p > 0.05).

Table 3 Center of Pressure Variables in Single-Right-Leg-Stance Position

Variable	Mean (SD)			
	Insole-A	Insole-B	Insole-C	Insole-D
Rng AP, cm	2.75 (0.60)	2.66 (0.63)	2.44 (0.66)	2.66 (0.93)
Rng ML, cm	2.28 (0.45)	2.28 (0.41)	2.47 (0.53)	2.51 (0.43)
TvlDist AP, cm	16.19 (4.86)	16.70 (6.31)	15.48 (4.90)	16.82 (4.73)
TvlDist ML, cm	17.77 (5.97)	18.45 (5.68)	19.04 (5.12)	19.46 (5.03)
TvlDist Total, cm	26.37 (8.07)	27.30 (8.79)	26.83 (7.29)	28.14 (7.17)
PkVel AP $^{\alpha}$, cm/s	8.87 (3.40)	9.55 $^{\beta}$(3.07)	7.72 $^{\beta}$(2.38)	9.06 (1.75)
PkVel ML, cm/s	9.64 (4.08)	10.06 (3.17)	10.22 (3.11)	10.57 (2.24)
PkVel Resultant, cm/s	12.29 (5.12)	12.86 (3.80)	12.19 (3.96)	12.89 (2.34)
PoSD AP, cm	0.63 (0.16)	0.61 (0.15)	0.54 (0.12)	0.61 (0.28)
PoSD ML, cm	0.50 (0.10)	0.51 (0.11)	0.54 (0.12)	0.56 (0.14)
PoSD Resultant, cm	0.81 (0.19)	0.81 (0.16)	0.78 (0.15)	0.85 (0.30)

$^{\alpha}$ indicates significant difference, $p < 0.05$ (Friedman Test)
$^{\beta}$ indicates significant difference, $p < 0.05$ (Post-hoc Wilcoxon Signed-Ranks Test)

C. Comparison of CoP within Single-Left-Leg-Stance

The results of the Friedman test indicate that no significant differences exist in Single-Left-Leg-Stance across all types of insole support regarding a) Rng AP ($\chi^2(3, N=10) = 1.061$, p > 0.05), b) Rng ML ($\chi^2(3, N=10) = 4.44$, p > 0.05), c) TvlDist AP ($\chi^2(3, N=10) = 4.44$, p > 0.05), d) TvlDist ML ($\chi^2(3, N=10) = 1.56$, p > 0.05), e) TvlDist Total ($\chi^2(3, N=10) = 0.96$, p > 0.05), f) PkVel AP ($\chi^2(3, N=10) = 1.364$, p > 0.05), g) PkVel ML ($\chi^2(3, N=10) = 4.44$, p >

0.05), h) PkVel Resultant ($\chi^2(3, N=10) = 0.84, p > 0.05$), i) PoSD AP ($\chi^2(3, N=10) = 1.08, p > 0.05$), j) PoSD ML ($\chi^2(3, N=10) = 0.96, p > 0.05$) and k) PoSD Resultant ($\chi^2(3, N=10) = 2.4, p > 0.05$).

The result of descriptive statistic (Table 4) indicates that Insole-C has lowest mean value of TvlDist ML. Moreover, Insole-D has lowest mean value of PkVel and PoSD in mediolateral direction.

Table 4 Center of Pressure Variables in Single-Left-Leg-Stance Position

Variable	Mean (SD)			
	Insole-A	Insole-B	Insole-C	Insole-D
Rng AP, cm	2.60 (0.52)	3.00 (1.04)	2.83 (0.76)	2.86 (0.77)
Rng ML, cm	2.27 (0.51)	2.39 (0.64)	2.47 (0.51)	2.32 (0.51)
TvlDist AP, cm	17.30 (6.61)	18.54 (8.37)	18.02 (6.01)	18.51 (7.23)
TvlDist ML, cm	18.79 (6.89)	19.50 (7.20)	18.73 (4.87)	18.56 (5.64)
TvlDist Total, cm	27.94 (9.97)	29.39 (11.39)	28.46 (7.95)	28.71 (9.55)
PkVel AP, cm/s	8.97 (3.37)	10.02 (4.23)	9.48 (3.66)	9.47 (3.58)
PkVel ML, cm/s	9.70 (3.83)	10.79 (3.66)	10.41 (3.00)	9.68 (2.50)
PkVel Resultant, cm/s	12.16 (4.76)	13.82 (4.89)	12.89 (4.34)	12.65 (3.81)
PoSD AP, cm	0.61 (0.14)	0.66 (0.21)	0.64 (0.16)	0.63 (0.12)
PoSD ML, cm	0.51 (0.10)	0.53 (0.16)	0.54 (0.11)	0.49 (0.12)
PoSD Resultant, cm	0.80 (0.15)	0.85 (0.26)	0.85 (0.18)	0.81 (0.16)

D. Comparison of CoP within Right-Tandem-Stance

The results of the Friedman test indicate that no significant differences exist in Right-Tandem-Stance across all types of insole support regarding a) Rng AP ($\chi^2(3, N=10) = 1.8, p > 0.05$), b) Rng ML ($\chi^2(3, N=10) = 3.12, p > 0.05$), c) TvlDist AP ($\chi^2(3, N=10) = 2.04, p > 0.05$), d) TvlDist ML ($\chi^2(3, N=10) = 6.12, p > 0.05$), e) TvlDist Total ($\chi^2(3, N=10) = 4.92, p > 0.05$), f) PkVel AP ($\chi^2(3, N=10) = 0.6, p > 0.05$), g) PkVel ML ($\chi^2(3, N=10) = 6.939, p > 0.05$), h) PkVel Resultant ($\chi^2(3, N=10) = 3.96, p > 0.05$), i) PoSD AP ($\chi^2(3, N=10) = 2.04, p > 0.05$), j) PoSD ML ($\chi^2(3, N=10) = 2.16, p > 0.05$) and k) PoSD Resultant ($\chi^2(3, N=10) = 1.44, p > 0.05$).

Table 5 Center of Pressure Variables in Right-Tandem-Stance Position

Variable	Mean (SD)			
	Insole-A	Insole-B	Insole-C	Insole-D
Rng AP, cm	2.75 (1.22)	3.00 (1.57)	3.46 (2.28)	3.52 (1.97)
Rng ML, cm	2.72 (0.97)	2.81 (0.93)	2.89 (0.77)	2.72 (0.78)
TvlDist AP, cm	22.12 (8.58)	20.87 (7.16)	24.34 (9.94)	22.68 (9.18)
TvlDist ML, cm	16.96 (5.30)	16.48 (6.18)	18.92 (5.42)	18.72 (4.84)
TvlDist Total, cm	30.50 (9.56)	29.15 (10.37)	33.76 (11.73)	32.30 (10.32)
PkVel AP, cm/s	13.26 (5.99)	11.67 (3.89)	16.44 (10.50)	16.13 (11.40)
PkVel ML, cm/s	8.87 (2.34)	9.72 (3.97)	10.91 (5.02)	10.81 (3.78)
PkVel Resultant, cm/s	15.27 (5.97)	13.44 (4.43)	18.91 (11.53)	19.12 (11.03)
PoSD AP, cm	0.56 (0.26)	0.69 (0.40)	0.70 (0.45)	0.84 (0.58)
PoSD ML, cm	0.63 (0.26)	0.64 (0.22)	0.65 (0.19)	0.62 (0.20)
PoSD Resultant, cm	0.87 (0.32)	0.96 (0.42)	0.99 (0.49)	1.07 (0.58)

However, the result of descriptive statistic (Table 5) indicates that Insole-B has lowest mean value of TvlDist AP, TvlDist ML, TvlDist Total, PkVel AP and PkVel Resultant. Insole-D has lowest mean value of Rng and PoSD in mediolateral direction.

E. Comparison of CoP within Left-Tandem-Stance

The results of the Friedman test indicate that significant differences do exist in Rng ML, TvlDist ML and PoSD ML across insole support types with chi square, $\chi^2(3, N=10) = 9.0$, $\chi^2(3, N=10) = 9.12$, $\chi^2(3, N=10) = 8.04$, $p < 0.05$ respectively. The post-hoc Wilcoxon test results indicate that two pairs of insoles are significant difference in Rng ML. First pair was the Insole-C and Insole-B, z = -2.191, $p < 0.05$, with lower Rng ML of center of pressure shift in Insole-B. Second pair was the Insole-D and Insole-C, z = -1.988, $p < 0.05$, with lower Rng ML of center of pressure shift in Insole-D. In TvlDist ML, post-hoc Wilcoxon test indicates that there are two pairs of insoles were found significantly difference. The Z-score of Insole-C paired with Insole-A was, z = -2.293, $p < 0.05$, and the Z-score of Insole-C paired with Insole-B was, z = -2.293, $p < 0.05$. Both pairs were also showing higher value in Insole-C. In PoSD ML, post-hoc results indicate that there were significant difference found in Insole-C paired with Insole-A and Insole-D paired with Insole-C, the Z-scores were, z = -1.988 and z = -2.395, $p < 0.05$ respectively with higher value in Insole-C.

The result of descriptive statistic (Table 6) indicates that Insole-B has lowest mean value of Rng, TvlDist and PkVel in mediolateral direction. Besides, Insole-B also has lowest mean value in TvlDist Total. Insole-D has lowest mean value of Rng, TvlDist, PkVel and PoSD in anteroposterior direction. Moreover, result also indicates that Insole-D has lowest mean value in PoSD ML, PoSD Resultant and PkVel Resultant.

No significant differences were found across insole support types regarding a) Rng AP ($\chi^2(3, N=10) = 6.6$, $p > 0.05$), b) TvlDist AP ($\chi^2(3, N=10) = 6.36$, $p > 0.05$), c) TvlDist Total ($\chi^2(3, N=10) = 6.48$, $p > 0.05$), d) PkVel AP ($\chi^2(3, N=10) = 2.28$, $p > 0.05$), e) PkVel ML ($\chi^2(3, N=10) = 5.204$, $p > 0.05$), f) PkVel Resultant ($\chi^2(3, N=10) = 4.44$, $p > 0.05$), g) PoSD AP ($\chi^2(3, N=10) = 1.8$, $p > 0.05$) and h) PoSD Resultant ($\chi^2(3, N=10) = 3.96$, $p > 0.05$).

Table 6 Center of Pressure Variables in Left-Tandem-Stance Position

Variable	Mean (SD)			
	Insole-A	Insole-B	Insole-C	Insole-D
Rng AP, cm	3.17 (1.16)	2.95 (1.15)	3.18 (1.46)	2.79 (1.00)
Rng ML $^{\alpha}$, cm	2.69 (0.55)	2.59 $^{\beta}$(0.91)	3.23 $^{\beta,\theta}$(1.03)	2.61 $^{\theta}$(0.71)
TvlDist AP, cm	23.07 (12.18)	21.16 (8.74)	25.66 (10.97)	21.15 (7.50)
TvlDist ML $^{\alpha}$, cm	16.79 $^{\beta}$(3.72)	16.60 $^{\theta}$(5.80)	19.72 $^{\beta,\theta}$(5.41)	16.81 (3.83)
TvlDist Total, cm	31.71 (11.95)	29.61 (10.85)	35.52 (12.52)	29.72 (8.29)
PkVel AP, cm/s	13.14 (8.39)	12.78 (6.18)	13.94 (6.40)	11.38 (3.99)
PkVel ML, cm/s	9.98 (3.56)	8.58 (2.56)	10.73 (3.61)	9.47 (2.26)
PkVel Resultant, cm/s	15.73 (8.14)	14.61 (5.99)	16.33 (6.69)	13.67 (3.91)
PoSD AP, cm	0.66 (0.21)	0.63 (0.27)	0.67 (0.29)	0.62 (0.25)
PoSD ML $^{\alpha}$, cm	0.60 $^{\beta}$(0.10)	0.59 (0.23)	0.74 $^{\beta,\theta}$(0.24)	0.58 $^{\theta}$(0.14)
PoSD Resultant, cm	0.91 (0.19)	0.89 (0.33)	1.01 (0.36)	0.87 (0.30)

$^{\alpha}$ indicates significant difference, *p < 0.05* (Friedman Test)
$^{\beta,\theta}$ indicates significant difference, *p < 0.05* (Post-hoc Wilcoxon Signed-Ranks Test)

IV. Discussion

It is important for a person to have good balance ability, since balance ability could reduce the risk of injury and enhance sports performance. Present study, exhibits instantaneous effects of generic foot arch support insole during static standing. In our analysis, the data showed there was significant difference found in anterior-posterior travelled distance as a whole. However there was no significant difference found in any combination pair of insoles used. This may be due to Friedman test which has less power than Wilcoxon signed-rank test. This means that Friedman test is more sensitive to detect the statistical differences [32]. Similar results were reported by Hatton et al. (2012) where the authors used textured insole as an intervention to investigate the immediate effects of CoP on elderly female population [13].

There were no significant differences among the selected variables in this study for the right-tandem-stance position. But in left-tandem-stance position, Insole-C was found to be significant in body sway compared to Insole-A. Insole-C was found to be significant higher value in medial-lateral travelled distance and medial-lateral PoSD. Insole-C showed highest mean value in all measured variables in our study. These imply that Insole-C had a poorest balance effects in left-tandem-stance.

Similar in single stance position, we only observed left side of position has significant different results. In single-right-leg-stance position, we found Insole-C was significantly lower in anterior-posterior peak velocity than

Insole-B. Based on descriptive analysis, Insole-C obtained the lowest CoP value for all the anterior-posterior variables. This suggests that use of Insole-C has a better control of postural sway in anterior-posterior direction. However, similar results were not seen in medial-lateral variables. All the lowest values of medial-lateral variables were obtained by without using any foot arch support insole (insole-A). In single-left-leg-stance position, no significant difference was found among the variables. We presume that, it may be due to the neuromuscular adaptation capability and different density of plantar mechanoreceptors between dominant leg and non-dominant leg. Thus, the time required for neuromuscular coordination and adaptation is a determining factor. To our best knowledge the time range required for neuromuscular adaptation may vary among individuals depending upon various factors such as foot arch and heel type.

V. Conclusion

Dominant leg may have an important role in neuromuscular adaptation time. Statistical analyses revealed that generic foot arch support insoles have no significant positive instantaneous effects on CoP measurement. Further research is needed to find out the varied effects which may be caused is due to inadequate period of neuromuscular adaptation or the insole's design itself was not significantly enriched to stimulate the plantar mechanoreceptors.

References

17. Nashner, L. M. (1993). Practical biomechanics and physiology of balance. Handbook of balance function testing.
18. Hrysomallis, C. (2007). Relationship between balance ability, training and sports injury risk. Sports Medicine, 37(6), 547–556. http://doi.org/10.2165/00007256-200737060-00007
19. Hrysomallis, C., McLaughlin, P., & Goodman, C. (2007). Balance and injury in elite Australian footballers. International Journal of Sports Medicine, 28(10), 844–7. http://doi.org/10.1055/s-2007-964897
20. Viswanathan, A., & Sudarsky, L. (2011). Balance and gait problems in the elderly. Handbook of Clinical Neurology (Vol. 103). http://doi.org/10.1016/B978-0-444-51892-7.00045-0
21. Muir, J. W., Kiel, D. P., Hannan, M., Magaziner, J., & Rubin, C. T. (2013). Dynamic parameters of balance which correlate to elderly persons with a aistory of falls. PLoS ONE, 8(8), 1–7. http://doi.org/10.1371/journal.pone.0070566
22. Bergland, A., & Wyller, T. B. (2004). Risk factors for serious fall related injury in elderly women living at home. Injury Prevention : Journal of the International Society for Child and Adolescent Injury Prevention, 10(5), 308–313. http://doi.org/10.1136/ip.2003.004721
23. Lamoth, C. J. C., van Lummel, R. C., & Beek, P. J. (2009). Athletic skill level is reflected in body sway: a test case for accelometry in combination with stochastic dynamics. Gait & Posture, 29(4), 546–51. http://doi.org/10.1016/j.gaitpost.2008.12.006
24. Kamen, G., Patten, C., Du, C. D., & Sison, S. (1998). An accelerometry-based system for the assessment of balance and postural sway. Gerontology, 44(1), 40–5. http://doi.org/10.1159/000021981
25. Nelson, B. L., & Johnson, J. K. (1979). Practical measurements for evaluation in physical education (4th ed.). Minneapolis: Burgess.
26. Rival, C., Ceyte, H., & Olivier, I. (2005). Developmental changes of static standing balance in children. Neuroscience Letters, 376(2), 133–136. http://doi.org/10.1016/j.neulet.2004.11.042
27. Kirshenbaum, N., Riach, C. L., & Starkes, J. L. (2001). Non-linear development of postural control and strategy use in young children: A longitudinal study. Experimental Brain Research, 140(4), 420–431. http://doi.org/10.1007/s002210100835
28. Mallau, S., Bollini, G., Jouve, J.-L., & Assaiante, C. (2007). Locomotor skills and balance strategies in adolescent's idiopathic scoliosis. Spine, 32(1), E14–22. http://doi.org/10.1097/01.brs.0000251069.58498.eb
29. Hatton, A. L., Dixon, J., Rome, K., Newton, J. L., & Martin, D. J. (2012). Altering gait by way of stimulation of the plantar surface of the foot: the immediate effect of wearing textured insoles in older fallers. Journal of Foot and Ankle Research, 5(Suppl 1), O21. http://doi.org/10.1186/1757-1146-5-S1-O21
30. Qiu, F., Cole, M. H., Davids, K. W., Hennig, E. M., Silburn, P. a., Netscher, H., & Kerr, G. K. (2012). Enhanced somatosensory information decreases postural sway in older people. Gait and Posture, 35(4), 630–635. http://doi.org/10.1016/j.gaitpost.2011.12.013
31. Davlin, C. D. (2004). Dynamic balance in high level athletes. Perceptual and Motor Skills, 98(1955), 1171–1176. http://doi.org/10.2466/PMS.98.4.1171-1176
32. Chapman, D. W., Needham, K. J., Allison, G. T., Lay, B., & Edwards, D. J. (2008). Effects of experience in a dynamic environment on postural control. British Journal of Sports Medicine, 42(2005), 16–21. http://doi.org/10.1136/bjsm.2006.033688
33. Mason, B. R., & Pelgrim, P. P. (1986). Body stability and performance in archery.pdf. Excel, 3(2), 17–20.
34. Ball, K. A., Best, R. J., & Wrigley, T. V. (2003). Body sway, aim point fluctuation and performance in rifle shooters: inter- and intra-individual analysis. Journal of Sports Sciences, 21(7), 559–566. http://doi.org/10.1080/0264041031000101881
35. Mononen, K., Konttinen, N., Viitasalo, J., & Era, P. (2007). Relationships between postural balance, rifle stability and shooting accuracy among novice rifle shooters. Scandinavian Journal of Medicine and Science in Sports, 17(2), 180–185. http://doi.org/10.1111/j.1600-0838.2006.00549.x
36. Behm, D. G., Wahl, M. J., Button, D. C., Power, K. E., & Anderson, K. G. (2005). Relationship between hockey skating speed and selected performance measures. Journal of Strength and Conditioning Research / National Strength & Conditioning Association, 19(2), 326–331. http://doi.org/10.1519/R-14043.1
37. Bressel, E., Yonker, J. C., Kras, J., & Heath, E. M. (2007). Comparison of static and dynamic balance in female collegiate soccer, basketball, and gymnastics athletes. Journal of Athletic Training, 4242(11), 42–46.

38. Matsuda, S., Demura, S., & Uchiyama, M. (2008). Centre of pressure sway characteristics during static one-legged stance of athletes from different sports. Journal of Sports Sciences, 26(7), 775–779. http://doi.org/10.1080/02640410701824099
39. Paillard, T., Noé, F., Rivière, T., Marion, V., Montoya, R., & Dupui, P. (2006). Postural performance and strategy in the unipedal stance of soccer players at different levels of competition. Journal of Athletic Training, 41(2), 172–176.
40. Sell, T. C., Tsai, Y. S., Smoliga, J. M., Myers, J. B., & Lephart, S. M. (2007). Strength, flexibility, and balance characteristics of highly proficient golfers. J Strength Cond Res, 21(4), 1166–1171. http://doi.org/10.1519/R-21826.1
41. Niinimaa, V., & McAvoy, T. (1983). Influence of exercise on body sway in the standing rifle shooting position. Canadian Journal of Applied Sport Sciences, 8(1), 30–33.
42. Era, P., Konttinen, N., Mehto, P., Saarela, P., & Lyytinen, H. (1996). Postural stability and skilled performance-a study On top-Level and naive rifle shooters. Journal of Biomechanics, 29, 301–306. http://doi.org/10.1016/0021-9290(95)00066-6
43. Konttinen, N., Lyytinen, H., & Era, P. (1999). Brain slow potentials and postural sway behavior during sharpshooting performance. Journal of Motor Behavior, 31(1), 11–20. http://doi.org/10.1080/00222899909601888
44. Paillard, T., Costes-Salon, C., Lafont, C., & Dupui, P. (2002). Are there differences in postural regulation according to the level of competition in judoists? British Journal of Sports Medicine, 36(4), 304–5. http://doi.org/10.1136/bjsm.36.4.304
45. Gross, M. T., Mercer, V. S., & Lin, F.-C. (2012). Effects of foot orthoses on balance in older adults. The Journal of Orthopaedic and Sports Physical Therapy, 42(7), 649–57. http://doi.org/10.2519/jospt.2012.3944
46. Palluel, E., Olivier, I., & Nougier, V. (2009). The lasting effects of spike insoles on postural control in the elderly. Behavioral Neuroscience, 123(5), 1141–7. http://doi.org/10.1037/a0017115
47. Wang, C. C., & Yang, W. H. (2012). Using detrended fluctuation analysis (DFA) to analyze whether vibratory insoles enhance balance stability for elderly fallers. Archives of Gerontology and Geriatrics, 55(3), 673–676. http://doi.org/10.1016/j.archger.2011.11.008
48. Zimmerman, D. W., & Zumbo, B. D. (1993). Relative power of the Wilcoxon test, the Friedman test, and repeated-measures ANOVA on Ranks. Journal of Experimental Education. http://doi.org/10.1080/00220973.1993.9943832

Author : Chok Chuen Leang
Institute : National Sports Institute of Malaysia
Street : National Sports Complex, Bukit Jalil, Sri Petaling
City : Kuala Lumpur
Country : Malaysia
Email : ccleang@isn.gov.my

Hadamard Transform Based PAPR Reduction for Telemedicine Applications Utilized for Epilepsy Classification

S. K. Prabhakar, and H. Rajaguru

Department of ECE, Bannari Amman Institute of Technology, Sathyamangalam, India

Abstract— **A set of long term associated neurological disorders which is characterized by seizures is epilepsy. Because of the abnormal electrical activities occurring in the brain, the occurrence of epileptic seizures become prominent. In epilepsy research, convulsions too are sometimes related to epileptic seizures. So to investigate and understand the brain's electrical activity, recording the scalp is important and that can be done with the help of Electroencephalograph (EEG). The time consumption with the long term measurements is enormous and so reviewing it becomes quite a hectic task. So the need for an automatic detection of seizures is required. As the long term measurements produce a huge amount of data, it would be pretty difficult to process and so with the help of Independent Component Analysis (ICA), the dimensions of the data are reduced. Then it is transmitted via a Space Time Block Coded Multiple Input Single Output (2 x 1) Orthogonal Frequency Division Multiplexing (STBC MISO-OFDM) System. As the system has a high Peak to Average Power Ratio (PAPR), it is reduced with the help of Hadamard Based PAPR Reduction and then at the receiver the Bit Error Rate (BER) is computed. Also at the receiver side, the classifier employed is Linear Kernel Support Vector Machines (L-SVM). Finally the epilepsy risk level classification from EEG Signals is measured in terms in Specificity, Sensitivity, Accuracy, Time Delay and Quality Values.**

Keywords— **EEG, Epilepsy, PAPR, STBC MISO-OFDM, L-SVM**

I. Introduction

Epilepsy is one of the serious disorders of the brain and has to be treated well [1]. The convulsions in epilepsy can be severe and prolonged and can vary from short lapses to long lapses of time accompanied by muscle jerks. When electrical discharges occur in the particular group of brain cells suddenly and in excess, then the seizures occur. Due to the intermittent abnormal firing of neurons in the brain, about 1 million of the entire population in the world suffers. During a seizure, the activity of the brain varies significantly from that during the normal state [2]. This is with respect to the frequency and pattern of neuronal firing. To understand the complex dynamics of the brain, EEG is used because it contains all the relevant information regarding the various physiological states of brain. Since EEG is non-invasive in most of the cases, it can be easily recorded for a very long time in order to monitor and discover the incidental disorders like epileptic seizures. These recordings are carefully observed, checked and monitored with the help of visual encephalographers. Only with this obtained information, the decision to exact diagnosis and followed by surgical treatment, medication and surgeries can be performed. But due to human error, improper diagnosis can occur which can prove fatal to the patient and so automated systems came into existence [3]. The block diagram is given in Figure 1.

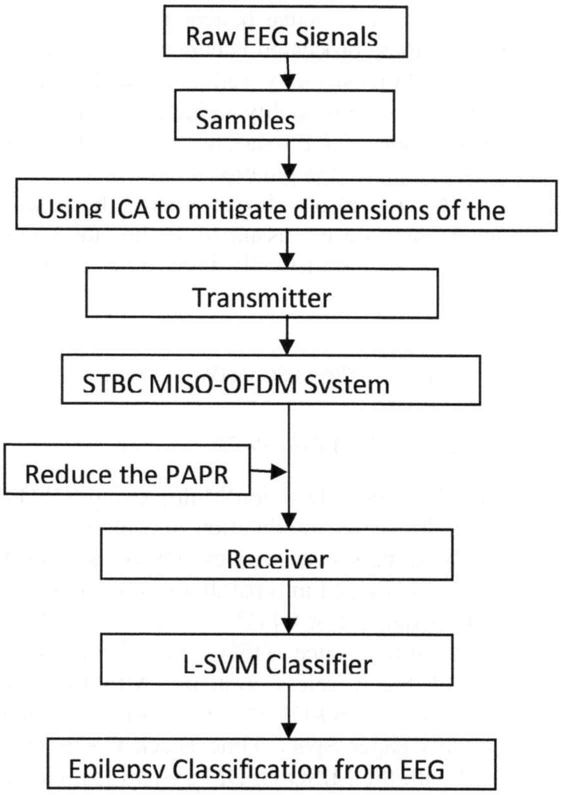

Figure 1 Proposed Block Diagram of the System

The structural organization of the paper is as follows. In section 2, the materials and methods along with the usage of ICA as a dimensionality reduction technique is discussed followed by the usage of STBC MISO-OFDM System and the need for a reduced PAPR and BER in section 3. Section

F. Ibrahim et al. (eds.), *3rd International Conference on Movement, Health and Exercise*,
IFMBE Proceedings 58,
DOI: 10.1007/978-981-10-3737-5_7

4 explains the application of L-SVM classifier at the receiver side followed by the results and discussion in section 5 and concluded finally.

II. Materials and Methods

For our analysis, the EEG data was taken from 20 epileptic patients in European Data format (EDF) who were under treatment for epilepsy in Sri Ramakrishna Hospital, Coimbatore, India. For all the 20 patients the recordings were taken and it was recorded for 30 minutes each continuously. The recorded signals were divided into 2 second epoch duration because to detect the essential changes happening in the signal it was more than sufficient. There are totally 16 channels measured over three different epochs for all patients. As the maximum frequency of the EEG is 50 Hz, based on the Nyquist criterion the sampling frequency is 200 Hz. There are totally 400 values for each and every epoch for a particular patient and therefore for 20 patients the data to be processed is enormous and so ICA is employed as a dimensionality reduction technique here. ICA is used as a dimensionality reduction technique here because to separate a multivariate signal into additive subcomponents, this computational method is quite efficient. ICA is also chosen here because the subcomponents are always assumed to be non-Gaussian signals and it is also statistically independent from each other.

III. STBC MISO-OFDM System

A. MISO-OFDM System and PAPR

Orthogonal Frequency Division Multiplexing (OFDM) is a versatile multi-carrier modulation technique used for transmission of signals over wireless channels. The high-rate data stream is divided into parallel lower rate data here and the main advantage of OFDM is that it can eliminate Inter Symbol Interference (ISI). For improving the performance of the wireless systems, MISO-OFDM is utilized [4]. With MISO-OFDM a transmit diversity is achieved. In this paper Space Time Block Codes (STBC) are combined with the MISO-OFDM system [5]. Generally a better performance is achieved by simply increasing the total number of transmit antennas and receive antennas. The modulation technique used here is Quadrature Phase Shift Keying (QPSK) and Rayleigh Channel mode is used here. As the system suffers from a high Peak to Average Power Ratio (PAPR), the PAPR has to be reduced and that is achieved with the help of Hadamard Transform. The performance results are analyzed in terms of probability value (symbol PAPR> value) versus PAPR and Bit Error Ratio (BER) Vs Signal Noise Ratio (SNR).

B. Hadamard Transform for PAPR Reduction

In OFDM with totally N number of sub-carriers, the discrete-time OFDM can be represented as

$$x_m = \frac{1}{\sqrt{N}} \sum_{n=0}^{N-1} X_n e^{j2\pi mn/QN}$$

where $m = 1,2,...N-1$ and $j = \sqrt{-1}$

In $X_n, n = 0,1,....N-1$ are the input symbols. It is modulated with Quadrature Phase Shift Keying (QPSK). Q is the oversampling factor which is an integer and it has to be always greater than 1. The data sequence is written as

$$X = [X(0), X(1),.... X(N-1)]^T$$

The PAPR of the transmitted signal x_m is defined as the ratio of the average signal power to the maximum signal power [6]. The PAPR of a complex pass band signal x_m is represented as

$$PAPR = \frac{Peak\ Power}{Average\ Power}$$

$$PAPR(x_m) = 10\log_{10} \frac{\max|x_m|^2}{E[|x_m|^2]}$$

In the above equation $|x_m|^2$ is the peak signal power and $E[.]$ denotes Expectation Operator.

The Hadamard Transform is simply a powerful and standard orthogonal matrix or order $'n'$ [7]. A Hadamard Transform of order n is nothing but an $N \times N$ matrix with the values of positive ones (+1's) and negative ones (-1's) such that

$$H_n H_n^t = nI_n$$

The Hadamard matrix (H) of order n can be modelled as

$$H_n = H_{n-1} \otimes H_{n-1}$$

$$H_n = \frac{1}{\sqrt{2}} \begin{bmatrix} H_{n-1} H_{n-1} \\ H_{n-1} H_{n-1} \end{bmatrix}$$

The vector X is represented as $X = [X_1, X_2,...., X_n]$ is simply transformed into a Hadamard matrix of order n, and therefore the transformed vector is given as $R = HX$. Ultimately, the signal with the minimum PAPR is selected for transmission purposes. The simulation parameters are shown in the Table 1.

Table 1. Simulation Parameters

Modulation Technique Used	QPSK
System analysed and designed	2 x1 STBC MISO-OFDM
Total number of subcarriers	128
Number of sub-blocks	4
Maximum symbols loaded	1e5
Symbol rate Utilized	250000
Number of time slots	2
Windowing function	Blackman-Haris
HPA Model	SSPA
Total Number of frames taken	10
No of OFDM symbols/ frame	4
Bandwidth	5 MHz
Oversampling factor	4

IV. L-SVM CLASSIFIER AT THE RECEIVER

At the receiver side, along with the computation of Bit Error Rate, L-SVM is also used as a post classifier. For binary classification, Linear SVM (L-SVM) was originally formulated [8]. The training data and its respective labels (x_n, y_n) are taken initially, where $n = 1,...., N$ and $x_n \in R^D$, $t_n \in \{-1,+1\}$

The learning of SVM consists of the constrained optimization as follows:

$$\min_{w,\xi_n} \frac{1}{2} w^T w + C \sum_{n=1}^{N} \xi_n$$

such that $w^T x_n t_n \geq 1 - \xi_n \quad \forall n$

$$\xi_n \geq 0 \quad \forall n$$

ξ_n is the slack variable where the data points are penalized which violates the original and actual margin requirement. The bias can be included by augmentation of the entire data vector x_n which has a scalar value of 1.

The L-SVM is used so that minimizing the square hinge loss is done as follows:

$$\min_{w} \frac{1}{2} w^T w + C \sum_{n=1}^{N} \max(1 - w^T x_n t_n, 0)^2$$

The above equation is easily differentiable and imposes a higher loss of point where the margin is violated. In order to assess the class label of a particular test data X, the following equation is used

$$\arg \max_t \ (w^T x) t$$

V. RESULTS AND DISCUSSION

In this part, the PAPR Results, BER Results and the Classification Results are explained elaborately.

A. PAPR Results

The PAPR results and the BER Analysis results are clearly elucidated here. It is evident from the careful analysis of the figure 2, it is apparent that when QPSK modulation is engaged the PAPR is reduced by 1.5 dB with the effective implementation of Hadamard transform. The Bit Error Rate is also analysed in Figure 3.

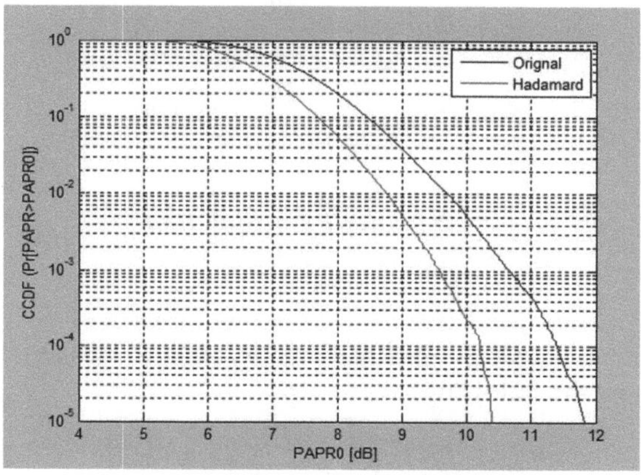

Figure 2 PAPR Reduction Using Hadamard Technique

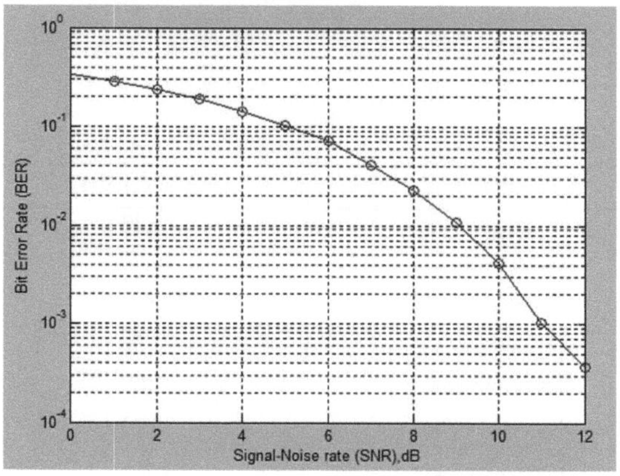

Figure 3 BER Analysis for the system

B. Classification Results

For ICA as dimensionality reduction technique and L-SVM as a Post Classifier, based on the terminologies like Performance Index, Quality values, Time Delay and Accuracy the results in the receiver side are computed in Table 2. The formulae for the Performance Index (PI), Sensitivity, Specificity and Accuracy are given as follows

$$PI = \frac{PC - MC - FA}{PC} \times 100$$

where PC – Perfect Classification, MC – Missed Classification, FA – False Alarm,

The Sensitivity, Specificity and Accuracy measures are stated by the following

$$Specificity = \frac{PC}{PC + MC} \times 100$$

$$Sensitivity = \frac{PC}{PC + FA} \times 100$$

$$Accuracy = \frac{Sensitivity + Specificity}{2} \times 100$$

The Time Delay and the Quality Value Measures are given by the following

$$Time\ Delay = \left[2 * \frac{PC}{100} + 6 * \frac{MC}{100} \right]$$

$$Quality\ \ Values = \frac{10}{\left[\frac{FA}{100} + 0.2 \right] * Time\ \ Delay}$$

Table 2 Average Performance Analysis Values of 20 Patients

Parameters	Obtained Values
PC (%)	79.23
MC (%)	20.76
FA (%)	0
PI (%)	72.85
Sensitivity (%)	100
Specificity (%)	79.23
Time Delay (sec)	2.83
Quality Values	17.79
Accuracy (%)	89.61

Conclusion

Thus in this paper, ICA was used as a dimensionality reduction technique and then it was transmitted using STBC MISO-OFDM System. As the system suffered a high PAPR it was reduced with the help of Hadamard Transform. At the receiver, the L-SVM is used as a post classifier for the classification of epilepsy risk levels from EEG Signals. Results prove that about 1.5 dB PAPR was reduced with the help of Hadamard transform and the overall average accuracy found here is about 89.61%. A time delay of about 2.83 seconds is produced and a quality value of about 17.79 is produced. The perfect classification rate is about 79.23% and the missed classification rate is about 20.76%. Future work plans to incorporate various other dimensionality reduction techniques, other types of MIMO-OFDM Systems and other types of post classifiers.

References

1. Tzallas AT, Tsipouras MG & Fotiadis DI (2009), 'Epileptic seizure detection in EEGs using time-frequency analysis', IEEE Trans. on Inf. Tech. in Biomed., vol. 13, pp. 703-710.
2. Srinivasan V, Eswaran C & Sriraam N, (2007), 'Approximate entropy-based epileptic EEG detection using artificial neural networks'. IEEE Trans. Inform. Technol. Biomed., vol.11, no.3, pp.288-295.
3. Sadasivam P.K, Narayana Dutt,(1996) "SVD Based Technique for noise reduction in Electroencephalographic signals", Elsevier Signal Processing, Vol 55, pp 179-189.
4. Harikumar Rajaguru, Sunil Kumar Prabhakar (2016), " Clinical Health Care for Long Distance using Matrix Factorization and Mahalanobis Based Sparse Representation Measures for Epilepsy Classification from EEG Signals", International Journal of Pharmaceutical Sciences Review and Research, Vol.38(1), Article 24, Pages:144-148.
5. H Yang (2005), "A Road to Future Broadband Wireless Access :MIMO-OFDM-Based Air Interface",IEEE Comm, Magazine, pp.53-60.
6. Han S.H and Lee J.H (2005), " An Overview of Peak-to-Average Power Ratio Reduction Techniques for Multicarrier Transmission", IEEE Wireless Comm, pp.56-65.
7. Myonghee Park et al (2000), "PAPR Reduction in OFDM transmission using Hadamard Transform", IEEE International Conference on Communication, Vol.1, pp. 430-433, New Orleans, LA
8. Harikumar Rajaguru, Sunil Kumar Prabhakar (2016), "LDA, GA and SVM's for Classification of Epilepsy From EEG Signals", Research Journal of Pharmaceutical, Biological and Chemical Sciences, Vol.7(3), pp:2044-2049.

Author: Sunil Kumar Prabhakar
Institute: Department of Electronics and Comunication Engineering, Bannari Amman Institute of Technology
City: Sathyamangalam
Country: India
Email: Sunilprabhakar22@gmail.com

Artificial Intelligence Techniques Used for Wheeze Sounds Analysis: Review

F.G. Nabi[1], K. Sundaraj[2], L.C. Kiang[1], R. Palaniappan[3], S. Sundaraj[4] and N.U Ahamed[5]

[1] School of Mechatronic Engineering, Universiti Malaysia Perlis (UniMAP), Arau, Perlis, Malaysia
[2] Faculty of Electronics and Computer Engineering, Universiti Teknikal Malaysia Melaka (UTeM), Durian Tunggal, Melaka, Malaysia
[3] School of Electronics Engineering (SENSE), Vellore Institute of Technology (VIT), Tamil Nadu, India
[4] Department of Anesthesiology, Hospital Tengku Ampuan Rahimah (HTAR), Klang, Selangor, Malaysia
[5] Faculty of Manufacturing Engineering, Universiti Malaysia Pahang, Pekan, Pahang, Malaysia

***Abstract*— Wheezes are acoustic, adventitious, continues and high pitch pulmonary sounds produce due to airway obstruction, these sounds mostly exist in pneumonia and asthma patients. Artificial intelligence techniques have been extensively used for wheeze sound analysis to diagnose patient. The available literature has not yet been reviewed. In this article most recent and relevant 12 studies, from different databases related to artificial inelegance techniques for wheeze detection has been selected for detailed review. It has been noticed that now trend is going to increase in this area, for personal assistance and continues monitoring of patient health. The literature reveals that 1) wheezes signals have enough information for the classification of patients according to disease severity level and type of disease, 2) significant work is required for identification of severity level of airway obstruction and pathology differentiation.**

***Keywords*— Wheeze, Wheeze Sounds, Respiratory Sounds, Airway Obstruction, Wheeze Analysis**

I. Introduction

Wheezes are continuous, adventitious and abnormal respiratory sounds. These are musical sounds super imposed on normal breath sounds mostly produce in asthma and pneumonia patients. The automatic wheezes recognition history started in 1980s. Physicians in the field of pulmonary medicine use a stethoscope to listen the respiratory sounds with the goal of diagnosing respiratory disorders and abnormalities. Oscillation of bronchi wall produces turbulent flow of breath which generates acoustic signal. Wheeze mostly produce in pneumonia and asthmatic patients. According to the WHO, 235 million individuals are suffering from asthma (2013), and 15% of children die due to pneumonia (*1*).

Wheezes are resembled to sinusoidal waves with different peaks in the spectrum greater than 100 Hz and length bigger than 100 ms (*2, 3*). The number of wheezes and their distribution on chest report allocation of pathology. Monophonic or polyphonic nature of wheezes defines exact type of related disease. Wheezes monitoring identifies asthma attack and results of therapies, and wheezes characteristics recognize severity level of obstruction (*4*). Wheezes have multiple tones, typically 4-6 and usually 1-6 in the spectrum 100–1200 Hz. The home medical personal assistance to asthma patients is important so research is being carried out, trying to explore low computational, low power consumption and user friendly devices to maintain the health (*5*).

Conventionally wheeze detection algorithms depend on wheeze amplitude (*4*). But wheeze amplitude dependent on flow rate, efforts are going to construct algorithms without sound attenuation. So currently authors are focusing to identify wheeze through artificial intelligence techniques related to classifiers and feature extraction (*4*).

II. Methodology

Search done with keywords "Wheeze classifiers" and "wheeze feature extraction". Only those studies included: (1) studies in English language, (2) studies discussing other than wheeze sound were discarded, (3) most relevant studies discussing wheezes detection through supportive machine learning process included, (4) most relevant articles in 2000 – 2016 meeting all criteria included, (5) studies with insufficient information or methodology discarded. Twelve studies selected for detailed review. Table 1 gives: author, data collection details, data collection environment, classifiers, feature extraction, results and key findings of selected studies.

III. Artificial intelligence techniques

Feature selection involves the extraction of parameters from the training data that employ to define the wheeze sounds, and classifiers discriminate data into various classes based on variety of learning algorithms.

F. Ibrahim et al. (eds.), *3rd International Conference on Movement, Health and Exercise*,
IFMBE Proceedings 58,
DOI: 10.1007/978-981-10-3737-5_8

Table 1: Wheeze classification analysis.

Author	Data collection Detail	Feature Extraction	Classifiers	Result
Stayline et al. 2010 (*4*)	21 adults: 10 patients with COPD and 11 patients with asthma	23 features, Local peaks (LP), Global peaks (GP), WBC, WBS etc.	Not mention	Pathologies can be differentiated based on sound signal information, using set of features.
Stayline et al. 2007 (*6*)	Asthma subjects	WBC, WBS	Not mention	Nonlinear characteristic of wheeze relate to the severity level of pathology.
Marcin et al. 2015 (*5*)	130 wheezy and 130 normal recordings	ASE and TI audio spectral	SVM	Accuracy 93.7 % ASE and TI features have low computational complexity and high accuracy.
Bahoura 2009 (*7*)	24 subjects; 12 normal and 12 wheezing	MFCC, FT , WT , LPC, SBC	FT, LPC, WT, MFCC	Sensitivity = 94.6%, Specificity = 91.9% GMM/MFCC combination performs well
Bahoura et al. 2004 (*8*)	24 subjects: 12 normal and 12 wheezing	MFCC and sub-band Based cepstral (SBC)	VQ, GMM and MLP	Accuracy = 95±5% Smoothing the score function increase accuracy.
Bahoura et al. 2003 (*9*)	24 subjects: 12 normal and 12 wheezing	MFCC	(VQ	Wheezing classification MFCC=77.5%, SBC= 76.6%
Mazic et al. 2015 (*10*)	16 children (age 1-6): wheezy patients	MFCC	Two cascaded SVM classifiers	100 % reliability , cascade classifier increase accuracy
Oud et al. 2000 (*11*)	10 subjects Induced Asthma	Box-Cox transformation	k-nearest neighbor	Average classification = 77±8 %
Oud 2003 (*12*)	10 subjects Induced Asthma	Box-Cox transformation	One-nearest-neighbor Feed-forward neural network	Interpolation of the degree of airway obstruction yielded a score of (26±2)% to (80.1±0.8)%
Marcin et al. 2012 (*13*)	140 normal recordings and 140 wheezy recordings	ASE, TI, CF, Kurtosis (K), energy ratio (ER)	SVM	Accuracy = 92.8%, sensitivity = 93%, Specificity = 100% (ASE, TI)
Marcin et al. 2012 (*14*)	Subjects not mention: 1024 samples of wheeze and normal sounds	Correlation based feature (CF), Energy ratio (ER), Linear prediction (LP) etc.	SVM	Efficiency = 84.2% (TI,CF)
Marcin et al. 2011 (*15*)	50 normal and 50 samples of artificial wheeze	Kurtosis (K), Spectral peak entropy (SPE), Spectral flatness (SF), TI	SVM	Efficiency = 84.2% with TI

A. Nonlinear Behavior

Stayline et al. investigated nonlinear behavior of wheeze musicality wave in time-frequency domain with wavelet transform. Wheeze wave's nonlinear behavior studied by wavelet bispectrum (WBS) and wavelet Bicoherence (WBC) after selecting 23 features. Local peaks behavior LPS calculated to investigate microscopic resolution of wheezes in time-bifrequency domain. Similarly global peaks (GPs) estimated to examine wavelet biamplitude in time-bifrequency phase (*4*).

Analysis done on 393 wheezes obtained from eleven asthma patients and ten from COPD patients. Wheezes segmented by two physicians in polyphonic and monophonic wheezes through audiovisual inspection for feature extraction (*4*). However parametric analysis of features done but validation of experiment has not been reported (*4*).

B. *Audio Spectral Behavior*

Marcin et al. in 2015 studied acoustic wheeze sounds features, tonality index (TI) and audio spectral envelope (ASE) adopted from MPEG-2 and MPEG-7 respectively, for wheezes identification. More than twenty tonality calculated based on signal statistics, Fourier spectrum energy, bi-spectrum and linear prediction. TI tonality measures when peaks appear due to wheezes, while ASE measure both normal and wheezy breaths. ASE measures the peaky of spectra due to normal and wheezy sounds i.e. spectrum fluctuation is measured. In ASE colored noise appears with the existence of multi-tones greater than one (*5*).

For experiment 130 wheezy and 130 normal recordings were collected, 30 % data collected from recording and other from online available systems. Data collected under the assistance of physician but the segmentation for feature extraction exhibit by researchers. SVM classifier use to classify wheezes and normal sounds, so 93% accuracy attained with the combination of ASE and TI. This algorithm has some limitations which include: limited number of samples for experiment, Number of subjects, age of patients not reported, ASE tonality is not efficient for single tone wheeze, only chest sounds collected for experiment (*5*).

Marcin et al. in 2012 calculated ASE by summing power spectrum energies in predefined frequency bands, which are further divided in lower and higher bands. Vector composition measured for spectrum analysis. Maximal relevance minimal redundancy (MRMR) applied for feature reduction (*13*).

C. *Frequency and Spectral Behavior*

In 2015 Mazic classify wheeze and non-wheeze in further classes by machine learning algorithm. The algorithm consists of two layers, first layer with two Cascaded SVM classifiers and second layer of threshold. Feature extracted by MFCC through energy distribution (kurtosis) and measure of disorder (entropy) (*10*).

In 2003 Bahoura classify normal and patients in two classes. Feature extracted through Mel Cepstrum Coefficient (MFCC) by cepstral analysis and vector quantization classifier for classification. Bahoura also did experiment on already defined features by other researchers with same data. MFCC and spectral base cepstrum shows high efficiency (*9*). In 2004 Multi-layer Perceptron (MLP) compared to Gaussian Mixture Model (GMM) and VQ classifiers results with the combination of MFCC and sub-band based cepstral (SBC) parameters (*9*). In 2009 total 17 combinations of classifiers and feature extraction methods studied by Bahora. Cepstral analysis (MFCC), wavelet transforms (WT), SBC, linear predictive coding (LPC), and Fourier Transform (FT) methods analyzed for feature extraction. Feature extracted from power spectra, mean square prediction error, amplitude spectra and energy spectra. GMM, MLP, and artificial neural network (ANN) classifiers (*7*). Finally best results reported by MFCC and GMM combination.

D. *Spectral Behavior*

In 2000 Oud classify asthma wheeze in classes according to airway obstruction using Box-Cox transformation and k-nearest neighbor classifier. Effect of discrete Fourier transforms (DFT) and Welch spectra on classification also discovered. It was conclude that (1) Welch spectra show better result of classification for small bins, while discrete Fourier transform (DFT) show better result for larger bins, (2) Detailed spectra information can classify individuals detailed airway obstruction in asthmatic patients. (3) 60% - 90% of sound spectra can be classified in class (*11*). In 2003 one-nearest neighbor classifier applied, and hidden layers feed forward neural network with back propagation of the error. Principal component analysis used for reduction of number of parameters and mapping data done for FEV_1 labels. It concluded that full spectral information is sufficient for the classification of asthma severity level of air way obstruction and to calculate lung function (FEV_1) (*12*).

IV. Conclusion

Judging from the increasing number of research papers on respiratory machine learning reveals that research is increasing in the area. However, the literature work reveals that classification of wheeze still need improvements. Mostly authors have not reported subject's details. Current study also reveals that, existing algorithms can only identify the wheeze for identification of patient health condition. Literature notice, mostly SVM classifier is used for classification of normal and wheezy patients. It has been conclude that lung sounds and lung function FEV_1 has correlation in asthma patients, so FEV_1 can be measured by respiratory sounds. It also has been concluded that respiratory sounds has sufficient information for severity level of airway obstruction and for differentiating pathologies of wheeze related patients.

Conflict of Interest

The authors declare that they have no conflict of interest.

References

1. WHO., W. H. O.
2. Meslier, N., Charbonneau, G., and Racineux, J. L. (1995) Wheezes, *Eur Respir J 8*, 1942-1948.
3. Palaniappan, R., Sundaraj, K., and Ahamed, N. U. (2013) Machine learning in lung sound analysis: A systematic review, *Biocybernetics and Biomedical Engineering 33*, 129-135.
4. Taplidou, S. A., and Hadjileontiadis, L. J. (2010) Analysis of wheezes using wavelet higher order spectral features, *IEEE Trans Biomed Eng 57*, 1596-1610.
5. Wisniewski, M., and Zielinski, T. P. (2015) Joint application of audio spectral envelope and tonality index in an e-asthma monitoring system, *IEEE J Biomed Health Inform 19*, 1009-1018.
6. Taplidou, S. A., and Hadjileontiadis, L. J. (2007) Nonlinear analysis of wheezes using wavelet bicoherence, *Comput Biol Med 37*, 563-570.
7. Bahoura, M. (2009) Pattern recognition methods applied to respiratory sounds classification into normal and wheeze classes, *Comput Biol Med 39*, 824-843.
8. Bahoura, M., and Pelletier, C. (2004) Respiratory sounds classification using Gaussian mixture models, *Electrical and Computer Engineering*, 2-5.
9. Bahoura, M., and Pelletier, C. (2003) New parameters for respiratory sound classification, In *Canadian Conference on Electrical and Computer Engineering, 2003. IEEE CCECE 2003.* , pp 1457-1460 vol.1453.
10. Mazić, I., Bonković, M., and Džaja, B. (2015) Two-level coarse-to-fine classification algorithm for asthma wheezing recognition in children's respiratory sounds, *Biomedical Signal Processing and Control 21*, 105-118.
11. Oud, M., Dooijes, E. H., and van der Zee, J. S. (2000) Asthmatic airways obstruction assessment based on detailed analysis of respiratory sound spectra, *IEEE Trans Biomed Eng 47*, 1450-1455.
12. Oud, M. (2003) Lung function interpolation by means of neural-network-supported analysis of respiration sounds, *Med Eng Phys 25*, 309-316.
13. Wisniewski, M., and Zielinski, T. P. (2012) MRMR-based feature selection for automatic asthma wheezes recognition, In *Signals and Electronic Systems (ICSES), 2012 International Conference on*, pp 1-5, IEEE.
14. Wisniewski, M., and Zielinski, T. P. (2012) Tonality detection methods for wheezes recognition system, In *2012 19th International Conference on Systems, Signals and Image Processing (IWSSIP)*, pp 472-475.
15. Wisniewski, M., and Zielinski, T. P. (2011) Application of Tonal Index to pulmonary wheezes detection in asthma monitoring, In *Signal Processing Conference, 2011 19th European*, pp 1544-1548, IEEE.

Author: Fizza Ghulam Nabi
Institute: Universiti Malaysia Perlis (UniMAP)
Street: School of Mechatronic Engineering
City: Pauh Putra
Country: Malaysia
Email: engr.fizza@yahoo.com

PCA Based Selective Mapping Technique for Reduced PAPR Implemented for Distributed Wireless Patient Monitoring Epilepsy Classification System

S. K. Prabhakar and H. Rajaguru

Department of ECE, Bannari Amman Institute of Technology, Sathyamangalam, India

Abstract— **To assess the neurological conditions, non-invasive techniques for recording activities of the brain are used widely. The recording of the brain activity helps us to understand the brain functions properly. The scalp Electroencephalography (EEG) recordings act as a fundamental tool to trace the pathological brain activity and so it is used widely to treat the patients affected with epilepsy. Since the recordings of the EEG are quite long and hectic to process, Linear Graph Embedding (LGE) is used to reduce the dimensions of the EEG data in this paper. It is then transmitted to the Space Time Trellis Coded Multiple Input Multiple Output Orthogonal Frequency Division Multiplexing (STTC MIMO OFDM) System. As the system suffers from a high PAPR, Principal Component Analysis Based Selective Mapping Technique (PCA-SLM) is used to reduce the PAPR. At the receiver the Gaussian Kernel Based Support Vector Machine (GK-SVM) is employed to act as a post classifier to classify the epilepsy risk levels from EEG signals. Along with the classification results, the Bit Error Rate (BER) is also analyzed in the receiver. The performance metrics here are Specificity, Sensitivity, Time Delay, Quality Values, Accuracy and Performance Index.**

Keywords— **Epilepsy, EEG, LGE, STTC MIMO-OFDM, PAPR, PCA-SLM, GK-SVM, BER**

I. Introduction

A number of temporary changes in the behaviour and perception are caused because of epileptic seizures [1]. Epileptic seizures are usually caused by excessive and synchronized activity of large group of neurons. In the EEG recordings of the human, they are often reflected by multiple ictal patterns. To initiate the in depth diagnostic analysis and procedures during the seizure period and to differentiate the seizure-like symptoms from original epileptic seizures, early detection of the onset of ictal patterns is to be observed in the EEG data [2]. Since the time consumption is more and quite hectic workload is involved in the detection of seizures by clinical experts and also it more prone to erroneous analysis, several attempts were made to develop and implement automatic seizure detection systems. Some of the reference papers which triggered us to perform research in epilepsy and provided an insight to the basics of EEG and epilepsy research as follows. Alarcon et.al explained the power spectrum and intracranial EEG patterns at seizure onset in the cases of partial epilepsy [3]. The statistical mapping of scalp-recorded ictal EEG records using wavelet analysis was done by Battiston et.al [4]. The detection of the sudden seizure onset in epileptic patients from intracranial EEG signals was studied in detail by Estellar [5]. The methodology of the paper is shown in the block diagram Figure 1.

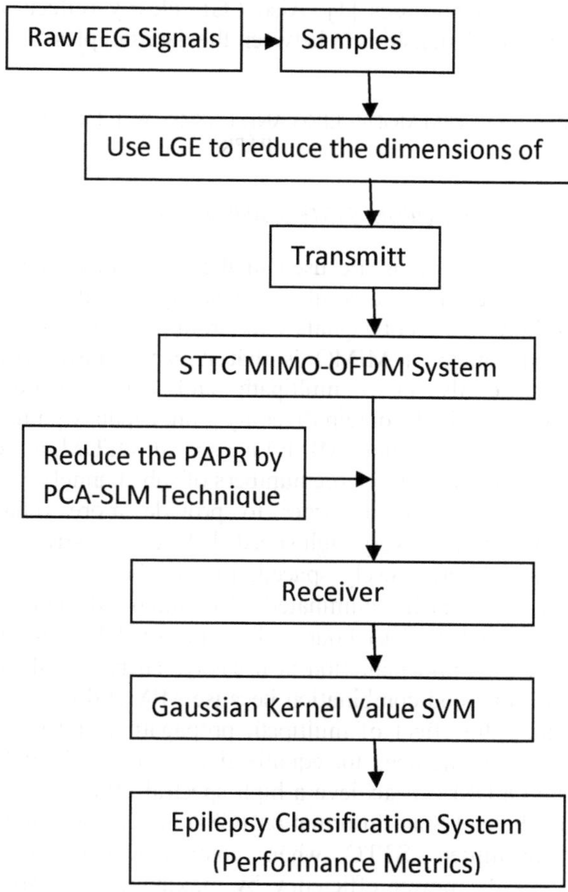

Figure 1 Block Diagram of the Work

The paper is organized as follows. In section 2, the materials and methods are discussed followed by the modeling and design of STTC MIMO-OFDM System with reduced PAPR with the help of PCA Based SLM in section 3. The Gaussian Kernel Value Based SVM is discussed in

F. Ibrahim et al. (eds.), *3rd International Conference on Movement, Health and Exercise*,
IFMBE Proceedings 58,
DOI: 10.1007/978-981-10-3737-5_9

section 4 and the results and discussion is discussed in section 5 and finally concluded.

II. MATERIALS AND METHODS

The EEG dataset was obtained from Sri Ramakrishna Hospital, Coimbatore from 20 epileptic patients in European Data Format (EDF). For all the 20 patients, the recordings were done continuously for thirty minutes. To identify the essential changes happening in the EEG signal, a 2 second epoch is required and so the recorded EEG signal was divided into 2 second epoch. There are totally 16 channels present and it measured for three different epochs. The Linear Graph Embedding is used as a dimensionality reduction technique because it is more robust to noise. Studies on LGE prove that sparse graphical representation can definitely convey highly valuable information for classification purposes [1]. Also LGE clearly reflects the underlying relationships in between the data samples.

III. SYSTEM MODELLING AND DESIGN WITH REDUCED PAPR

A. STTC MIMO-OFDM SYSTEM AND PAPR

Multiple antennas are used in the transmitter side and receiver side and as a result the capacity is multiplied by transmitting different signals over multiple antennas. The main advantage of MIMO is that it can achieve spatial diversity easily in a multipath and dense scattering environment [6]. To obtain diversity gain, capacity gain and to combat signal fading, MIMO is used. In OFDM, a radio channel is divided into large numbers of sub channels which are closely spaced in order to provide more reliable communications at very high speeds [7]. As the narrowband sub channels are closely spaced, Intersymbol Interference (ISI) can be easily eliminated. The main advantage of OFDM is that the Fast Fourier Transform (FFT) is used to simplify the implementation in it easily. Therefore MIMO-OFDM is a good combination because MIMO does not try to reduce the effect of multipath propagation and OFDM easily avoids the need for equalization of signal. Therefore MIMO-OFDM can achieve a high spectral efficiency even if the Channel State Information (CSI) is not possessed by the transmitter. STTC when combined with MIMO improves the power efficiency by maximizing the spatial diversity. In order to improve the data rate and the reliability of wireless communication, STTC MIMO-OFDM is used [8].

The PAPR is nothing but the peak amplitude squared (which gives the peak power) divided by the RMS value squared (which gives the average power) [9]. It is the square of the Crest factor C

$$PAPR=\frac{|x|_{peak}^{2}}{x_{rms}^{2}}=C^2$$

$$PAPR_{dB}=10\log_{10}\frac{|x|_{peak}^{2}}{x_{rms}^{2}}=C_{dB}^2$$

The PAPR is widely used in a lot of signal processing applications.

B. PCA BASED SLM FOR PAPR REDUCTION

The input signals are applied with Principal Component Analysis to find out the exact values of linearly uncorrelated variables called principal components [10]. Then the input signals are multiplied with various phase sequences in order to compute the IFFT operations [11]. The shifting of the sparse matrix is done and the input signals are modulated using QPSK modulation. Then IFFT is used to compute the mapped sequences. The calculation of the threshold value is easily done and it is checked to find whether the PAPR exceeds the threshold value. Then the CCDF plot versus probability of PAPR is computed and the BER is accepted. If each and every mapping is considered as statistically independent, then the Complementary Cumulative Distributive Function (CCDF) of PAPR is represented as

$$P(PAPR>Z)=F(Z)^N=\left(1-(1-e^{-Z})^N\right)^U$$

where U is the number of phase sequences, N is the number of subcarriers and Z is the threshold function. The simulation parameters are shown in Table 1.

I. GAUSSIAN KERNEL VALUE BASED SVM

The principle of structural risk minimization is utilized in support vector machines. If a given set of positive and negative examples are present in a scenario, then the SVM learns an optimal method to separate the hyper plane from it [12]. Support Vector Machine is unique because it is quite difficult from the old and traditional pattern recognition techniques. SVM's can be applied to both linearly separable data and linearly inseparable data. SVM's tries and traces the separating hyperplane in linearly separable data so that the data is separated from the largest margin. SVM maps the entire data in the input space into a very high dimensional space for linearly inseparable data and is mathematically represented as follows:

$$x\in R' \rightarrow \phi(x)R^H$$

Table 1 Analysis of Simulation Parameters

Modulation Technique used	QPSK
System Designed	2 x2 STTC MIMO-OFDM
Number of subcarriers	512
Number of sub-blocks	4
Maximum symbols loaded	1e5
Symbol rate considered	250000
Total no. of time slots	2
Windowing function	Blackman-Haris
HPA Model	SSPA
Number of frames used	10
No of OFDM symbols/ frame	4
Bandwidth	5 MHz
Oversampling factor	4

where $\phi(x)$ is nothing but the kernel function and it is used to find the separating hyper plane. There are a lot of kernels like Polynomial, Gaussian, Sigmoidal etc...In this paper; the type of Kernel used is Gaussian [13] and given below.

Type of Kernel	**Inner Product of Kernel** $K(x^t, x_i)$
Gaussian	$Exp\left[-\left\|x^t - x_i\right\|^2 \Big/ 2\sigma^2\right]$

where x is the input vector, x_i is the support vector and σ^2 is the variance. The range of i should be in between $1 \leq i \leq N_s$, where N_s is the total number of support vectors. Thus SVM is a highly versatile tool for learning purpose from a given data. With the effective combination of a feature selection procedure along with a conventional classifier, SVM's can perform more effectively.

II. Results and Discussion

In this part, the PAPR Results, BER Results and the Classification Results are illustrated in detail.

A. PAPR Results

The PAPR results and the BER Analysis results are shown here. It is very obvious from the careful analysis of the figure 2, that when QPSK modulation is engaged the PAPR is mitigated by 2 dB with the effective utilization of PCA Based Selective Mapping Scheme. The Bit Error Rate is also analyzed in Table 2.

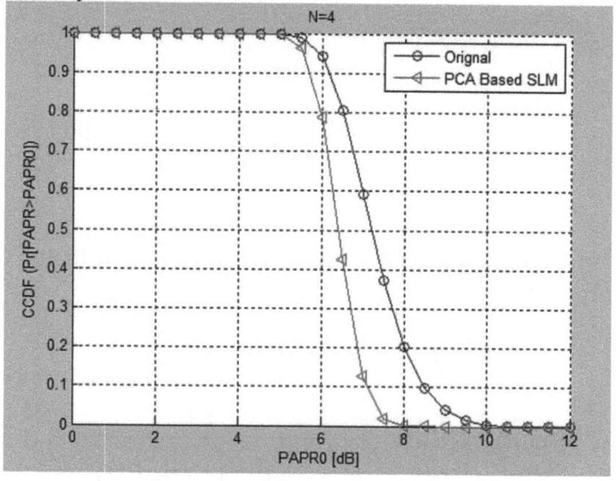

Figure 2 PAPR Reduction Using PCA-SLM Technique

Table 2 BER Analysis for the system

SNR	BER
0	0.1688
2	0.1298
4	0.0629
6	0.0389
8	0.0109
10	0.0004
12	0.0001

B. Classification Results

For LGE as dimensionality reduction technique and GK-SVM as an effective Post Classifier, based on the Performance Index, Quality values, Time Delay and Accuracy the results in the receiver side are calculated and shown in Table 3. The formulae for the Performance Index (PI), Sensitivity, Specificity and Accuracy are given as follows

$$PI = \frac{PC - MC - FA}{PC} \times 100$$

where PC denotes Perfect Classification, MC denotes Missed Classification and FA denotes False Alarm.

The Sensitivity, Specificity and Accuracy measures are stated by the following

$$Specificity = \frac{PC}{PC + MC} \times 100$$

$$Sensitivity = \frac{PC}{PC + FA} \times 100$$

$$Accuracy = \frac{Sensitivity + Specificity}{2} \times 100$$

The Time Delay and the Quality Value Measures are given by the following

$$Time\ Delay = \left[2 * \frac{PC}{100} + 6 * \frac{MC}{100}\right]$$

$$Quality\ Values = \frac{10}{\left[\frac{FA}{100} + 0.2\right] * Time\ Delay}$$

Table 3 Average Performance Analysis Values of 20 Patients

Parameters	Obtained Values
PC (%)	89.02
MC (%)	5.41
FA (%)	5.55
PI (%)	86.93
Sensitivity (%)	94.44
Specificity (%)	94.58
Time Delay (sec)	2.105
Quality Values	19.89
Accuracy (%)	94.51

III. Conclusion

This paper proposes the usage of LGE as a dimensionality reduction technique along with the usage of Gaussian Kernel SVM as a post classifier for the classification of epilepsy risk levels from EEG signals. For the distributed patient monitoring system and telemedicine applications, the Space Time Trellis Coded MIMO OFDM System is utilized completely in this paper. The Perfect Classification of about 89.02% is obtained, with a time delay of about 2.105 seconds, quality value of 19.89. Also, the Performance Index of about 86.93% is found out with an overall average accuracy of 94.51%. The sensitivity and specificity rates obtained are 94.44% and 94.58% respectively. Future works incorporate the usage of various other dimensionality reduction techniques, other types of PAPR Reduction techniques, other types of system design and modeling and finally other types of post classifiers.

References

1. R.Harikumar, P.Sunil Kumar (2015), "Dimensionality Reduction Techniques for Processing Epileptic Encephalographic Signals", Biomedical and Pharmacology Journal, Vol.8, No.1, pp:103-106
2. R.Harikumar, P.Sunil Kumar (2015), "Dimensionality Reduction with Linear Graph Embedding Technique for Electroencephalography Signals of an Epileptic Patient" Research Journal of Pharmacy and Technology, Vol.8, No.5, pp: 554-556
3. Alarcon G, Binnie C, Elwes R, Polkey C (1994) Power spectrum and intracranial EEG patterns at seizure onset in partial epilepsy. *Electroencephalogr Clin Neurophysiol*,94:326 –337.
4. Battiston JJ, Darcey TM, Siegel AM, et al (2003). Statistical mapping of scalp recorded ictal EEG records using wavelet analysis. *Epilepsia* 44:664–672.
5. Esteller, R (1999) *Detection of Seizure onset in Epileptic Patients from Intracranial EEG Signals*. PhD thesis, School of Electrical and Computer Engineering.
6. Lozano A et al (2005), "High-SNR Power Offset in Multi-antenna Communication. *IEEE Tr.Comm*", pp.4134-4151
7. Seyran Khademi et al (2013), " Constant Modulus Algorithm for Peak-to-Average Power Ratio (PAPR) Reduction in MIMO OFDM", IEEE Signal Processing Letters, Vol.20, No.5.
8. Z. Chen, J. Yuan, B. Vucetic (2001), "Improved space-time trellis coded modulation scheme on slow Rayleigh fading channels," Eletronics Lett., vol.37, pp.440-441
9. Xiaodong Zhu et al (2013), "Simplified Approach to Optimized Iterative Clipping and Filtering for PAPR Reduction of OFDM Signals", IEEE Transactions on Communications, Vol.61, No.5.
10. Sunil Kumar Prabhakar, Harikumar Rajaguru (2016), " Soft Thresholding Techniques with PCA as Post Classifier for Epilepsy Risk Level Classification", International Journal of Pharma and Bio Sciences, Vol.7, Issue 2, pg: 687- 693
11. Tao Jiang, Chunxing Ni, and Lili Guan (2013), "A Novel Phase Offset SLM Scheme for PAPR Reduction in Alamouti MIMO-OFDM Systems Without Side Information, " IEEE Signal Processing Letters, Vol. 20, No 4.
12. Harikumar Rajaguru, Sunil Kumar Prabhakar (2016), "LDA, GA and SVM's for Classification of Epilepsy From EEG Signals", Research Journal of Pharmaceutical, Biological and Chemical Sciences, Vol.7(3), pp:2044-2049.
13. R.Harikumar,A. Keerthi Vasan , M Logesh Kumar (2008), "Performance Analysis of Support Vector Machine (SVM) for Optimization of Fuzzy Based Epilepsy Risk Level Classifications from EEG Signal Parameters" International Conference ICBME 2008, IFMBE & BES, Singapore.

Author:	Sunil Kumar Prabhakar
Institute:	Department of Electronics and Comunication Engineering , Bannari Amman Institute of Technology
City:	Sathyamangalam
Country:	India
Email:	sunilprabhakar22@gmail.com

The Effect of Single Bout of 15 Minutes of 15-degree Celsius Cold Water Immersion on Delayed-Onset Muscle Soreness Indicators

N.H. Amir[1], H.A. Hashim[2] and S. Saha[2]

[1] Faculty of Sport Science and Recreation, Universiti Teknologi MARA Cawangan Perlis, Malaysia.
[2] Sport Science Unit, School of Medical Science, Universiti Sains Malaysia, Malaysia.

Abstract— **The aim of this study is to investigate the effectiveness of a single bout of 15 minutes Cold Water Immersion (CWI) to alleviate DOMS symptoms following the plyometric exercise protocol. Sixteen physically healthy young males with the mean age of 21.6±2.3 years old, weight of 65.7±13.1 kg, height of 170.5±6.9 cm, BMI 22.4±3.3 kg.m-1 and fat percentage 20.5±6.2% were required to complete 10 x 10 counter-movement jumps (CMJ) to induce muscle damage. They were randomly assigned into control group (n = 8) and CWI group (n = 8). The CWI group was given a single bout of 15 minutes lower limb of CWI therapy at 15°C following damage-inducing exercise with room temperature was maintained at 16°C. Indicators of DOMS such as perceived muscle soreness, maximal voluntary contraction, range of motion (ROM), thigh circumference, creatine kinase (CK) and lactate dehydrogenase (LDH) were assessed prior to the commencement of CMJ, immediately after the jumps and at 24, 48, 72 and 96 hours post CMJ. The results of mixed-factorial ANOVA revealed a significant (p < 0.05) interaction between groups across the experimental sessions in perceived muscle soreness, ROM and LDH for participants in the CWI group. In conclusion, a single bout of CWI at 15°C for 15 minutes is effective to elicit beneficial effects in some DOMS indicators.**

Keywords— **DOMS; Muscle damage; Cold Water Immersion**

I. Introduction

Participation in physical and sporting activities provides numerous health benefits [1]. However, following exercise and physical activities, the sensation of discomfort are often experienced. It is widely known as "Delayed onset muscle soreness" (DOMS). It is one of the factors that discourage individuals from maintaining exercise participation.

DOMS is commonly experienced by an individual who performed any unacquainted exercise or high intensity eccentric exercise [2][3][4][5]. It can be categorized as one of the most common soft tissue injury occurring due to repetitive and high intensity contraction [6]. The associated symptoms of DOMS include shortening of muscle, increment of passive stiffness within muscle, decrement in strength and power, localized soreness and proprioception disturbance [7]. It usually occurs between eight to 24 hours of post exercise, peaking at 24 – 48 hours and typically diminishes within 96 hours [3][8][9][10]. The effects of DOMS include impairment of athletic performance due to its functional damage [11]. Thus, the strategies to minimize the effects of DOMS are recommended in order to hasten the return of the muscle to its pre-exercise state following exercise [12]. Among various recovery methods, cold water immersion (CWI) has received a growing attention [4][5][13][14][15]. Despite of the popularity of CWI since the time of ancient Greek, research findings in relation to CWI have been inconsistent. The beneficial effects appeared when the participants immersed for 10 minutes at temperature 10°C [16] while in other studies the effects seems to appear at 14 – 15 minutes at temperature of 15°C [17][5]. On a contrary, another study has found there was no beneficial effect observed after single session of 10 minutes and multiple session of 12 minutes immersion at 10°C and 15°C respectively [18][19].

Despite many evidences on the benefits of CWI, inconsistencies exist in relation to methodological variations. It was suggested the important factors that lead to such inconsistencies are frequency of the session (number of immersion either single bout or multiple bouts), duration of each session and temperatures of cold water [20]. It was suggested that CWI treatment should not be time consuming and difficult to set up. [19]. Therefore, the present study aim to examine the effects of single bout of 15 minutes' cold water immersion of 15°C on DOMS indicators following eccentric exercise. It was hypothesized that the prescribed protocol is effective in regulating DOMS markers.

II. Methods and Materials

A. Subjects

Sixteen physically healthy male with age ranging from 19 – 25 years old were recruited from Universiti Sains Malaysia (Health Campus), Kubang Kerian. Inclusion criteria such as having no previous history of lower limb injury in the past three months, untrained and unfamiliar with plyometric exercise protocol were included in this study. Those with absolute and relative contraindications

F. Ibrahim et al. (eds.), *3rd International Conference on Movement, Health and Exercise*,
IFMBE Proceedings 58,
DOI: 10.1007/978-981-10-3737-5_10

were excluded from the study. Absolute contraindications to cryotherapy includes individual with Raynaud's Disease, cold allergy, cryoglubulinaemia and paroxysmal cold hemoglobunuria; whereas individual with arthritic conditions, pheochromocytoma, anaesthetic skin and cardiovascular disease patients are regarded as relative contraindications. Subjects were randomly assigned into two groups; Cold Water Immersion (CWI), and Control group (C) with each group consisted of eight persons.

B. Materials and procedures

Permission to conduct the study was obtained from the Human Research Ethics Committee (HREC), Universiti Sains Malaysia. Subjects were recruited through fliers. Potential subjects were briefed regarding the nature of the study and potential risks and benefits. Prior to the start of the study, signed informed consent form was obtained from the subjects. On the testing day, subjects' height and weight were measured using a stadiometer (Seca 220, Germany) whereas BMI and body fat percentage were analysed using a body composition analyzer machine (Tanita ® TBF- 410, Japan) in shoeless and light clothing condition. All values were recorded to the nearest 0.1 cm and 0.1 kg respectively.

C. Blood parameters measurement.

A 2 mL of venous blood sample was collected prior to exercise protocol from superficial forearm vein using standard venipuncture technique for subsequently four days to determine the Creatine Phosphokinase (CK) and Lactate Dehydrogenase (LDH) serum concentration levels. The blood serum was refrigerated at -20°C until it was analysed using standard Creatine Phosphokinase and Lactate Dehydrogenase kits (BP Clinical Lab Sdn. Bhd, Perak, Malaysia). The testing parameters were repeatedly measured at every 24 hours for four consecutive days.

D. Perceived muscle soreness

Perceived muscle soreness was measured following a 2-seconds body weight squat at 90 degree of knee flexion and muscle soreness was rate whilst adopting 90 degree angle of knee flexion using isokinetic dynamometry machine and repeatedly measured after 24, 48, 72 and 96 hours. Perception of muscle soreness was assessed using 100mm Visual Analog Scale (VAS) with 0 was indicated as "no pain/soreness" and 10 was indicated as "extremely pain/soreness" [21][18][9][22].

E. Maximal Voluntarily Contraction (MVC) measurement

The dominant lower body strength was measured using isokinetic dynamometry machine (Multi-Joint System 3 Pro; Biodex Medical Systems, Shirley, New York, USA) following 5 to 10 minutes warm up and gentle static stretching.

Before assessment begins, positional adjustment of knee flexion and extension were made to ensure the movement was restricted to the sagittal plane and the axis of rotation was parallel to the femoral condyles. The axis of the knee was aligned with the axis of Biodex dynamometer exercise arm. The dynamometer was slid and positioned outside of the leg followed by the adjustment of the knee attachment adjacent to the medial malleoli. The lateral epicondyle was marked as bony landmark to align the axis rotation of the knee joint with the axis rotation of the dynamometer shaft. Two-finger-width distance was used to position the resistant pad above the malleoli on the lower leg to ensure exercise was performed at correct angle. Both knee flexion and extension range of motion was set at 90°.

Subjects were required to complete five maximal voluntary concentric contractions at 45°.s-1 (0.91 rad/s) of the dominant leg through 90° range of movement. The face should facing forward while performing the test in order to avoid any muscular substitution. The angular velocities were chosen in accordance to a similar previous reference study [18]. The angular velocities of 45°.s-1 was the slowest velocity thus appropriate to produce maximum concentric contraction of quadriceps muscle group. Subjects were encouraged verbally as it would motivate them to complete the testing protocol particularly when performing exercise to measure maximum muscle strength. The MVC was expressed as a value of uncorrected peak torque and was used as indicator of maximal quadriceps contractile ability.

F. Knee Flexion (Range of Motion) measurement

Knee flexion joint angle of participants was assessed using a stainless steel goniometer (JAMAR stainless steel 180° 14 inch goniometer, Greendale, Wisconsin, USA). Subjects were instructed to lie down on a bed provided in a prone position with both knees are fully extended. From this position, participants were asked to perform full active knee flexion on the dominant leg and the knee joint angle was determined by using the universal landmarks between the lateral epicondyle of the femur, lateral malleolus of fibula and greater trochanter to ensure alignment [23][19][5][14]. These landmarks were marked on during the first day of measurement with semi-permanent pen to provide re-test reliability and ensure consistency during repeated measurements (post-exercise, 24, 48, 72 and 96 hours post-intervention). Three measurements were taken and only mean value was reported.

G. Thigh circumference (swelling)

Thigh circumference was assessed by using the anthropometric tape which acted as indicator of changes of thigh volume that were likely to occur from osmotic fluid shifts or inflammation [17]. Mid-thigh area was determined between the greater trochanter and lateral epicondyle of femur [19][5][14]. Similar to the previous measurement, the mid-thigh area was marked during the first day of measurement with semi-permanent pen to ensure consistency on the subsequent measurements (post-exercise, 24, 48, 72 and 96 hours post-intervention).

H. Cold Water Immersion Protocol

Subjects in the CWI group were required to immerse their lower limbs (fully submerged up to iliac crest level) within 15 minutes after the completion of plyometric exercise in cold water. Crushed ices were added in water. The temperature was maintained at 15°C ± 1°C. Participants who were assigned into control group were required to sit on a chair to allow normal recovery for 15 minutes following plyometric exercise protocol.

I. Plyometric Exercise Protocol

Subjects were asked to perform plyometric exercise protocol to induce DOMS using 10 sets x 10 repetition of countermovement jump [18]. Subjects were first coached to perform maximal vertical jumps to ensure proper technique to reduce the risk of injury; and adopting 90° knee angle upon landing to induce muscle damage. Precaution steps were taken by separating each set with 30 seconds rest to reduce the risk of injury. This plyometric exercise protocol was performed only once. Participants were verbally encouraged throughout the plyometric exercise protocol to provide continuous support and prevent dropped out possibility before they can reach minimum 80% of the completed jumps.

J. Statistical Analysis

All statistical analyses were performed using the Statistical Package for Social Sciences (IBM SPSS version 20). Descriptive statistics were used to calculate age, height, weight, body mass index (BMI) and body fat percentage. Mixed factorial ANOVA was used to compute the present data as the data has met the assumption for parametric test. Significant value was set at $p < 0.05$ and all data were reported in mean ± standard deviation.

III. Results

Table 1 presented on physical characteristic of participants. A significant interaction was found between group across the testing sessions for perceived muscle soreness (F = 3.05, df = 5, P = 0.02, ETA squared = 0.18). Control group had significantly higher soreness post exercise compared to the CWI group (Table 2). Range of motion (ROM) was found significant between group across the testing sessions for (F = 13.32, df = 5, P = 0.001, ETA squared = 0.49). Control group had significantly higher ROM at pre and post exercise. However, it is worth noting that CWI group improved significantly at 24h to 96h evidence from non-significant group difference (Table 3).

A significant interaction was found in thigh circumference (F = 4.23, df = 5, P = 0.02, ETA squared = 0.23). However, comparison between groups at each testing session revealed no difference. Instead, comparison between measures for each group revealed differences between each testing session for both groups (Table 4). Meanwhile, no significant interaction was found between group across the testing sessions for peak torque (F = 2.32, df = 5, P = 0.05, ETA squared = 0.14) (Table 5).

As for the blood marker parameters, a significant interaction was found between group across the testing sessions for LDH (F = 2.58, df = 5, P = 0.03, ETA squared = 0.16) as presented in Table 6. On the other hand, a nonsignificant interaction was found between group across the testing sessions for CK level (F = 1.59, df = 5, P = 0.17, ETA Squared = 0.10) (Table 7).

Table 1 Physical Characteristic of Participants

Variables	CWI group (n = 8) *M* ± *SD*	Control group (n = 8) *M* ± *SD*
Age (years)	21.13 ± 2.75	22.75 ± 2.32
Weight (kg)	64.30 ± 13.87	58.91 ± 7.37
Height (cm)	168.25 ± 7.61	166.06 ± 4.93
Body mass index [BMI ($kg.m^{-1}$)]	22.43 ± 2.79	21.36 ± 2.73
Fat percentage (%)	18.40 ± 7.56	20.00 ± 5.54

Table 2 Descriptive Statistic for Perceived Muscle Soreness

	Group	Perceived Muscle Soreness	
		M	*SD*
Pre	Control	1.88	1.96
	CWI	2.25	1.49
Post	Control	6.38*	1.41
	CWI	4.00	2.20
24h	Control	6.63	1.06
	CWI	4.87	2.95
48h	Control	6.63	0.92
	CWI	5.50	3.21
72h	Control	3.75	1.28
	CWI	3.63	2.45

96h	Control	2.13	0.84
	CWI	2.50	1.60

*Significant ($p < 0.05$) between group

Table 3 Descriptive Statistic for Range of Motion

	Group	Range of Motion (°)	
		M	*SD*
Pre	Control	133.64*	2.61
	CWI	124.10	3.71
Post	Control	128.28*	4.21
	CWI	118.04	4.49
24h	Control	123.50	7.04
	CWI	117.65	6.51
48h	Control	120.11	3.95
	CWI	120.98	5.23
72h	Control	121.94	4.14
	CWI	123.06	4.59
96h	Control	123.48	4.15
	CWI	124.38	3.95

Table 4 Descriptive Statistic for Thigh Circumference

	Group	Thigh Circumference (cm)	
		M	*SD*
Pre	Control	50.56	5.04
	CWI	52.29	5.24
Post	Control	51.80	4.40
	CWI	53.79	4.73
24h	Control	51.34	4.37
	CWI	53.84	5.35
48h	Control	52.53	4.46
	CWI	53.03	5.19
72h	Control	52.08	4.51
	CWI	52.35	5.56
96h	Control	51.30	4.09
	CWI	52.13	5.27

Table 5 Descriptive Statistic for Peak Torque

	Group	Peak Torque (N·m)	
		M	*SD*
Pre	Control	144.46	36.49
	CWI	172.86	51.26
Post	Control	118.14	33.19
	CWI	131.10	35.06
24h	Control	134.43	44.57
	CWI	116.14	58.05
48h	Control	127.71	48.67
	CWI	131.15	58.95
72h	Control	141.84	46.74
	CWI	148.86	56.04
96h	Control	144.28	46.87
	CWI	174.58	49.49

Table 6 Descriptive Statistic for Lactate Dehydrogenase

	Group	Lactate Dehydrogenase (U/L)	
		M	*SD*
Pre	Control	176.00	40.50
	CWI	133.88	28.36
Post	Control	179.13	40.64
	CWI	133.75*	48.66
24h	Control	174.38	55.43
	CWI	138.88	66.34
48h	Control	183.63	61.16
	CWI	113.25*	37.22
72h	Control	136.38	18.04
	CWI	142.50	27.48
96h	Control	157.63	81.19
	CWI	185.38	79.93

Table 7 Descriptive Statistic for Creatine Kinase

	Group	Creatine Kinase (U/L)	
		M	*SD*
Pre	Control	131.63	32.12
	CWI	166.00	31.56
Post	Control	149.25	37.39
	CWI	195.50	37.70
24h	Control	351.25	236.46
	CWI	256.13	105.25
48h	Control	403.25	261.58
	CWI	298.00	171.81
72h	Control	530.13	374.82
	CWI	316.25	262.41
96h	Control	658.13	496.92
	CWI	396.13	293.31

IV. Discussion

The main findings of this study indicated that CWI induced beneficial effects on perceived muscle soreness, range of motion, and lactate dehydrogenase (LDH) serum. Specifically, muscle soreness began to subside 72 hours onwards after receiving CWI treatment and greater range of motion was recorded at 48 hours post-intervention. Meanwhile, LDH was reported to be significantly lower at post-test and 48 hours post-intervention.

A number of previous studies had suggested that reduction in the pain perception in the first few hours could be resulted from the analgesic effects of cooling rather than inhibition of muscle damage [24[25[26]. However, the duration of analgesic effects is limited to 1-3 hours [26]. Thus, the attenuation in DOMS which might had occurred at one hour post-intervention was unable to be observed as the soreness measurements were taken 24 hours after intervention sessions. Therefore, analgesic effect experienced by the participants was probably diminished after 24 hours post-recovery intervention. This explained the possible increment pattern of soreness at 24 hours and 48 hours' time point. The observed reduction pattern in perceived muscle soreness of CWI group in the present study is consistent with previous investigations at 24 hours, 48 hours and 72 hours [5][15].

Intracellular release of CK and LDH in plasma had been used as indirect biochemical markers of exercise-induced muscle damage (EIMD) to characterize muscle membrane disruption [27][28][8]. It was reported that altered chemical environment within muscles were created due to damage at the sarcolemma and extracellular matrix thus releasing proteins and ions to plasma indicating inflammation [29]. This explained the elevation of LDH, protein metabolites and myoglobin in plasma up to 48 hours following eccentric exercise. The discharge of these substances begins approximately 24 hours post-exercise [30] just before phagocytic cells invaded injury sites [29].

The marked trend of CK in this study is contrary to the previous studies with similar damaging protocol [18]. In the previous study, CK typically peaks at 24 hours [19][5][14][15][16][18] and 48 hours [31] using similar damaging exercise in the lower limb. In addition, previous study has observed that CK peaks beyond 24 hours post-exercise following exercise in the upper limb [21][32]. The reason of these discrepancies are unclear but it was speculated the upper limb muscle are more unaccustomed to eccentric loading hence highly susceptible to greater damage compared to lower limb resulting longer inflammatory response and greater CK efflux in the plasma [33]. It is important to note that lower CK concentration was observed from trained individuals compared to untrained individuals [31][16].

Lactate had been widely known to quickly subside during active recovery due to elevated blood flows. Active recovery is believed to enhanced lactate removal from the muscle cell [34]. It was reported that LDH enzymatic values remained elevated for at least two days after exercise-induced muscle damage [29]. Previous study also reported that CWI was found not significantly beneficial on LDH [5]. However, LDH in the present study was found to have significantly decreased at 48 and 72 hours post-exercise.

It was speculated that CWI could possibly increase the re-absorption of interstitial fluids thus reducing edema [5][22] to improve not only contractile functions but also increased chances to reduce secondary damage as a result of cellular infiltration [20]. However, there was no significant difference found in the present study between recovery interventions although there were reductions observed in the thigh circumference trend throughout time points particularly in CWI groups. This finding was parallel with previous studies that observed no noticeable difference was found between intervention groups in thigh circumference [19][17]. It was suggested the compressive effects resulted from the immersion was thought to facilitate fluid displacements from the periphery to the central cavity thus results in physiological changes such as increases in substrate transport, cardiac output and peripheral resistance [20]. However, these changes apparently were not observed in the present study.

Previous finding have reported no significant difference in range of motion (ROM) of knee [19][14]. However, the present study reported a significant difference was found. ROM observed in CWI group was begun to increase at 48 hours after intervention session compared to control group. The results suggested CWI was effective to reduce the state in which muscle and connective tissue become shortened or stiffed after plyometric exercise. The trend marked in the present study suggested recovery occurred much faster at 48 hours compared to a study by Eston and Peters, (1999) demonstrated more pronounced increment after 72 hours despite no significant difference was found at this time course [17].

Maximal voluntary contraction (MVC) of quadriceps peak torque in the present study was reported insignificant between all groups. The peak torque results in this study followed similar pattern marked by previous studies which indicated a distinct decline immediately post-exercise followed by gradual returned to baseline level throughout 96 hours [19][18][14][35][17][36][37].

There were several study limitations of the present study that need to be highlighted. Firstly, subjects recruited in this study were not matched based on weight, age, height, body mass index (BMI) and percentage body fat (%). Therefore, different results are expected if the participants were matched prior study. Secondly, the study was conducted among sedentary healthy individuals, thus, it is possible that CWI may have different acute and chronic effects in well-trained athletes, which still remains to be elucidated to date.

V. Conclusion

As a conclusion, the present study found that CWI provide beneficial effects on perceived muscle soreness, range of motion, and lactate dehydrogenase (LDH) serum when compared to the control group. Therefore, a single bout of 15 minutes CWI at 15° C can be used as a treatment after completing an exercise induced muscle damage.

Acknowledgment

We would like to thank all the subjects for their participation in the present study.

Conflict of Interest

The authors declare that they have no conflict of interest.

References

1. Warburton, D E, Nicol, C W & Bredin S S. (2006) Health benefits of physical activity: the evidence. Canadian Medical Association Journal 174(6), 801-809.
2. Cheung, K, Hume, P A & Maxwell, L. (2003) Delayed onset muscle soreness. Sports Medicine, 33(2), 145-164.
3. Connolly, D A, Sayers, S E & McHugh, M P. (2003) Treatment and prevention of delayed onset muscle soreness. The Journal of Strength & Conditioning Research, 17(1), 197-208.
4. Sellwood, K L, Brukner, P, Williams, D et al. (2007) Ice-water immersion and delayed-onset muscle soreness: a randomised controlled trial. British Journal of Sports Medicine, 41(6), 392-397.
5. Vaile, J, Halson, S, Gill, N et al. (2008) Effect of hydrotherapy on the signs and symptoms of delayed onset muscle soreness. European journal of applied physiology, 102(4), 447-455
6. Allen, T J, Dumont, T L & MacIntyre, D L. (2004) Exercise-induced muscle damage: mechanisms, prevention, and treatment. Physiotherapy Canada, 56(2), 67-79.
7. Proske, U & Morgan, D. (2001) Muscle damage from eccentric exercise: mechanism, mechanical signs, adaptation and clinical applications. The Journal of physiology, 537(2), 333-345.
8. Clarkson, P M & Sayers, S P. (1999) Etiology of exercise-induced muscle damage. Canadian journal of applied physiology, 24(3), 234-248.
9. Cleak, M & Eston, R. (1992) Muscle soreness, swelling, stiffness and strength loss after intense eccentric exercise. British Journal of Sports Medicine, 26(4), 267-272.
10. Howell, J, Chleboun, G & Conatser, R. (1993) Muscle stiffness, strength loss, swelling and soreness following exercise-induced injury in humans. The Journal of physiology, 464(1), 183-196.
11. Byrne, C, Twist, C & Eston, R. (2004) Neuromuscular function after exercise-induced muscle damage. Sports Medicine, 34(1), 49-69.
12. Cochrane, D J. (2004) Alternating hot and cold water immersion for athlete recovery: A review. Physical Therapy in Sport, 5(1), 26-32.
13. Barnett, A. (2006) Using recovery modalities between training sessions in elite athletes. Sports Medicine, 36(9), 781-796.
14. Howatson, G, Goodall, S & Van Someren, K. (2009) The influence of cold water immersions on adaptation following a single bout of damaging exercise. European journal of applied physiology, 105(4), 615-621.
15. Ingram, J, Dawson, B, Goodman, C et al. (2009) Effect of water immersion methods on post-exercise recovery from simulated team sport exercise. Journal of Science and Medicine in Sport, 12(3), 417-421.
16. Bailey, D, Erith, S, Griffin, P et al. (2007) Influence of cold-water immersion on indices of muscle damage following prolonged intermittent shuttle running. Journal of Sports Sciences, 25(11), 1163-1170.
17. Eston, R & Peters, D. (1999) Effects of cold water immersion on the symptoms of exercise-induced muscle damage. Journal of Sports Sciences, 17(3), 231-238.
18. Jakeman, J, Macrae, R & Eston, R. (2009) A single 10-min bout of cold-water immersion therapy after strenuous plyometric exercise has no beneficial effect on recovery from the symptoms of exercise-induced muscle damage. Ergonomics, 52(4), 456-460.
19. Goodall, S & Howatson, G. (2008) The effects of multiple cold water immersions on indices of muscle damage. Journal of Sports Science and Medicine, 7, 235-241.
20. Wilcock, I M, Cronin, J B & Hing, W A. (2006) Physiological response to water immersion. Sports Medicine, 36(9), 747-765.
21. Howatson, G, Van Someren, K & Hortobagyi, T. (2007) Repeated bout effect after maximal eccentric exercise. International journal of sports medicine, 28(07), 557-563.
22. Vaile, J M, Gill, N D & Blazevich, A J. (2007) The effect of contrast water therapy on symptoms of delayed onset muscle soreness. The Journal of Strength & Conditioning Research, 21(3), 697-702.
23. Tokmakidis, S, Kokkinidis, E, Smilios, I et al. (2003) The effects of ibuprofen on delayed muscle soreness and muscular performance after eccentric exercise. The Journal of Strength & Conditioning Research, 17(1), 53-59.
24. Denegar, C R & Perrin, D H. (1992) Effect of transcutaneous electrical nerve stimulation, cold, and a combination treatment on pain, decreased range of motion, and strength loss associated with delayed onset muscle soreness. Journal of Athletic Training, 27(3), 200.
25. Gulick, D T, Kimura, I F, Sitler, M et al. (1996) Various treatment techniques on signs and symptoms of delayed onset muscle soreness. Journal of Athletic Training, 31(2), 145.
26. Meeusen, R & Lievens, P. (1986) The use of cryotherapy in sports injuries. Sports Medicine, 3(6), 398-414.
27. Howatson, G, Gaze, D & Van Someren, K. (2005) The efficacy of ice massage in the treatment of exercise-induced muscle damage. Scandinavian journal of medicine & science in sports, 15(6), 416-422.
28. Clarkson, P M & Hubal, M J. (2002) Exercise-induced muscle damage in humans. American Journal of Physical Medicine & Rehabilitation, 81(11), S52-S69.
29. Sudhakar, S & Naiya, A. (2012) Biochemical markers an indirect method for evaluating delayed onset muscle soreness among recreational athletes. Unpublished Manuscript.
30. Clarkson, P M, Nosaka, K & Braun, B. (1992) Muscle function after exercise-induced muscle damage and rapid adaptation. Medicine and science in sports and exercise, 24(5), 512-520.
31. Ascensão, A, Leite, M, Rebelo, A et al. (2011) Effects of cold water immersion on the recovery of physical performance and muscle damage following a one-off soccer match. Journal of Sports Sciences, 29(3), 217-225.
32. Nosaka, K, Newton, M & Sacco, P. (2002) Delayed-onset muscle soreness does not reflect the magnitude of eccentric exercise-induced muscle damage. Scandinavian journal of medicine & science in sports, 12(6), 337-346.
33. Jamurtas, A Z, Theocharis, V, Tofas, T et al. (2005) Comparison between leg and arm eccentric exercises of the same relative intensity on indices of muscle damage. European journal of applied physiology, 95(2-3), 179-185.
34. Nichols, W W, Estrada, J C, Braith, R W et al. (2006) Enhanced external counterpulsation treatment improves arterial wall properties and wave reflection characteristics in patients with refractory angina. Journal of the American College of Cardiology, 48(6), 1208-1214.

35. Isabell, W K, Durrant, E, Myrer, W et al. (1992) The effects of ice massage, ice massage with exercise, and exercise on the prevention and treatment of delayed onset muscle soreness. Journal of Athletic Training, 27(3), 208.
36. Miyama, M & Nosaka, K. (2004) Muscle damage and soreness following repeated bouts of consecutive drop jumps. Adv Exerc Sports Physiol, 10(3), 63-69.
37. Miyama, M & Nosaka, K. (2004) Influence of surface on muscle damage and soreness induced by consecutive drop jumps. The Journal of Strength & Conditioning Research, 18(2), 206-211

Author: Nurul Hidayah Amir
Institute: Universiti Teknologi MARA Cawangan Perlis
Street: Kampus Arau
City: 02600 Arau, Perlis
Country: Malaysia
Email: hidayah131@perlis.uitm.edu.my

Conceptual Design of Orthosis for Clubfoot Model Developed Using Image Reconstruction Techniques

E. Vijayaragavan[1], T. V. Gopal[1] and S. Viswanath[2]

[1] Department of Mechanical Engineering, SRM University, Tamil Nadu, India
[2] National Institute of Sports, Kuala Lumpur, Malaysia

Abstract— **Clubfoot is one of the most common deformities distributed worldwide in the ratio of 1to 2 per live births. The subjects with clubfoot have their foot twisted at the ankle, which makes them walk on the lateral surface of the foot. This study involves the design of an orthosis, which is an externally applied device that regulates the musculoskeletal functions of the foot. An infant with clubfoot deformity is selected for this study. Computed tomography (CT) images of the clubfoot with 1 mm slice thickness were captured and fed in MIMICS, an image processing software, where in 3D clubfoot model was developed. Rapid prototyping (RP) model of the clubfoot was prepared using Stratasys FORTUS 360mc. This prototype is taken as reference for designing the orthosis for clubfoot treatment. Considering, the level of deformity, its nature, bean ratio, and age of the subject different conceptual models are developed. The deep heel cup (Model I) orthosis, provides required support for the heel to regulate bone growth and to counteract unbalanced forces. Spiral encloser (Model II) orthosis is supported from the thigh and wounds around the tibia to promote bone growth in the required profile. A mechanical orthosis (Model III) with adjustable supports was developed to cater subject to attain normal posture. Considering the complex profile of the foot and orthosis assembly, Solid 45 – tetrahedral elements are used for meshing. By applying necessary material properties and loading conditions, the structural analysis of the assembly was performed to predict the stress distribution and the pattern of bone growth. The finite element analysis on mechanical orthosis (Model III) shows a maximum displacement of 1.45 mm and the maximum stress of 0.79 N/mm². This study can be further extended by manufacturing and testing the orthosis on different subjects.**

Keywords— **Clubfoot, Orthosis, Image reconstruction, Rapid prototyping.**

I. HUMAN FOOT

The most complex part of the human body is the foot. It serves three important functions: (i) Weight transmission, (ii) Maintaining body balance and (iii) Assists ambulation. The foot and ankle of a healthy individual comprise of 26 bones (about ¼th of the bones in the human body), 107 ligaments, 19 muscles 33 joints, and tendons [1]. These components must work in unison for the proper functioning of the foot. The foot takes the entire body weight while standing and gradually rises to 1.5 times during walking and about eight times during running. The foot provides mechanical and structural strength. It also functions as a shock absorber and transmits forces to the ground during regular activities. The human foot can be divided into three regions as forefoot – comprises of five toes (phalanges) and five metatarsals, midfoot – forms pyramid-like structure responsible for foot arches, consists of cuneiforms, cuboid and the navicular bones and hind foot – forms the heel and ankle. It contains load carrying members the talus and the calcaneus [2]. Any structural deformity or malfunction of the foot inhibits problems elsewhere in the body; similarly, an abnormality in any other parts leads to problems in the feet [3]. The common foot abnormalities in humans are a bunion, plantar fasciitis, flat foot, dropfoot, talipes, pescavus, clubfoot, hump foot, hollow claw foot [4]. Study on the effects of structural characteristics of the bones in the foot on plantar pressure distribution and stress concentration in the bones are very much required for individualized treatment for various foot deformities [5]. The structural arrangement of the human foot with all the bones is shown in Fig. 1.

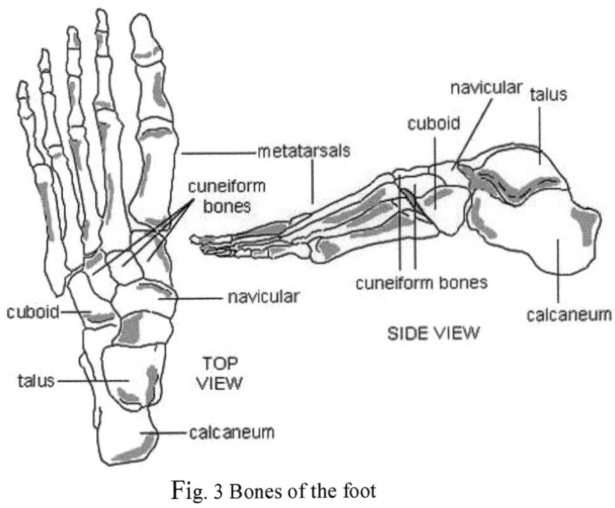

Fig. 3 Bones of the foot

II. CLUBFOOT

Any structural deformity or malfunction of the foot inhibits problems elsewhere in the body; similarly, an abnormality in any other parts leads to problems in the feet [3]. The common foot abnormalities in humans are bunion, plantar fasciitis, flat foot, dropfoot, talipes, pescavus, clubfoot, hump foot, hollow claw foot [4]. Study on the effects of structural characteristics of the bones in the foot on plantar pressure distribution and stress concentration in the bones are very much required for individualized treatment for

F. Ibrahim et al. (eds.), *3rd International Conference on Movement, Health and Exercise*,
IFMBE Proceedings 58,
DOI: 10.1007/978-981-10-3737-5_11

various foot deformities [5]. Clubfoot is medically termed as Congenital talipes equinovarus (CTEV), is a

relatively common developmental deformity affecting one or both feet. The statistics shows an incidence of 0.3–7 per 1000 live births [6]. The affected foot would be inversed, plantar flexed and adducted that makes the subject to walk with the lateral surface of the foot [7]. Treatment for clubfoot involves foot manipulation, serial casting followed by bracing. Nowadays surgical procedures involving Achilles tenotomy is more common. Aetiology of CTEV is unknown, and it may be due to impairment of ontogenesis, chondrogenesis, neurogenesis, angiogenesis, and myogenesis; but the precise mechanisms remain unclear. A child affected with clubfoot and the structural arrangement of the foot is shown in Fig. 2.

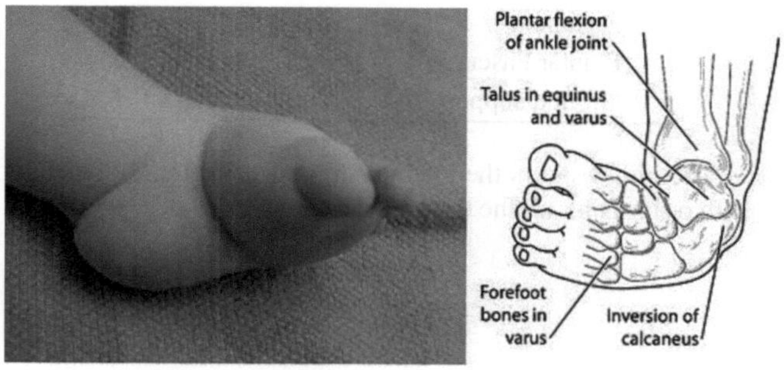

Fig. 2 Child affected with clubfoot

III. BIOMECHANICS OF CLUBFOOT

For a healthy foot, the angle between the foot axis and bimolar axis is to be between 85° to 90°. Similarly, the range of plantar flexion and dorsiflexion must be around 40° and 20° respectively [8]. The misalignment of bony structures in clubfoot is shown in Fig. 3.

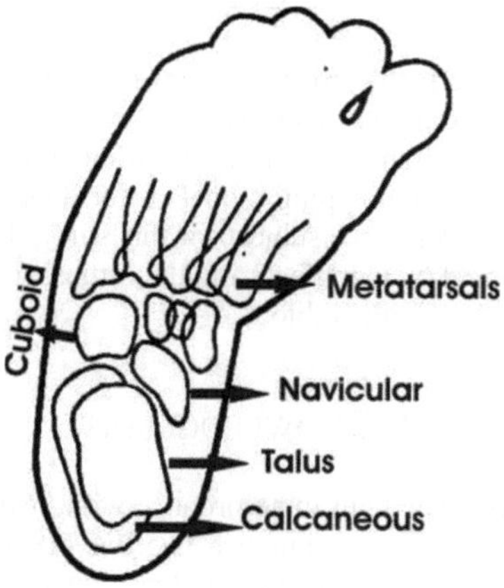

Fig. 3 Arrangement of bones in clubfoot

In this study, it is planned to correct the foot parameters by using an external device, which would regulate the growth pattern to the standard profile.

IV. MODELING OF CLUBFOOT

A child of 6 months old, is selected for this study after obtaining the necessary permission from the parents concerning the ethical committee of SRM University. The Computed tomography (CT) of the affected clubfoot were captured using SIEMENS Somatom Spirit with a slice thickness of 1 mm. These images were exported to image processing software MIMICS 12.1 for a surface generation as shown in Fig. 4.

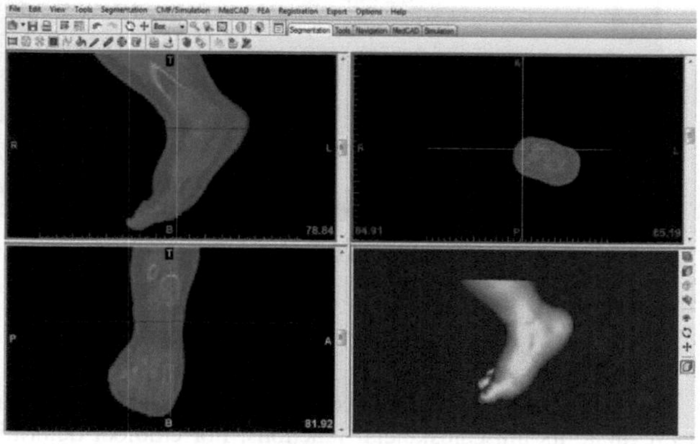

Fig. 4 CT images exported to MIMICS

Segmentation is the process in which the 2D images of anatomical data are stacked to form 3D models. In this process, the region of interest (soft tissues) is separated from other entities by thresholding. The threshold frequency ranging from -700 to 225 was selected for developing the 3D model of the clubfoot. After thresholding, any mask can be divided into a finite number of objects by Region Growing in which unwanted pixels were removed. For further segmentation, masking is done and finally calculate 3D is used for 2D to 3D transformations. The *.STL file was generated and fed to INSIGHT software of the FORTUS 360mc as shown in Fig. 5.

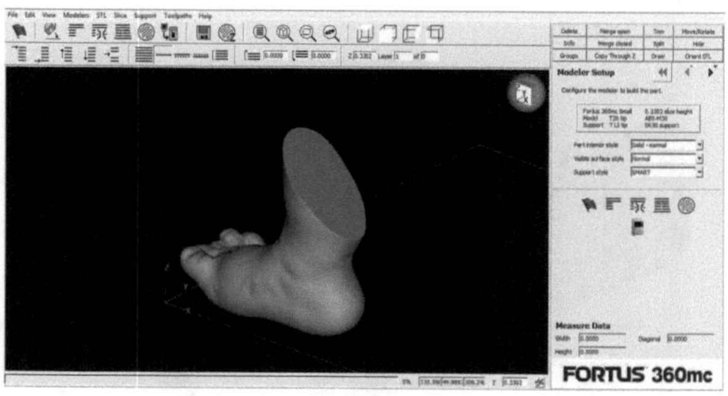

Fig. 5 CT images exported to MIMICS

ABS–M30 material was selected to make this model, and the quantity of the support material and base material were calculated. The RP model of clubfoot thus generated is shown in Fig. 6.

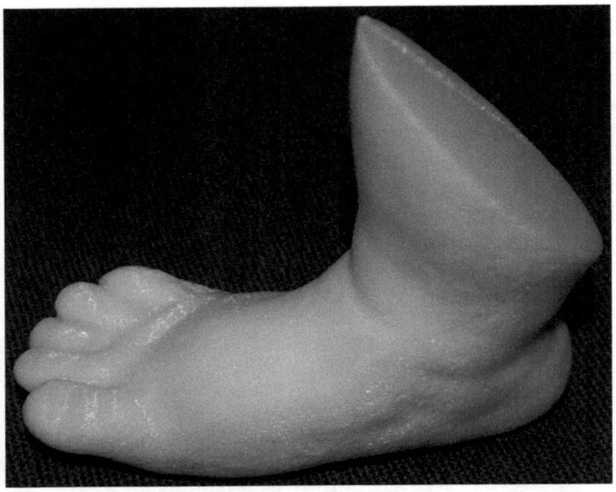

Fig. 6 RP model of clubfoot

V. MODELING OF ORTHOSIS

Orthosis is an external device that assists the anatomical parts to regulate its musculoskeletal functions. For clubfoot deformity, the orthosis is designed based on various factors [9] including (i) age, (ii) level of deformity, (iii) bean ratio – length to width ratio, (iv) nature of deformity.

Model 1 – Deep heel cup: This model forces the heel to grow straight and counteracts any deviations in the foot as shown in Fig. 7(a)

Model II – Spiral encloser: This model consists of three regions: shoe, spiral and thigh as shown in Fig. 7(b). This generates counter forces which regulate bone growth.

Model III – Mechanical orthosis: It consists of cylindrical joints and hinge joints that can be adjusted to cater the needs of the subject as shown in Fig. 7(c).

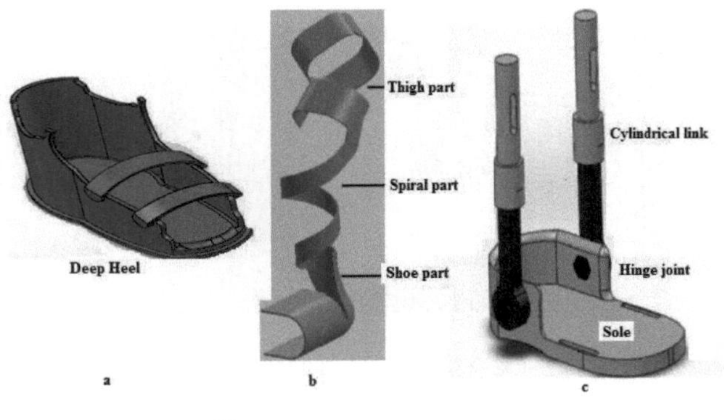

Fig. 7 Model of orthosis

VI. ANALYSIS OF ORTHOSIS

The orthosis model assembled with clubfoot is meshed using solid 45 and prepared for analysis in ANSYS 14.5. The material properties and loading conditions [10-14] are specified in Table 1.

Table 4 Material properties

Component	Modulus of elasticity (MPa)	Poisson's ratio
Bone	7300	0.3
Soft tissue	0.15	0.4
Cartilage	1	0.4
Ligament	260	0.4
Plantar Fascia	350	0.4
Ground support	20000	0.1

The Fig. 8 shows the internal stress distribution in the clubfoot with orthosis for all the three models thus developed.

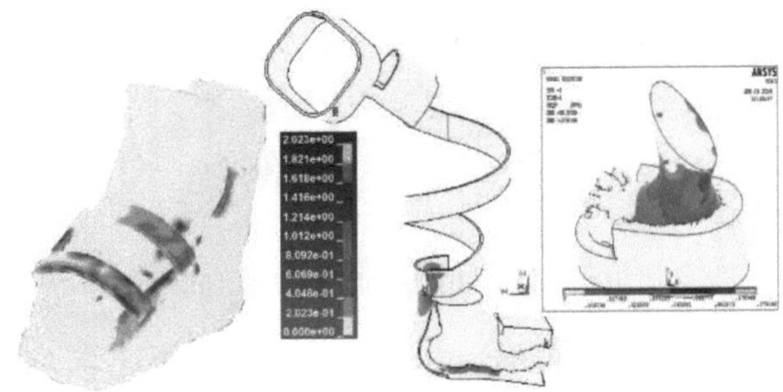

Fig. 8 Analysis of orthosis

VII. CONCLUSION

This paper reports the process of converting the CT images into solid model of the human foot. Three different kinds of orthosis were developed and analyzed in a virtual environment to predict its function. Rapid prototyping model of the foot was developed which serves as a reference subject for this study. This project can be further prepared by manufacturing the orthosis and testing it on the patients affected with clubfoot.

ACKNOWLEDGMENT

This work is carried out in accordance with Institutional Ethics Committee vide clearance number 965/IEC/2016. The Authors thank Dr. Shantanu Patil, Head of the Department of Translational Medicine and Research, SRM University for his guidance in ethical approval and offering valuable suggestions and feedback for this study. Authors also acknowledge the considerable work done by Ms. Leya Miriam Kurain, Mr. D. Somasundaram and Mr. R. Raj Bharath towards this project.

Conflict of Interest

The authors declare that they have no conflict of interest.

References

1. Susan J Hall, (2012) Basic Biomechanics. 6th ed. McGraw-Hill, 251–257
2. Cheryl LR, (1988) Anatomy of the ankle and foot. Physical Therapy, 68(12):1802–1814
3. Diane L France, (2009) Human and Nonhuman Bone Identification: A Color Atlas. CRC Press, pp. 477–504
4. William M Scholl, (1916) The Human Foot - Anatomy, Deformities and Treatment, Foot Specialist Publishing Co., 1916
5. Jason TC, Ming Z, Aaron KL et al. (2005) Three-dimensional finite element analysis of the foot during standing–a material sensitivity study. Journal of Biomechanics, 38(5):1045–1054
6. Miedzybrodzka Z (2003) Congenital talipes equinovarus (clubfoot): a disorder of the foot but not the hand. J Anat 202: 37–42.
7. Ponseti IV (1992) Treatment of congenital club foot. J Bone Joint Surg Am 74:448–454
8. Vijayaragavan E, Leya MK, Sulayman H et al. (2014) Application of Rapid Prototyping in the treatment of clubfoot in children. Procedia Engineering 97:2298–2305
9. Herd F, Ramanathan AK, Cochrane et al. (2008) Foot pressure in clubfoot—The development of an objective assessment tool. The Foot 18:99–105
10. Liu XC, Channing T, Robert R et al. (2014) Newly designed foot orthosis for children with residual clubfoot after Ponseti casting. Journal of prosthetics and orthotics 26(1):38-442
11. Jason TC, Ming Z, Aaron KL et al. (2005) Three-dimensional finite element analysis of the foot during standing - a material sensitivity study. Journal of Biomechanics 38:1045–1054
12. Camacho DL, Ledoux WR, Rohr DS et al. (2002) A three-dimensional, anatomically detailed foot model: a foundation for a finite element simulation and means of quantifyingfoot-bone position. Journal of Rehabilitation Research and Development 39(3):401-410
13. Song HR, Carroll NC, Neyt J et al. (1999) Clubfoot analysis with three-dimensional foot models. Journal of Pediatric Orthopaedics 8(1):5-11
14. Herzenberg JE, Carroll NC, Christofersen MR et al. (1998) Clubfoot analysis with three-dimensional computer modeling. Journal of Pediatric Orthopaedics 8(3):257-262

Author: S.Viswanath
Institute: National Institute of Sports
Street: Bukit Jalil
City: Kuala Lumpur 57000
Country: Malaysia
Email: viswabdu@gmail.com

The Control of an Upper-Limb Exoskeleton by Means of a Particle Swarm Optimized Active Force Control for Motor Recovery

A. P.P. Abdul Majeed, Z. Taha, I. Mohd Khairuddin, M. Y. Wong, M. A. Abdullah and M. A. Mohd Razman

Innovative Manufacturing, Mechatronics & Sports Laboratory (iMAMS),
Faculty of Manufacturing Engineering, Universiti Malaysia Pahang, Pahang, Malaysia

Abstract—**The modelling and control of a two degree of freedom upper extremity exoskeleton for motor recovery is presented in this paper. The dynamic modelling of the upper arm and the forearm for both the anthropometric based human upper limb as well as the exoskeleton was attained via the Euler-Lagrange formulation. A proportional-derivative (PD) architecture is employed to assess its effectiveness in performing joint-space control objectives namely the forward adduction/abduction on the shoulder joint and the flexion/extension of the elbow joint. An intelligent active force control (AFC) optimised by means of the Particle Swarm Optimisation (PSO) algorithm is also integrated into the aforesaid controller to examine its efficacy in compensating disturbances. It was established from the simulation study that the PD- PSOAFC performed notably well in catering the disturbances introduced to the system whilst maintaining its excellent tracking performance as compared to its pure classical PD counterpart.**

Keywords— **Active force control, Particle swarm optimisation, Robust, rehabilitation, Trajectory tracking control.**

I. Introduction

Over the past couple of decades, the life expectancy amongst the ageing has increased steadily around the globe [1]. Approximately 8.3% of Malaysia's population is well over 60 years old [1]-[2]. The number of stroke patients as reported in the 2013 Malaysian Ministry of Health's annual report, demonstrated an average increase of threefold annually [3]. Furthermore, it was also reported in the 2011 report, that 11% and 7.2% of Malaysians between the age of an infant to 18 years old are affected by physical and cerebral palsy disabilities [2]. It is not uncommon for people that fall into the aforesaid figures are affected by complete or partial loss of motor control of the upper-limb which essentially affects their activities of daily living [4].

It has been established from the literature that through continuous and repetitive rehabilitation activities, these individuals may regain their mobility [4]–[7]. Nonetheless, it is worth to note that conventional rehabilitation therapy is often deemed to be laborious and costly which in turn limiting the patient's rehabilitation activities [4]–[6]. Owing to the increasing demand for rehabilitation coupled by the shortcomings of conventional rehabilitation therapy has led the research community to address the aforementioned issues through the engagement of robotics [4], [5], [8], [9]. It is hypothesised that the use of exoskeletons may progressively eliminate the long hours of consultation as well as rehabilitation sessions and subsequently accommodate more patients.

The control strategies developed over the years with respect to rehabilitation robotics reported in the literature can be classified into four main classes, viz. position tracking control, bio-signal, impedance and force control based control as well as adaptive control [10]. As previously mentioned, one's mobility may well be further developed through repetitive and continuous exercise on the impaired limb. This form of training is of particular importance, especially in the early stage of rehabilitation whereby passive mode is required, and this therapy may be achieved through positional or joint based trajectory tracking control.

Hitherto, a number of control strategies have been employed with regard to positional or joint based trajectory tracking control for the upper limb exoskeleton system i.e. Proportional-derivative (PD) controller [11]-[12], nonlinear sliding mode control (SMC) [13]-[14], modified non-linear computed-torque control [15] as well as intelligent based controller such as fuzzy-based PD [16] amongst others. The objective of the study is to examine the tracking performance of a robust control scheme viz. a hybrid proportional-derivative particle swarm optimised active force control (PD-PSOAFC) that is to some extent oblivious to the presence of disturbances of a two DOF upper limb exoskeleton system. The performance of the proposed controller is then compared to a classical pure PD controller by exciting the same form of disturbance into both systems. To the best of the authors' knowledge, the study is novel as the proposed controller has yet been utilised in any upper limb exoskeleton system.

F. Ibrahim et al. (eds.), *3rd International Conference on Movement, Health and Exercise*,
IFMBE Proceedings 58,
DOI: 10.1007/978-981-10-3737-5_12

II. Upper limb dynamics

Figure 1 illustrates the upper limb dynamics of the human limb and exoskeleton are modelled as rigid links joined by joints (bones). The two-link model restricted to the sagittal plane is a rather conservative system as the human-machine interaction is presumed smooth. Furthermore, it is noteworthy to mention that the frictional elements, as well as other unmodelled variables, are disregarded.

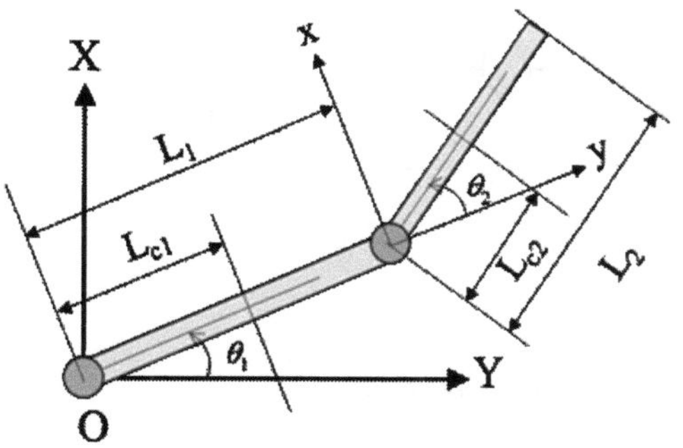

Fig. 4 The two-link manipulator that mimics the human upper limb

The parameters of the upper arm and the forearm illustrated in Figure 1 are represented by the subscripts 1 and 2, respectively. The joint position of the shoulder and elbow is also depicted in the figure. L is the length segments of the limb; L_c is the length segments of the limb about its centroidal axis and θ is the angular position of the limbs. The equation of motions for the non-linear dynamic system is derived by means of the Lagrangian. The following vector form expresses the upper-extremity dynamics of the coupled nonlinear differential equations

$$\begin{bmatrix}\tau_1\\ \tau_2\end{bmatrix}=\begin{bmatrix}D_{11} & D_{12}\\ D_{21} & D_{22}\end{bmatrix}\begin{bmatrix}\ddot{\theta}_1\\ \ddot{\theta}_2\end{bmatrix}+\begin{bmatrix}C_{11} & C_{12}\\ C_{21} & C_{22}\end{bmatrix}\begin{bmatrix}\dot{\theta}_1\\ \dot{\theta}_2\end{bmatrix}+\begin{bmatrix}G_1\\ G_2\end{bmatrix}+\begin{bmatrix}\tau_{d1}\\ \tau_{d2}\end{bmatrix} \quad (1)$$

where **τ, D, C, G** and $\boldsymbol{\tau}_d$ denote the actuated torque vector, the 2 × 2 inertia matrix of the human limb and the exoskeleton, the Coriolis and centripetal torque vector, the gravitational torque vector and the external disturbance torque vector, respectively. The readers are referred to [17] for the detail components of the above vector formulation.

III. Control architecture

A. Active Force Control

Hewit and Burdess initially proposed the concept of AFC in the early eighties, and it is established on the principle of invariance and Newton's second law of motion [18]. Mailah et al. have improved the effectiveness of the aforementioned control scheme by integrating artificial intelligent techniques in estimating the inertial matrix of the dynamic system that essentially activates the compensation effect of the controller [19]–[28]. Furthermore, the robustness of this method has also been well demonstrated in various applications both numerically as well as experimentally [22],[23],[26]-[28].

The torque generated is governed by the classical PD control law, typically expressed as [17]

$$\tau = K_p(\theta_d - \theta) + K_v(\dot{\theta}_d - \dot{\theta}) \quad (2)$$

where, $\dot{\theta}_d$ and $\dot{\theta}$ are the desired and present angular velocities, respectively, θ_d and θ are the desired and present positions, respectively, whilst K_p and K_d are the proportional and derivative constants, respectively. In the study, the controller gains were assumed to be appropriately tuned by heuristic means.

To eliminate the actual disturbances τ_d, the estimated disturbance torque τ_d* has to be calculated and is given by the following equation

$$\tau_d^* = \tau - \mathbf{IN}\,\ddot{\theta} \quad (3)$$

where **IN** is the estimated inertial matrix, whilst τ and $\ddot{\theta}$are the measured applied control torque acceleration signal, respectively. The value of **IN** may be expressed in the following form

$$[\mathbf{IN}] = [\mathbf{D}] \quad (4)$$

where only the diagonal terms of **D** are considered. The off-diagonal terms of the matrix are ignored as it has been demonstrated that the cross-coupling term may be disregarded [18]. The acceleration, as well as the actuated torque of the lower limbs, were also assumed to be perfectly modelled (i.e. the noises from sensors are also totally neglected) in this particular study. A graphical representation of the PSOAFC scheme with the PD controller applied to the exoskeleton is illustrated in Figure 2. The PD-PSOAFC control scheme is only activated upon the engagement of the AFC loop. The system is driven by the classical pure PD architecture without the initiation of the AFC loop.

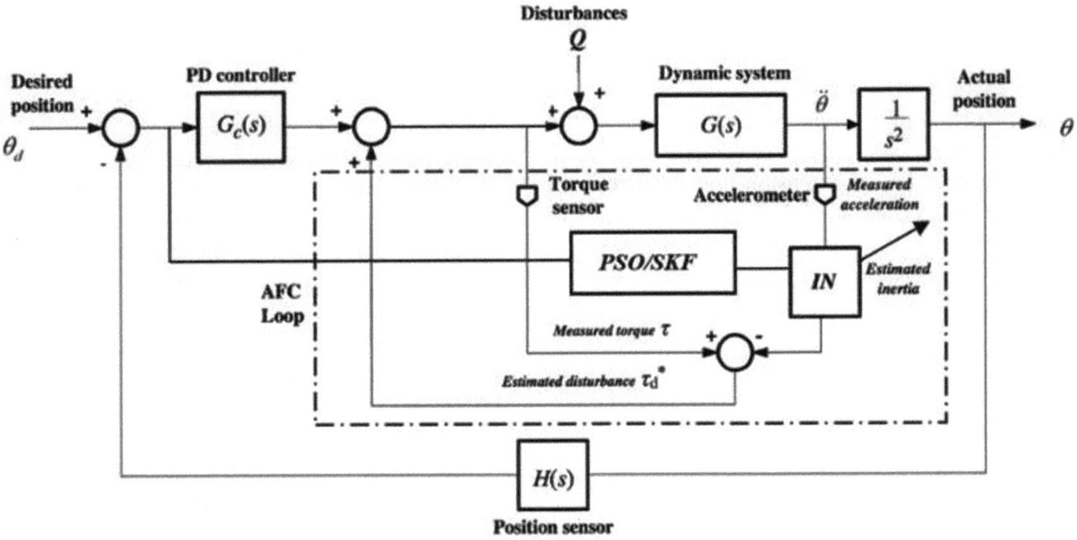

Figure 1 PD with PSOAFC scheme for the control of the exoskeleton

B. Particle Swarm Optimisation

PSO is an evolutionary computational technique developed by Kennedy and Eberhart in the mid-nineties [29]. It is a robust stochastic optimisation approach motivated by the social behaviour of a bird flock. Owing to the limited number of parameter adjustments that is required as well as its robustness, PSO has been demonstrated to work well over a broad range of both linear and non-linear applications [29]–
[32]. It is based on a notion that a swarm may be represented by a population, whilst a particle is represented by an individual. This particle is treated as a point in a multidimensional space, in which the particle adjusts its flying route based on its own flying experience as well as the experience of other particles. The adjustment is driven by the success of a particle itself and the success of its neighbouring particles. Each particle will produce the position of the particles as well as its velocity. The particle position for the next generation is given by the following equation

$$s_i^{k+1} = s_i^k + v_i^{k+1} \tag{5}$$

From (5), the next particle position, s_i^{k+1} is governed by its current position, s_i^k and the velocity for the next placement. This velocity, v_i^{k+1} may be computed through

$$v_i^{k+1} = \omega v_i^k + c_1 r_1 \left(pbest_i - s_i^k\right) + c_2 r_2 \left(gbest_i - s_i^k\right) \tag{6}$$

Referring to (5) and (6), v_i^k and s_i^k are the particle *i*'s velocity and position at the *k*-th generation. r_1 and r_2 are random numbers, whilst c_1 and c_2 are the cognitive and social learning factors taken as 1.42 respectively. In this study the decrease inertia weight, ω is used and it may be expressed as

$$\omega = \omega_{\max} - \left(\frac{\omega_{\max} - \omega_{\min}}{k_{\max}}\right) * k \tag{7}$$

Where $\omega_{\max}$ and $\omega_{\min}$ are the maximum and minimum value of the inertia weight selected as 0.9 and 0.4 respectively. Whilst, k and $k_{\max}$ are the particle at *k*-th generation and its maximum generation, respectively. Figure 3 illustrates the process flow of the PSO algorithm.

In this study, PSO is applied to estimate a set of appropriate **IN** value based on the track error information. The sum of the root mean square error (RMSE) will be used as the performance index in which the lower value of RMSE represents the best set of values for **IN**.

IV. SIMULATION

MATLAB and Simulink software packages are utilised in performing the simulation work for the study. The Simulink block diagram for the proposed scheme consists of a number of subsystems namely the trajectory planner, the PD controller; the main PSO optimised AFC loop, the upper limb dynamics model as well as the disturbance model. The following are the simulation parameters employed in the simulation study

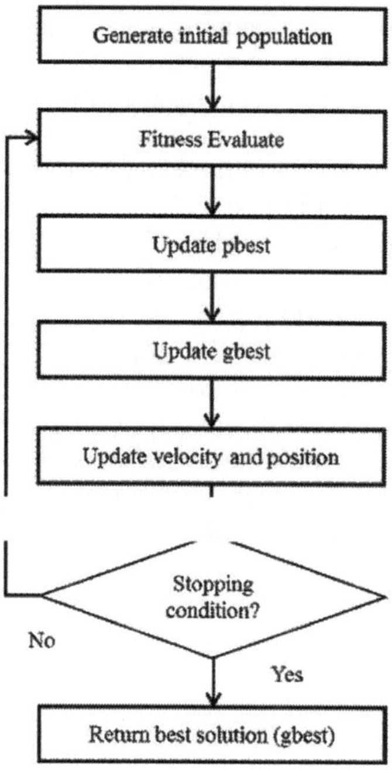

Fig. 3 The PSO algorithm

Upper-limb parameters:
Limb and exoskeleton robot lengths: $L_1 = 0.34$ m, $L_2 = 0.25$ m;
Centre of mass: $L_{c1} = 0.17$ m, $L_{c2} = 0.125$ m;
Forearm and upper arm masses:
$m_{upperarm} = 1.91$ kg, $m_{forearm} = 1.22$ kg;
Exoskeleton masses: $m_{exo1} = 0.34$ kg, $m_{exo2} = 0.25$ kg;
Mass moment of inertia of limb:
$J_{limb1} = 0.2374$ kg.m^2, $J_{limb2} = 0.0873$ kg.m^2;
Mass moment of inertia of exoskeleton:
$J_{exo1} = 0.0131$ kg.m^2, $J_{exo2} = 0.0052$ kg.m^2;
Controller parameters:
Controller gains (heuristically acquired):
$K_{p1} = 1\,000$, $K_{d1} = 90$;
$K_{p2} = 1\,000$, $K_{d2} = 90$;
PSO parameters:
The range of the diagonal elements of estimated inertia matrix:
$0 \leq \mathbf{IN}_1 \leq 0.02$ kg.m^2 $0 \leq \mathbf{IN}_2 \leq 0.005$ kg.m^2
Population size: 20
Number of generation: 100
Fitness function:

$$RMSE = \sqrt{\frac{1}{N}\sum_{i=1}^{N} e^2} \tag{8}$$

Simulation parameters:
Simulation start time: 0.0
Simulation stop time: 10 sec

The upper limb model is prescribed to perform a predefined trajectory that corresponds to a typical rehabilitation exercise for each individual joint i.e. shoulder and elbow respectively. The trajectories simulate the flexion/extension of the elbow joint and the forward adduction/abduction of the shoulder joint within the range of 0° and 45°. The trajectories (in radian) are as follows:

$$\theta_{shoulder} = \theta_{elbow} = 0.3927\sin(2\pi t - 1.3319) \tag{9}$$

V. Results and Discussion

Initially, the classical PD controller gains were heuristically tuned to demonstrate desirable stable joint space tracking response as per the prescribed trajectory. These tuned gains are not required to be retuned upon the incorporation of the AFC algorithm into the PD controller in the event that disturbances are introduced, unlike the conventional pure PD scheme where the gains are required to be retuned depending on the type of disturbances it is subjected to. A number of successive trials were implemented by means of PSO in order to ensure that the appropriate **IN** values were obtained through minimising the fitness function defined namely the sum of root mean square of the tracking error obtained by each joint. The number of population size was varied from 10 to 20 with an increment of 10 particles, whilst for each population size, the number of iteration or generation was varied from 20 to 100 with an increment of 20 iterations. The optimised values obtained by means of PSO for IN_1, and IN_2 are 0.0012 kg.m^2 and 0.6969 g.m^2, respectively. Figure 4 illustrates the convergence rate of the optimised estimated inertial matrices for a population size of 20 iterated 100 times through the PSO algorithm, respectively.

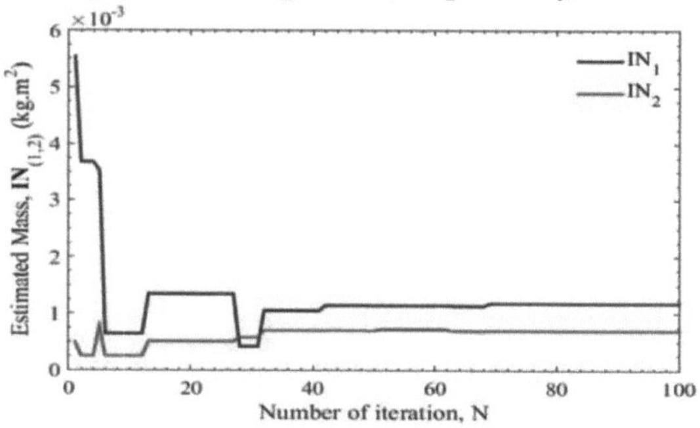

Fig. 4 Estimated inertial parameters of the actuated joints by means of PSO algorithm

Figures 5 to 6 shows the results acquired from the simulation study. The results illustrate the tracking performance of the classical PD and PD-PSOAFC algorithm under three different conditions to investigate its robustness viz. no disturbance, constant disturbance of 100 N.m. and harmonic disturbance with an amplitude of 200 N.m. and a frequency of 10 rad/sec applied to each joint respectively (shoulder and elbow). Table 1 lists the tracking error performance in terms of root mean square (RMS) of each joint under the aforementioned scenarios.

Table 5 Tracking error performance

Disturbance Type	Shoulder joint, θ_1 error$_{RMS}$ (mrad)		Elbow joint, θ_2 error$_{RMS}$ (mrad)	
	PD	PD-PSOAFC	PD	PD-PSOAFC
None	3.825	0.794	1.136	4.249
Constant (100 N.m.)	96.142	0.794	98.632	4.249
Harmonic (200 N.m.)	110.477	0.794	107.212	4.249

The results shown in Figures 5 (a) to 5 (b) suggest that the proposed control scheme (PD-PSOAFC) performs reasonably well as compared to its conventional counterpart as well as the PD in terms of tracking performance under the no disturbance scenario. The accumulated RMS error of the PD-PSOAFC was found to be 4.961 mrad whilst the PD scheme acquired an accumulated RMS error of 5.068 mrad. Although the tracking error of the elbow joint of the PD control scheme is better compared to the PD-PSOAFC, the error is rather permissible considering the actual magnitude of the trajectory.

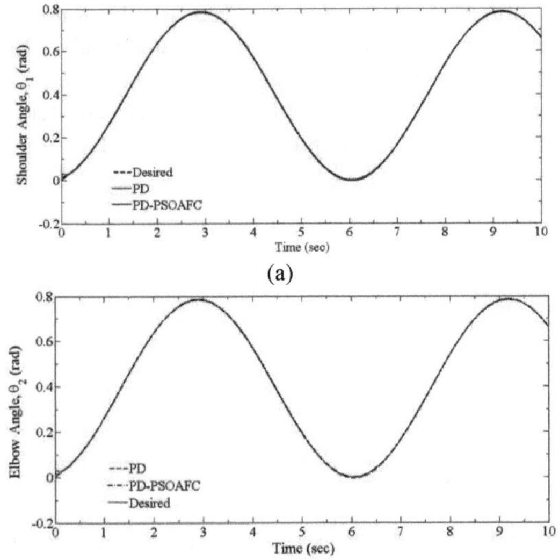

Fig. 5 (a) Shoulder trajectory tracking without disturbance (b) Elbow trajectory tracking without disturbance

Upon the introduction of a constant disturbance on each individual joints as depicted in Figures 6 (a) to 6 (b) thePSO AFC-based schemes appear to fare well in compensating the disturbance whilst maintaining good trajectory tracking as compared to its PD counterpart.

The RMS error accumulated by the PD-PSOAFC schemes were found to remain the same, which is innately in stark contrast to that of the PD scheme whereby the RMS error sum was found to be 194.774 mrad. A similar pattern is observed upon the wake of harmonic disturbance as illustrated in Figures 7 (a) to 7 (b). It is also evident from the trajectory tracking path as well as the RMS error readings, that in this scenario, the PD–PSOAFC method is deemed more superior in compensating disturbances than the conventional PD control approach scheme. It could be readily inferred from the cases presented; the proposed control scheme is much more robust and accurate in rejecting the disturbance effects.

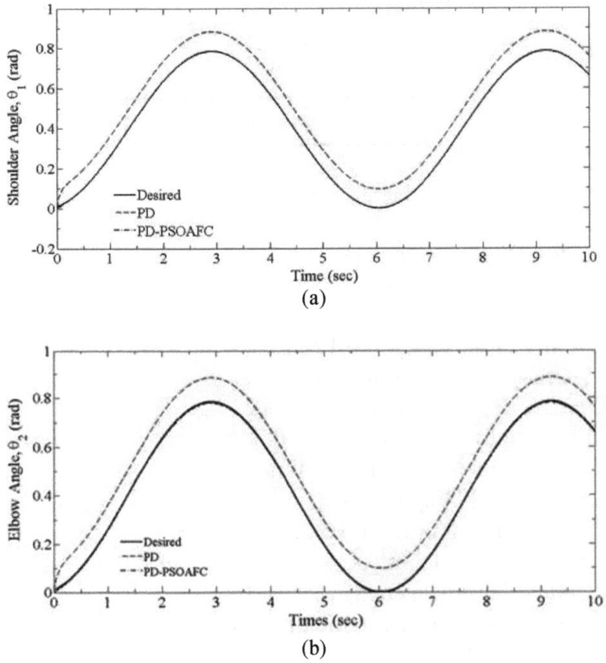

Fig. 6 (a) Shoulder trajectory tracking with constant disturbance (b) Elbow trajectory tracking with constant disturbance

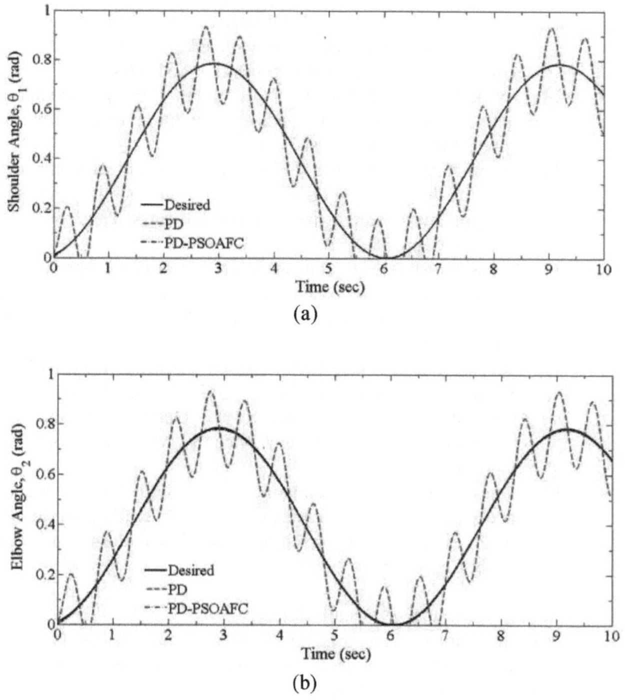

Fig. 7 (a) Shoulder trajectory tracking with harmonic disturbance (b) Elbow trajectory tracking with harmonic disturbance

VI. Conclusions

It is apparent from the simulation study performed, that the AFC-based control scheme has been shown to demonstrate its robustness in the event that disturbances and uncertainties are introduced as compared to its conventional counterpart. The PD control scheme provides satisfactory tracking performance without the presence of any form of disturbances acting on the system, nonetheless suffers considerably upon the onset of disturbances. Furthermore, the exceptional joint tracking performance achieved under various conditions by the PD-PSOAFC, suggest its applicability in the early stage of rehabilitation.

Acknowledgment

The authors would like to thank Universiti Malaysia Pahang for funding this research (Grant No. RDU160375).

Conflict of Interest

The authors declare that they have no conflict of interest.

References

1. World Health Organization. World Health Statistics 2014, http://scholar.google.com/scholar?hl=en&btnG=Search&q=intitle:No+Title#0
2. Taha Z, Majeed APPA, Tze MYWP, et al. (2015) Preliminary Investigation on the Development of a Lower Extremity Exoskeleton for Gait Rehabilitation: A Clinical Consideration. J Med Bioeng Vol 4.
3. Ministry of Health Malaysia HICPD. Health Facts 2013. Heal Facts 22.
4. Lo HS, Xie SQ. (2012) Exoskeleton robots for upper-limb rehabilitation: state of the art and future prospects. Med Eng Phys; 34: 261–8, http://www.ncbi.nlm.nih.gov/pubmed/22051085 (2012, accessed 21 January 2014).
5. Loureiro RC V, Harwin WS, Nagai K, et al. (2011) Advances in upper limb stroke rehabilitation: A technology push. Med Biol Eng Comput 49: 1103–1118.
6. Volpe BT, Ferraro M, Krebs HI, et al. (2002) Robotics in the rehabilitation treatment of patients with stroke. Curr Atheroscler Rep 4: 270–276.
7. Maciejasz P, Eschweiler J, Gerlach-Hahn K, et al. (2014) A survey on robotic devices for upper limb rehabilitation. J Neuroeng Rehabil 11:3, http://www.jneuroengrehab.com/content/11/1/3
8. Marchal-Crespo L, Reinkensmeyer DJ. (2009) Review of control strategies for robotic movement training after neurologic injury. J Neuroeng Rehabil; 6: 20, http://www.scopus.com/inward/record.url?eid=2-s2.0-68249158648&partnerID=tZOtx3y1
9. Meng W, Liu Q, Zhou Z, et al. (2015) Recent development of mechanisms and control strategies for robot-assisted lower limb rehabilitation.Mechatronics, http://linkinghub.elsevier.com/retrieve/pii/S0957415815000501
10. Carignan C, Liszka M, Roderick S. (2005) Design of an arm exoskeleton with scapula motion for shoulder rehabilitation. In: 2005 International Conference on Advanced Robotics, ICAR '05, Proceedings. 2005, pp. 524–531.
11. Kawasaki H, Ito S, Ishigure Y, et al. (2007) Development of a hand motion assist robot for rehabilitation therapy by patient self-motion control. In: 2007 IEEE 10th International Conference on Rehabilitation Robotics, ICORR'07. 2007, pp. 234–240.
12. Rahman MH, Saad M, Kenné JP, et al. (2010) Exoskeleton robot for rehabilitation of elbow and forearm movements. In: Control & Automation (MED), 2010 18th Mediterranean Conference on. IEEE, 2010, pp. 1567–1572.
13. Rahman MH, Kittel-Ouimet T, Saad M, et al. (2012) Development and control of a robotic exoskeleton for shoulder, elbow and forearm movement assistance. Appl Bionics Biomech 9: 275–292.
14. Rahman MH, Kittel-Ouimet T, Saad M, et al. (2011) Robot assisted rehabilitation for elbow and forearm movements. Int J Biomechatronics Biomed Robot 1: 206–218.
15. Xu G, Song A, Li H. (2011) Control System Design for an Upper-Limb Rehabilitation Robot. Adv Robot 25: 229–251.
16. Veeger HEJ, Yu B, An KN, et al. (1997) Parameters for modeling the upper extremity. J Biomech 30: 647–652.
17. Taha Z, Majeed APPA, Tze MYWP, et al. (2016) Modelling and control of an upper extremity exoskeleton for rehabilitation. IOP Conf Ser Mater Sci Eng; 114: 12134, http://stacks.iop.org/1757-899X/114/i=1/a=012134

18. Hewit J., Burdess J. (1981) Fast dynamic decoupled control for robotics, using active force control. Mech Mach Theory 16: 535–542.
19. Mailah M, Hooi HM, Kazi S, et al. (2012) Practical active force control with iterative learning scheme applied to a pneumatic artificial muscle actuated robotic arm. Int J Mech 6: 88–96.
20. Mailah M, Jahanabadi H, Zain MZM, et al. (2009) Modelling and control of a human-like arm incorporating muscle models. Proc Inst Mech Eng Part C J Mech Eng Sci; 223: 1569–1577, http://pic.sagepub.com/content/223/7/1569.abstract (2009, accessed 11 June 2015).
21. Mailah M. (1998) Intelligent active force control of a rigid robot arm using neural network and iterative learning algorithms.
22. Mailah M, Hewit JR, Meeran S. (1996) Active force control applied to a rigid robot arm. J Mek 2: 52–68.
23. As'arry A, Md Zain MZ, Mailah M, et al. (2013) Hybrid learning control for improving suppression of hand tremor. Proc Inst Mech Eng H; 227: 1171–80, http://www.ncbi.nlm.nih.gov/pubmed/23901066
24. Mailah M, Pitowarno E, Jamaluddin H. (2006) Robust motion control for mobile manipulator using resolved acceleration and proportional-integral active force control. Int J Adv Robot Syst; 2: 125–134, http://arxiv.org/abs/cs/0601057
25. Noshadi A., Mailah M. (2012) Active disturbance rejection control of a parallel manipulator with self learning algorithm for a pulsating trajectory tracking task. Sci Iran; 19: 132–141, http://dx.doi.org/10.1016/j.scient.2011.11.040
26. Noshadi A, Mailah M, Zolfagharian A. (2012) Intelligent active force control of a 3-RRR parallel manipulator incorporating fuzzy resolved acceleration control. Appl Math Model; 36: 2370–2383, http://www.sciencedirect.com/science/article/pii/S0307904X11005336 (2012, accessed 29 June 2015).
27. Priyandoko G, Mailah M, Jamaluddin H. (2009) Vehicle active suspension system using skyhook adaptive neuro active force control. Mech Syst Signal Process 23: 855–868.
28. Kwek LC, Wong EK, Loo CK, et al. (2003) Application of active force control and iterative learning in a 5-link biped robot. J Intell Robot Syst 37: 143–162.
29. Eberhart RC, Kennedy J. (1995) A new optimizer using particle swarm theory. In: Proceedings of the sixth international symposium on micro machine and human science. New York, NY, 1995, pp. 39–43.
30. Khassetarash A, Hassannejad R. (2015) Towards optimal design of sport footwear based on muscle activity and minimum loading rate using simplified model. Proc Inst Mech Eng Part H J Eng Med 229: 537–548.
31. Bahgaat NK, El-Sayed MI, Hassan MAM, et al. (2015) Load Frequency Control in Power System via Improving PID Controller Based on Particle Swarm Optimization and ANFIS Techniques. Res Methods Concepts, Methodol Tools, Appl Concepts 462.
32. Hamza MF, Yap HJ, Choudhury IA. (2015) Genetic algorithm and particle swarm optimization based cascade interval type 2 fuzzy PD controller for rotary inverted pendulum system. Math Probl Eng 2015.

Author: Anwar P. P. Abdul Majeed
Institute: Innovative Manufacturing, Mechatronics & Sports Laboratory (iMAMS), Faculty of Manufacturing Engineering, Universiti Malaysia Pahang
City: Pekan
Country: Malaysia
Email: anwarmajeed@imamslab.com / anwarpp@ump.edu.my

A Depth Image Approach to Classify Daily Activities of Human Life for Fall Detection Based on Height and Velocity of the Subject

Y. Nizam[1], M. Mahadi Abdul Jamil[1] and M. Norzali Mohd[2]

[1] Modeling and Simulation (BIOMEMS) Research Group, Department of Electronic Engineering, Faculty of Electrical and Electronic Engineering, Universiti Tun Hussein Onn Malaysia, Johor, Malaysia

[2] Embedded Computing Systems (EmbCos), Department of Computer Engineering, Faculty of Electrical and Electronic Engineering, Universiti Tun Hussein Onn Malaysia, Johor, Malaysia

Abstract— **Human fall detection systems can be categorized according to the approaches used such as some sort of wearable devices, ambient based devices or non-invasive vision based devices using live cameras. Wearable and ambient based devices are very often rejected by users due to the high false alarm and difficulties in carrying them during their daily life activities. This work proposes a fall detection system using depth information from Microsoft Kinect sensor. Classification of human fall from other activities of daily life is accomplished using height and velocity of the subject extracted from the depth information. Results of the simulated activities showed that the proposed system is able to gain an accuracy of 93.75% with 100% sensitivity and a specificity of 92.5%.**

Keywords— **Human fall, Kinect sensor, Depth images, Non-invasive, Depth sensor.**

I. Introduction

Due to the development in assistive technologies and the emerging of numerous applications requiring human posture recognition, classification of daily activities of human life has become a major research topic. Among such application is human fall detection systems, which also requires to distinguish human activities. Fall detection system are important especially in today's aging population, since it is the major problem for aged people to live individually. In addition to this most of deaths for seniors aged 79 from injuries are revealed [1, 2] to be caused from falls. Various approaches are used for developing human fall detection systems. Such as wearable sensors, non-wearable sensors (floor sensors, pressure sensors) and vision sensor (cameras). Thus the three basic approaches used in developing such systems are wearable sensor based device, vision sensor based and ambience sensor based devices. Fall detection systems using cameras (vision based) are more widespread because it can recognize and classify human movements more accurately than the other two approaches. As a result, vision based devices and such sensors used [3, 4] for human recognition are common sensors among the researchers, particularly when they want to base their system to be non-invasive [5].

Most of the human fall detection systems developed so far employed either wearable sensors or ambient sensors. Because of the amounts of false alarms generated by these devices, many users and their health supporters are not willing to use and rely on it. As a result, recent researches are going towards non-invasive based approaches for the development of human fall detection systems. Such systems using live cameras are very accurate in sensing human falls and therefore it generates less false alarms. But such cameras need sufficient lighting and also due to privacy concern people are not ready have such systems with their private life. Furthermore, a regular color camera cannot achieve the accuracy than that of a depth sensor, in extracting and tracking the subject. Thus, a depth sensor would be a preferable option over color camera, since it can identify human even in dark and at the same time preserve the privacy of users. Microsoft Kinect for windows is one of such sensors that generates depth images which can be used to track human. This work proposes a depth image based human fall detection system, which uses depth information from Microsoft Kinect sensor to classify human fall from other activities of daily life.

A. Related Work

Since the proposed system is based on depth information to categorize human activities for fall detection, this section will highlight only selected papers that had used depth sensors for developing human fall detection systems.

One of the related work had used a wearable wireless accelerometer and a depth sensor for fall detection. They used distance between the person centre of gravity and floor to confirm fall. Confirmation of fall after potential fall activity from the accelerometer is accomplished using Microsoft Kinect sensor [6].

A Spatio temporal context (STC) tracking of depth images using Kinect sensor was proposed by Yang et al. They used an estimate of Single Gauss Model (SGM) parameter and coefficients of the floor plane which is extracted in pre-processing. The position head is than extracted using foreground coefficient of ellipses and it is tracked using STC algorithm. Than the distance between

F. Ibrahim et al. (eds.), *3rd International Conference on Movement, Health and Exercise*,
IFMBE Proceedings 58,
DOI: 10.1007/978-981-10-3737-5_13

head to floor plane is calculated from every consecutive frame and fall detection is accomplished using an adaptive threshold [7].

Two feature based method was presented by Bian et al which uses distance between subject and the floor, and velocity. A fall is detected if the distance between the joints and the floor is close. Then the velocity of the joint hitting the floor is used to distinguish the event from a fall accident or a sitting/lying down on the floor [5].

Kawatsu et al presented a fall detection and reporting system using two algorithms with Microsoft Kinect sensor. The first uses only a single frame to determine a fall and the second uses time series data to distinguish between fall and slow lying down on the floor. They use the position and the velocities of joints. The reporting can be sent as an email or text messages and can include pictures during & after the fall [8].

In another study, introduced a mobile robot system which follows a person and detects when the person has fallen using a Kinect sensor. They also used the distance between the skeleton joints and the floor plane to detect fall [9].

Rougier et al presented a fall detection which uses human centroid height relative to the ground and subject's velocity. This work also dealt with occlusions, which was a weakness of previous works and claimed to have a really good fall detection results with an overall success rate of 98.7% [10].

A privacy-preserving fall detection method for indoor environment using Microsoft Kinect depth sensor was proposed by Gasparrini et at. They used Ad-Hoc segmentation algorithm on a ceiling mount configuration. The raw data directly from the sensor were analyzed and the system extracts the elements to categorize every blobs through the implemented solutions. A tracking algorithm between frames follows the subject recognized from the blobs. The distance between the blobs associated to the subject and floor is the threshold for human fall confirmation [11].

The related works either uses height of subject and the velocity or position of subject to identify fall. Many algorithms were used in the related works to identify human from the depth images. The algorithm applied in the proposed system uses velocity and height of the subject to classify human fall from other activities of daily life.

II. Methodology

The proposed methodology uses the depth information generated from Microsoft Kinect v1 sensor to compute the subject's height and velocity of body within the viewing angle of the sensor. The changes in height and velocity are observed to identify daily activities for the classification of human fall. Fall detection in the proposed system is accomplished by considering the changes in these two parameters over time. The height of the subject in this proposed system is meant for the distance from subject's head to floor and velocity is the speed of the subject with direction or position changes with respective to the previous location, which is measured from shoulder centre. The consideration of shoulder centre for velocity calculation is due to the fact that for a human fall or any movement which is similar to a fall such as lying on floor, the position of this joint will be on the floor or close too floor and the drop of height with respective to the velocity can give a clear gap for shoulder centre rather than other joints such as hip centre. The following section will discuss on the computation of the parameters used in the algorithm and the next section highlight the algorithm applied to classify human fall from other activities of daily life.

A. Fall detection parameters

The parameters required for the proposed algorithm are the height and velocity of the subject. Height which is the distance between head and floor and velocity is computed from shoulder centre to measure the movement of subject. To compute the height of the subject the skeleton frame generated from depth image and the containing floor-clipping plane vector is used. This skeleton data contains all detected joints coordinates stored as (x, y and z) expressed in meters and the floor-clipping vector, which has the coefficients of an estimated floor-plane equation as shown in equation 1. In the x, y and z joint coordinate system, the positive y-axis extends vertical upwards from the sensor, the positive x-axis extends to the left placing the sensor on a surface level and the positive z-axis extending in the direction in which the sensor is pointed. The z-axis gives the distance of the subject to the sensor (those close to sensor will have a small depth value and those far away will have larger depth value). With the help of this joint coordinate system, the movement of any joint and velocity (speed and direction) can be computed.

$$Ax + By + Cz + D = 0 \quad (1)$$

Here: A = vFloorClipPlane.x   C = vFloorClipPlane.z
B = vFloorClipPlane.y   D = vFloorClipPlane.w

This equation is normalized such that D is the height of the camera from the floor in meters. Floor plane or even stair plane can be identified at the same time using this equation. To calculate the distance from between head and the floor, the joint coordinates of head and floor plane equation in 1, has to be applied to the following Equation 2.

$$D\,(\text{distance}) = \frac{|Ax + By + Cz + D|}{\sqrt{(A^2 + B^2 + C^2)}} \quad (2)$$

Here x, y, z are the coordinates of the mentioned joint.

Velocity is calculated by taking the difference of the shoulder center position in between three frames (skipping one frame) from the respective axis after considering the direction of the movement (direction component of velocity). The direction is found using the joint coordinate system of the Kinect sensor as shown in fig. 1e, where any movement to the right or top or going far (to any axis) gives a positive value for the distance difference from current position to previous position. Likewise any movement to left or down or coming close gives a negative value for the distance difference from current position to previous position. Using this idea, the direction of all the movements are computed as presented in fig. 1e.

If the direction is straight (right or left), x-axis is consider and if it is horizontal (up or down) y-axis and z-axis is used if the movement is coming close or going far as shown in fig. 1a to 1c respectively. For the magnitude component of velocity, the difference of distance between two frames is divided by the time taken for the movement which is 1/15 seconds, because the sensor generates 30 frames per second and the joint position is taken after skipping one frame (time for two frames). The equation 3, shows the formula used for the calculation of magnitude component of velocity.

$$\text{Velocity (magnitude component)} = \frac{D_c - D_p}{t_c - t_p}\ \text{m/s} \quad (3)$$

Here Current Distance (D_c) and Previous Distance (D_p) is the current and previous joint coordinates respectively. Current time (t_c) and previous time (t_p) is in second.

If the direction is vertically (irregular) to any side (any axis), the distance travelled, cannot be simple calculated by subtracting the position between two frame on any axis, because the changes is not on the axis and if the changes is considered as on the axis, than the distance will be less than the actual distance travelled. This distance can be calculated by assuming the distance travelled as the hypotenuse of a right-angled triangle whose opposite and adjacent are form by joining the current and previous position of shoulder centre to the x-axis and y-axis. Here the actual distance travelled is the hypotenuse and the other two sides (opposite and adjacent) of the triangle will be formed by taking the straight movement to the two axis (x and y) depending on the direction of the movement. For an example if the direction is vertically down or up than the opposite and adjacent of the triangle will be formed from movement straight onto x and y axis as shown in fig. 1d. The formula used to calculate this distance (hypotenuse of the triangle formed) is shown in equation 4. Once distance is calculated the magnitude part of the velocity is calculated by using the equation in 3.

$$\text{Irregular Distance (d)} = \sqrt{((y - y`)^2 + (x - x`)^2)} \quad (4)$$

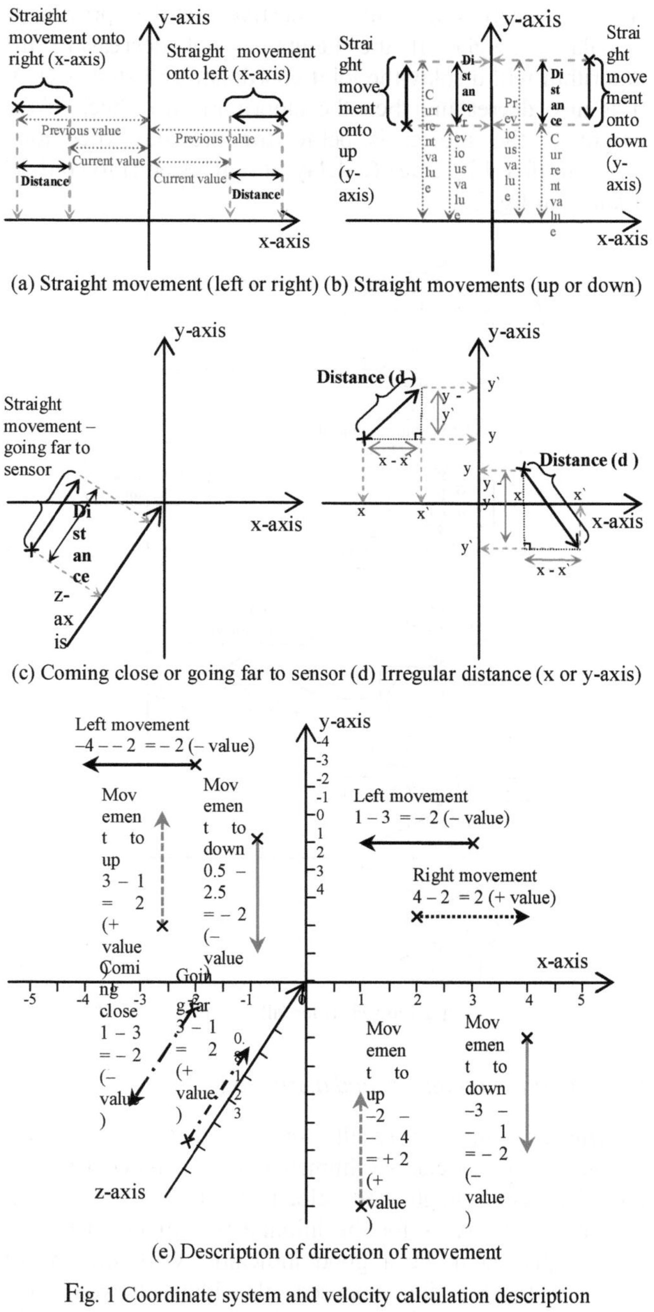

(a) Straight movement (left or right) (b) Straight movements (up or down)

(c) Coming close or going far to sensor (d) Irregular distance (x or y-axis)

(e) Description of direction of movement

Fig. 1 Coordinate system and velocity calculation description

B. Fall detection algorithm

For classification of human fall from other activities of daily life, the proposed algorithm principally depends on the velocity of the subject. If there is an increased velocity between 10 frames delay to vertically down to floor, the algorithm will calculate the height of the subject and check if it is decreasing with respective to the previously calculated height. If the height is not decreasing the algorithm will display the joint coordinate and start over. If it is also decreasing then the algorithm will check if the height of the subject is below the previous knee joint position after 15 frames (a delay of half second) to confirm a fall as in fig. 2.

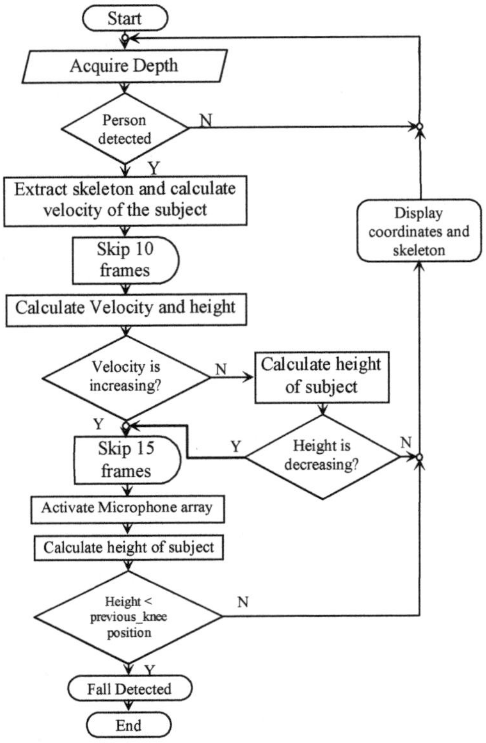

Fig. 2 Flow chart for fall detection

C. Experimental results and discussion

The experimental results showed that the proposed system is able to classify human fall from other activities daily life using height and velocity of the subject. Use of velocity as the basis for the initialization of fall detection process proved to be a good indicator. With the use of subject's height, the proposed algorithm showed good performance in eliminating activities that are not closely analogous to fall and differentiating human falls from activities that are similar like fall. The following figure 3, shows the distance pattern for some of activities like walking, bend while standing, and sitting on floor from standing, lying on floor from standing, sit on chair, and fall while sitting on chair. The activities shown in figure 3 are combination of different movements from one sample.

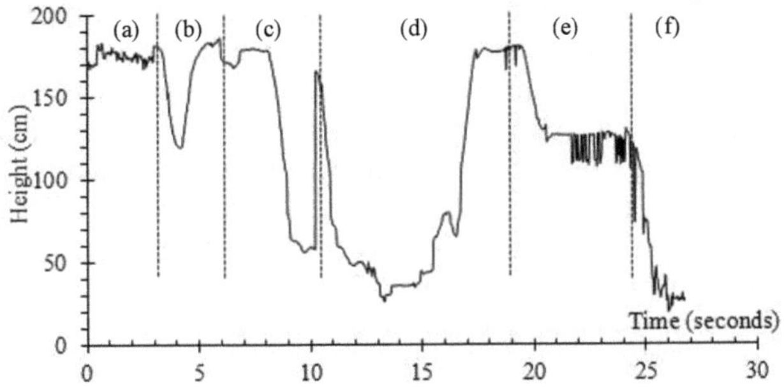

Fig. 3 Height change pattern for some activities. (a) walking; (b) bend while standing; (c) sitting on floor from standing; (d) lying on floor from standing; (e) sit on chair; (f) fall while sitting on chair.

As mentioned the proposed algorithm was able to eliminate activities that are dissimilar to fall and thus the challenge remain on classifying human activities that are analogous to fall. The changes of height over time showed to be a good variable to distinguish such movements. The following figure 4 shows the drop of height over time during a fall and lying down on floor plotted with the changes of height with respective to previous location. As seen from the figure 4, the rate of changes of height between frames shows clear gap for falling and lying on floor. The usual height change pattern during lying on floor goes slowly and spreads over time but during falling the height drops rapidly. This pattern of change was clearly distinguishable using the rate of change of height between frames as seen in fig. 4.

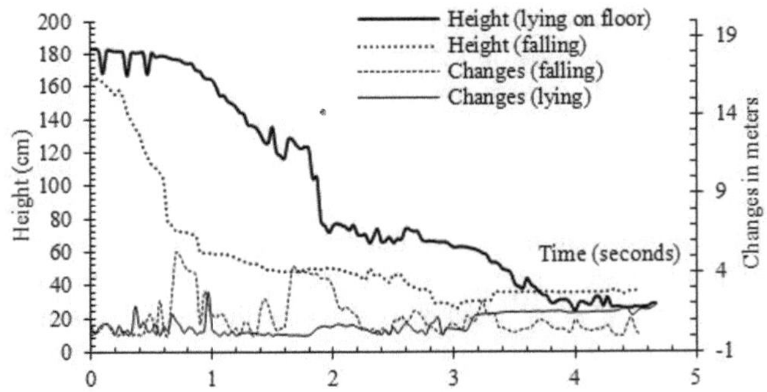

Fig.4 Height change pattern with rate of change between frames

The other indicator, which is the velocity also showed unique differences between lying on floor and falling as seen in fig. 5. The rate of change for lying on floor was slower compared with time and for falling the rate of change is higher. Most importantly the changes of magnitude component of velocity for falling is higher than lying on floor. Thus making it easy to distinguish such similar activities.

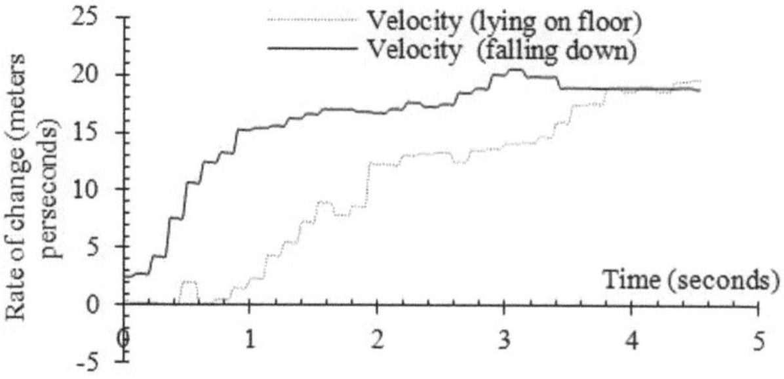

Fig. 5 Rate of change of velocity

Table 1 shows the confusion matrix for falls and non-falls for a sample 48 activities. It shows the overall number of trials conducted with the predicted value for each of the activity. Accuracy is calculated by dividing the sum of TP and TN over the total sample (TP + TN + FP + FN). Sensitivity is calculated by dividing the actual falls that are correctly predicted as fall (TP) over actual falls (TP + FN). The ratio of actual non-falls that are predicted correctly (TN) with actual non-fall (FP + TN) gives specificity. Using this, the system was able gained an average accuracy of 93.75% with specificity of 92.5% and sensitivity of 100%. The reason for the100% sensitivity, is the fact that the developed system was able to classify all fall movements conducted in the trial. The reduced specificity is due to the inability of the system to recognize some lying down on floor from fall.

Table 1 The matrix for average of fall and non-fall values

		Predicted	
	Total	Falls	None-falls
Actual	Falls	True Positive (TP) 8	False Negative (FN) 0
	None-falls	False Positive (FP) 3	True Negative (TN) 37

III. Conclusions

This paper proposed a human fall detection system based on human height and velocity of body computed from depth images generated by the Microsoft Kinect sensor. The experimental results showed that the algorithm used on the system can accurately distinguish fall movements from other daily activities with an average accuracy of 93.75%. The system was also able to gain a sensitivity of 100% with a specificity of 92.5%. The proposed system was able to distinguish all fall movements from other activities of daily life accurately, even though it failed to identify some lying down on floor from fall. The velocity of joints greatly help to classify certain movements were the distance changes possess similar variation. The proposed system could be further improved by incorporating posture identification.

Acknowledgment

The Author Yoosuf Nizam would like to thank the Universiti Tun Hussein Onn Malaysia for providing lab components and GPPS sponsor for his studies.

Conflict of Interest

The authors declare that they have no conflict of interest.

References

1. Griffiths, C.; Rooney, C.; Brock, A. Leading causes of death in england and wales–how should we group causes. Health Statistics Quarterly 2005, 28.
2. Baker, S.P.; Harvey, A. Fall injuries in the elderly. Clinics in geriatric medicine 1985, 1, 501-512.
3. Mohd, M. N. H., Kashima M., K. Sato, K., Watanabe, M. 2015. Internal state measurement from facial stereo thermal and visible sensors through SVM classification. ARPN Journal of Engineering and Applied Sciences, 10(18): 8363-8371.
4. Mohd, M. N. H., Kashima M., K. Sato, K., Watanabe, M. 2015. A Non-invasive Facial Visual-Infrared Stereo Vision Based Measurement as an Alternative for Physiological Measurement. In Lecture Notes in Computer Science (including subseries Lecture Notes in Artificial Intelligence and Lecture Notes in Bioinformatics), 9009: 684-697.
5. Bian, Z.-P.; Chau, L.-P.; Magnenat-Thalmann, N. A depth video approach for fall detection based on human joints height and falling velocity. International Coference on Computer Animation and Social Agents 2012.
6. Kwolek, B.; Kepski, M. Human fall detection on embedded platform using depth maps and wireless accelerometer. Computer methods and programs in biomedicine 2014, 117, 489-501.
7. Yang, L.; Ren, Y.; Hu, H.; Tian, B. New fast fall detection method based on spatio-temporal context tracking of head by using depth images. Sensors 2015, 15, 23004-23019.
8. Kawatsu, C.; Li, J.; Chung, C. Development of a fall detection system with microsoft kinect. In Robot intelligence technology and applications 2012, Springer: 2013; pp 623-630.
9. Mundher, Z.A.; Zhong, J. A real-time fall detection system in elderly care using mobile robot and kinect sensor.
10. Rougier, C.; Auvinet, E.; Rousseau, J.; Mignotte, M.; Meunier, J. Fall detection from depth map video sequences. In Toward useful services for elderly and people with disabilities, Springer: 2011; pp 121-128.

11. Gasparrini, S.; Cippitelli, E.; Spinsante, S.; Gambi, E. A depth-based fall detection system using a kinect® sensor. Sensors 2014, 14, 2756-2775

Author: M. Mahadi Abdul Jamil
Institute: Universiti Tun Hussein Onn Malaysia
Street:
City: 86400 Parit Raja, Batu Pahat, Johor
Country: Malaysia
Email: mahadi@uthm.edu.my

Effect of the Long Term 'Training and Competitive' Cycle on Urinary Protein and Creatinine in Elite Male Triathletes in Malaysia – A Pilot Study

H. Zainudin[1], B. Caszo[2], V.F. Knight[2], S. Lumley[2], A. Subramanium[3] and J. Gnanou[2]

[1]Centre for Research and Innovation Management, National Defence University of Malaysia, Kuala Lumpur, Malaysia;
[2] Faculty of Medicine and Defence Health, National Defence University of Malaysia, Kuala Lumpur, Malaysia;
[3]Head Coach, Team Time Triathlon, Kuala Lumpur, Malaysia;

Abstract— **Background: Strenuous exercise may cause DNA, skeletal muscle as well as renal damage. Triathletes experience strenuous muscular activity both during competition and training. Studies have shown changes in renal function after competition which returns to normal after few days. However, these finding have only focused on a single event. There is no literature studying the cumulative effect of "training and competition" cycles over a season of triathlon competitions. The objective of this study was to evaluate the cumulative effect of "training and competition" on renal function in elite male triathletes using urine protein and creatinine.**

Methods: Seven male elite triathletes were recruited for the study. They were on a standardized training regimen and competed in at least one endurance event every month for the past 3-4 years. They were followed up for nine months. Urine samples were collected at the beginning (Phase 1) and at the end of the triathlon season (Phase 2). Urine protein and creatinine levels were estimated using spectrophotometric methods and compared with a Student's t test ($p<0.05$ was considered significant).

Results: The urine protein was 5.26 ± 3.99 mg/dl and 11.48 ± 6.87 mg/dl and urine creatinine was 11.67 ± 5.16 mmol/L and 19.09 ± 7.15 mmol/L for phase 1 and phase 2 respectively. There was a statistically significant difference in urine protein and creatinine ($p<0.05$) between phase 1 and phase 2.

Discussion: Urine protein and creatinine are considered markers of renal function. Our results show that at the end of the "training and competition" cycle of an elite triathlete, there is two-fold increase in urinary protein and creatinine. This finding provides evidence to the cumulative effect of training and competition over a period of 9 months.

Keywords— **DNA damage, Triathlon, Urinary protein, Urinary creatinine.**

I. Introduction

Triathlon is an endurance sport which involves three different disciplines (swimming, running and cycling) which are continuous and sequential. Triathlon races vary in distance and a typical standard or intermediate race involves 1.5 kilometers (0.93 mi) swimming, 40 kilometers (25 mi) biking and 10 kilometers (6.2 mi) running to the finish line. There are events which have longer race distances called the ultra-distance, most commonly called the ironman triathlons [1]. The race typically starts in the morning (7.00 am) and ends around midnight lasting approximately 17 hours. Triathlon training also involves strenuous exercise and it requires the athlete to divide training time between the three disciplines. Training usually involves reaching similar endurance levels as a race but of lesser duration, however more importantly on day to day basis. Hence, for a competitive elite triathlete it would mean periods of "high endurance training" and "high endurance competitions" throughout one's career. Triathletes are also busy the whole year around with a constant "competition and training" cycles every 2 months.

The race on its own is a strenuous exercise [2] and it is well documented in literature that at the end of a triathlon race, there is evidence of muscle damage [3]. A strong body evidence exists which shows marked DNA damage [4] due to increased oxidative stress [5]. High intensity training which every triathlete undergoes training too contributes to the effects of strenuous exercise on the body. This also may lead to excessive urinary excretion of protein and this may vary from 18 to 100% [6]. Proteinuria after exercise is usually depended on the type and intensity of the exercise rather than the duration of the exercise and is caused due to changes in the glomerular permeability and filtration ratio. [7,8] However proteinuria after exercise is usually transient and it returns to physiological levels following few hours of resting [9]. Another marker of renal function and hemodynamics is urinary creatinine. Creatinine is an enzymatic breakdown product of creatine mainly from the skeletal muscle following exercise induced skeletal muscle degradation. Creatinine is excreted through urine and its increased excretion indicates compromised renal function [10].

Though many studies have established the effects of participation in a triathlon event on the renal function and hemodynamics, the contribution of training as a factor has not been investigated. Thus the aim of this study was to investigate the cumulative effect of "training and competition" cycle on the renal function using urinary protein and creatinine measures.

F. Ibrahim et al. (eds.), *3rd International Conference on Movement, Health and Exercise*,
IFMBE Proceedings 58,
DOI: 10.1007/978-981-10-3737-5_14

II. Methods

A. Subjects

Seven male triathletes with an average age of 17.71 ± 3.59 years volunteered to take part in this study. They are members of a professional triathlon team and undergo a standardised structured and rigorous training program. Table 1 provides the anthropometric characteristics of the subjects. The subjects were informed of the potential benefits and risks involved in this study and informed consent was obtained. The study was conducted in accordance with the Declaration of Helsinki and the guidelines of Resolution on 198/96 of the National Health Council, and was approved by the Institutional Ethics Committee of the National Defence University of Malaysia.

B. Training schedule

The triathletes have on average consistently trained for 4.43 ± 1.27 years. They trained regularly between 9-15 hours of training per week. For 9 months this included 3-5 swim sessions, 2-3 bike sessions and 3-4 running sessions. The training sessions were periodised into 4 x 4 training blocks, with the 4th week being a lower volume 'recovery' week. The average distance completed by the triathlete was 144.8 km/week (including swimming, biking and running). Apart from this, strength and conditioning sessions were also carried out.

C. Experimental Design

The subjects were recruited at the beginning of the triathlon season (February). They were asked to report to the lab for the baseline anthropometric measurement and urine sample collection (Phase 1). The subjects were advised not to take part in vigorous physical activity for at least 24 hours prior to the day of the sample collection. They were also advised to avoid caffeinated and alcoholic drinks 48 hours prior to the sample collection. Body composition measurements were carried out using a N2O segmental body composition analyser (U. Healthcare System, Singapore). Urine samples were collected in sterile containers and stored at -80°C until analysis. After the baseline collection, the athletes were asked to begin their training. The training was monitored under the supervision of a professional coach and the athletes were encouraged to take part in at least 6 triathlon events during this period. At the end of the triathlon season, the triathletes were once again asked to report to the lab (Phase 2). All samples were collected 14 days after the subject had participated in the last triathlon. Anthropometry and urine collections were repeated at this phase of the study.

D. Urinary Protein and Creatinine Assay

Urinary protein was estimated using the protein dye-binding spectrophotometric method. In this method, an improved pyrogallol red-molybdate protein dye-binding was used (BioAssay Systems, Hayward, CA 94545, USA). In this single step method, the color developed at the end of the assay was measured at 600nm and using a calibration curve, urine protein was estimated and expressed as mg/dl. Urinary creatinine was measured using a spectrophotometric assay. In this assay, creatinine is converted to creatine by creatininase, creatine is converted to sarcosine, which is specifically oxidized to produce a product which reacts with a probe to generate red color (Biovision, Milpitas, CA 95035 USA). The intensity of the red color which was measured at 570nm was proportional to the creatinine concentration. Using a calibration curve, the concentration of creatinine in urine was estimated and was expressed as nmol/L. In order to eliminate the effect of hydration on these assays, urine specific gravity was measured. The specific gravity of all the urine samples were within normal limits.

E. Statistical Analysis

All data were expressed as means ± standard deviation. To study the effect of 'training and competition' cycle on renal function, student's t-test was used to determine the difference between baseline (Phase 1) and end of triathlon season (Phase 2) for both urine protein and creatinine measurements. Statistical significance was accepted at $p < 0.05$.

III. Results

Figure 1 and 2 shows the change in urine protein and urine creatinine between phase 1 and phase 2. We found a statistically significant increase in urinary protein. There was almost a 2-fold increase in urinary protein between phase 1 and phase 2. Urinary protein is an indicator of changes in glomerular permeability and filtration. This is usually a transient process seen in athletes after strenuous exercise; however in our study we find urinary protein showing a cumulative increase. Similarly, urinary creatinine also showed a significant increase between phase 1 and phase 2. Urinary creatinine is measure of skeletal muscle breakdown. Like urinary protein, increase in urinary creatinine after strenuous exercise is a transient phenomenon which should return to baseline in few days. However similar to urinary protein, we find a cumulative increase in urinary creatinine in our triathletes.

Table 1: Anthropometric characteristics of triathletes
All data are presented in mean ± SD

Item	Triathletes
Height(cm)	166.09±10.83
Weight (Kg)	55.13±10.65
BMI (Kg/m^2)	19.80±2.21
Fat free Mass(Kg)	45.96±9.23
Percent Body Fat (%)	16.71±3.88
Skeletal Muscle Mass(Kg)	24.77±5.47

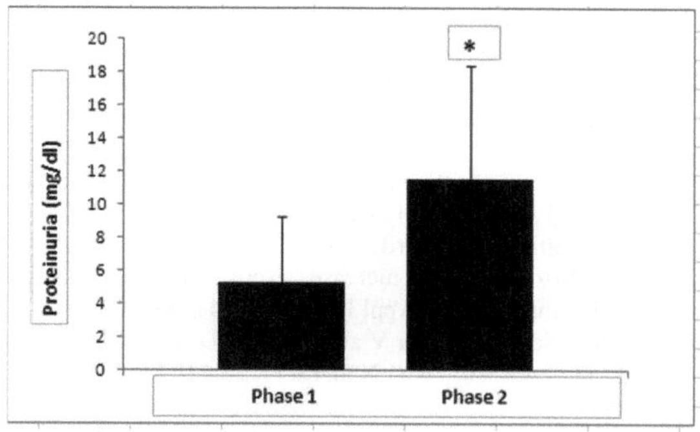

Fig. 1: Urinary protein (mg/dl) in Triathletes at phase 1 and Phase 2. *= $p<0.05$

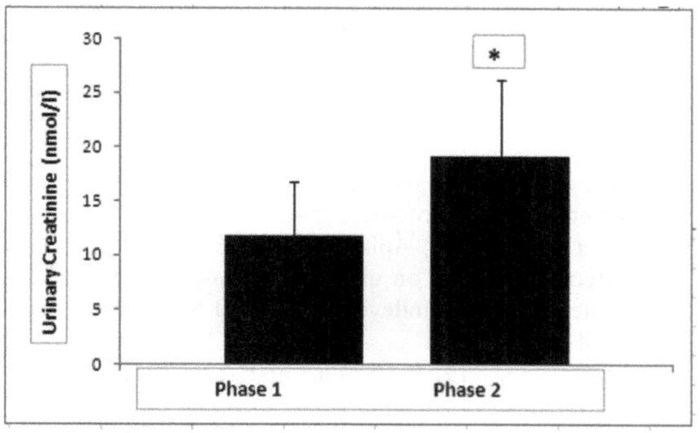

Fig. 2: Urinary creatinine (nmol/l) in Triathletes at phase 1 and Phase 2. *= $p<0.05$

IV. Discussion

The main aim of this study was to establish whether a 'training and competitive' cycle for one complete triathlon season (9 months) has a cumulative effect on renal function. This was achieved by measuring renal function markers such as urinary protein and creatinine in triathletes before and after a triathlon season. During this season, the triathletes underwent a structured systematic training session as well as took part in at least 6 triathlon events at regular intervals. In this study we used proteinuria as a measure of renal function. Strenuous physical activity causes increased excretion of protein in the urine and this has been termed sport proteinuria. This is due to the changes in the glomerular membrane permeability. In normal conditions, glomerular membrane acts as filter and allows selective permeability of molecules of different sizes [7]. This prevents filtration of plasma proteins. However, strenuous physical activity causes changes in the glomerular membrane permeability as well as decrease in protein re-absorption in proximal tubules. This causes increase in protein excretion leading to proteinuria. One of the important determining factors for proteinuria is the intensity of exercise rather than the duration. Presence of proteinuria immediately after a triathlon race has been very well documented. In a study on triathletes before and after a half-ironman triathlon, Puggina et al showed an 8 fold increase in urinary protein levels [8]. In a study on Japanese triathletes, a significant increase in urinary protein was noted [9]. However, most of these studies were conducted to show the effect of a single race or event on the urinary protein excretion. The long term effect of training and competition was not accessed. In one study, effect of 12 week of training on urinary protein in triathletes showed no significant increase in urinary protein [8]. This study looked at only one training and competition cycle. We have not come across any literature to show the effect of one complete season of triathlon on urinary protein excretion. In our study we find a significant increase in protein excretion (Fig.1). We feel there is a cumulative effect of long term 'training and competition' cycle on the glomerular permeability which causes increased proteinuria. This could be an indication of the strenuous nature of the training and competition as well as insufficient recovery period for the triathletes.

Urinary creatinine excretion is a good measure of glomerular filtration ratio since creatinine is excreted at a constant rate in a day. Strenuous physical activity causes increase in blood flow to the kidney, increasing the glomerular filtration ratio and thereby resulting in increased excretion of creatinine. Studies have shown this phenomenon in athletes from different sports [11,12,13]. Increased urinary creatinine excretion in triathletes after a triathlon race/event has also been very well documented [8,14] . Similar to proteinuria, few studies have looked into the effect of training on urinary creatinine excretion. In the same study, Puggina et al did not find a increase in creatinine excretion after 12 weeks of training, but found a significant increase after the triathlon race [8]. Again this study was limited to only one 'training and competition cycle'. In our study we found a statistically significant increase in urinary creatinine excretion (Fig.2) similar to urinary protein, showing a cumulative effect of multiple training and competition cycles.

The present results confirm that multiple 'training and competition' cycles have a cumulative effect on the urinary protein and creatinine excretion, thus providing evidence for changes in renal hemodynamics following long term strenuous physical activity. Our study did have certain limitations. The sample size of this study was low mainly due to the following factors: Our study required for the athlete to be followed up for an entire season (9 months) and also for the athletes to take part in triathlon events at a regular intervals, which caused a drop out of at-least 50% of our subjects. Secondly, since we needed all the athletes to follow a similar training regimen, we had to limit ourselves to one coach and his team. All these factors lead to the low sample size. The other main drawback of this study was non-inclusion of control subjects. Since in this study we only wanted to see the pre and post effect of training and competition of triathletes on renal function, we used the "one group pre-test and post test" experimental design. Since our measurement parameters were urine protein and creatinine, we presumed that there we be will no change in these parameters in non- triathlete subjects and hence did not include control subjects. We were also not able to collect 24 hour urine protein excretion mainly due to logistical reasons. However studies have shown that spot urine protein measurements and comparable to 24 urine protein measurement and can be used to assess renal marker for glomerular permeability.

V. Conclusion

In conclusion, it appears that there is a cumulative effect on continuous training and competition on the renal function as assessed by urinary protein and creatinine excretions. Thus for coaches, the findings from our study could provide good justification for proper recovery periods for the triathletes, in order to avoid a permanent renal damage.

Acknowledgment

The authors thank all the technical staff at the Faculty of Medicine and Defence Health, National Defence University of Malaysia for their excellent technical assistance. The authors would also like to thank all triathletes and their parents for their sincere involvement and dedication in this study. This research was funded by the Ministry of Higher Education, Malaysia under the *Geran Penyelidikan Sukan IPT* 2013 (SG/2014/SP/UPNM/1). The authors declare no conflict of interest.

Conflict of Interest

The authors declare that they have no conflict of interest.

References

1. Wu SS, Peiffer JJ, Brisswalter J, Nosaka K, Abbiss CR (2014) Factors influencing pacing in triathlon. Open Access J Sports Med. 5:223-234
2. O'Toole ML, Douglas PS. (1995) Applied physiology of triathlon. Sports Med. 19(4):251-67.
3. Baur DA, Bach CW, Hyder WJ, Ormsbee MJ (2016) Fluid retention, muscle damage, and altered body composition at the ultraman triathlon. Eur J Appl Physiol. 116(3):447-458.
4. Wagner KH, Reichhold S, Neubauer O (2011) Impact of endurance and ultraendurance exercise on DNA damage. Ann N Y Acad Sci. 1229:115-123.
5. Palazzetti S, Richard MJ, Favier A, Margaritis I (2003) Overloaded training increases exercise-induced oxidative stress and damage. Can J Appl Physiol. 28(4):588-604.
6. Belinghieri G, Savica V and Santoro D (2008) Renal alterations during exercise. J Ren Nutr 18(1): 158-164.
7. Poortmans JR, Jeannaud F, Baudry S, Carpentier A (2015) Changes in Kidney Functions During Middle-Distance Triathlon In Male Athletes. Int J Sports Med. 36(12):979-983.
8. Puggina EF, Machado DR, TourinhoFilho H, Barbanti VJ (2014) Half-ironman induces changes in the kidney function of triathletes. An Acad Bras Cienc. 86(1):429-36.
9. Yaguchi H, Ishigooka M, Hayami S, Kobayashi T, Kubota Y, Nakada T, Mitobe K (1998) The effect of triathlon on urinary excretion of enzymes and proteins. IntUrolNephrol. 30(2):107-112
10. Levey A, Perone R, Madias N (1988) Serum creatinine and renal function. Annu Rev Med 39: 465-490
11. Ayca B, Sener A, ApikogluRabus S and Oba R (2006). The effect of exercise on urinary gamma-glutamyl transferase and protein levels of volleyball players. J Sports Med Phys Fitness 46(4): 623-627
12. Poortmans J (1984) Exercise and renal function. Sports Med 1(2): 125-153
13. Levey A, Perone R and Madias N (1988). Serum creatinine and renal function. Annu Rev Med 39: 465-490.
14. Edes TE, Shah JH, Thornton WH Jr (1990) Spontaneous decline in exercise-induced proteinuria during a 100-mile triathlon. South Med J 83(9):1044-6, 1052.

Author: Assoc. Prof. Dr. Justin Vijay Gnanou
Institute: Faculty of Medicine and Defence Health, National Defence University of Malaysia
Street: Kem Sg. Besi
City: Kuala Lumpur
Country: Malaysia
Email: justin@upnm.edu.my

Classification of Respiratory Sounds in Smokers and Non-smokers using k-NN Classifier

N.S. Abdullah[1], C.K. Lam[1], K. Sundaraj[2] and R. Palaniappan[3]

[1] School of Mechatronic Engineering, Universiti Malaysia Perlis (UniMAP), Kampus Pauh Putra, Perlis. Malaysia.
[2] Faculty of Electronics and Computer Engineering, Universiti Teknikal Malaysia Melaka (UTeM), Hang Tuah Jaya, Melaka, Malaysia.
[3]School of Electronics Engineering (SENSE), VIT University, Vellore, India.

Abstract— **Respiratory sounds contain significant information on physiology and pathology of the lung and the airways. Its analysis provides vital information of the present condition of the lungs. Pulmonary disease is a major cause of ill-health throughout the world. The frequency spectrum and the amplitude of sound, i.e. tracheal or lung sounds without adventitious sound components (wheeze), may reflect airway dimension and other their pathologic changes (airways obstruction). Wheezes may have acoustic features indicating not only the presence of abnormality in the respiratory system but also the severity and locations of airway obstruction most frequently found in asthma and also found in smoker but not all smokers have airway obstruction. The significance of this study is to develop a classification system to distinguish between normal and smoker from respiratory sounds. 15 smokers and 15 non-smokers are recruited to collect respiratory sounds using Wireless Digital Stethoscope. The performance analysis of the K-Nearest Neighbor (k-NN) classifier, which uses entropy as the suitable feature, revealed that the classification accuracy on non-smokers and smokers are 89.33% and 78.67% respectively.**

Keywords— **Respiratory sounds, Airways Obstruction, Fourier Transform, K-Nearest Neighbor**

I. Introduction

Lung sounds are part of the sound of breathing or commonly called respiratory sound. Respiratory sounds include the sounds from the mouth, trachea, and the lung sounds occur in the chest (chest wall) [1]. As informed before, abnormal breath sounds like wheeze and rhonchi at times manifest similar morphologies and pathological features of lung airways obstruction that occur in smokers [2]. This may pose problems to proper diagnosis and evaluation of the underlying respiratory condition by human auscultation. Based on the spectral features (frequency domain), it is able to detect and classify normal and abnormal breath sounds in smokers using WISE Digital Stethoscope. The significance of this study is to develop a classification system using K-Nearest Neighbor (k-NN) classifier to distinguish between smoker and non-smoker from respiratory sounds.

F. Related works

In recent years, researchers are developing and proposing a numerous method to classify lung sounds into two categories: normal and abnormal respiratory sounds (RS). As the main focus of the research, the process must be independently of different breath sounds signal of the subjects. In early 1980's, respiratory pathologies were recognized using computerized respiratory sound analysis [3]. This section will discuss a few related works that have been done by other researchers in recent years about the computerized respiratory sound analysis.

A. Respiratory sound recording

The sensors that are used most regularly in computer based respiratory sound analysis are piezoelectric microphones, accelerometers, contact microphones, and electret microphones. Some of the popular electret microphones used in the interpret studies are ECM 44 (Sony), which the sound wave was captured using 14 air-coupled electret microphones [4-6]. Other than that, single electret condenser microphone (ECM 77B, Sony, Tokyo, Japan) is used in [7,8].

In Palaniappan et al. [9], all pathological respiratory sounds were recorded using electret microphone with the subjects in a sitting position from two positions on the chest: right basilar (RB) and left basilar (LB). In [10,11], the respiratory sounds were recorded from the patients using electronic stethoscope. Moreover, the respiratory sounds taken from the R.A.L.E® database is the only commercially available respiratory sound database and EMT 25C (Siemens) is a well-known accelerometer used in respiratory sound analysis [3,12]. There are also a several commercially available multichannel instruments for respiratory sound analysis. In addition, the latest electronic stethoscopes are able to record digital respiratory sound recordings that make easier in filtering the heart sounds and ambient noise from the respiratory sounds. Data is then transmitted to a computer for display and analysis [10].

F. Ibrahim et al. (eds.), *3rd International Conference on Movement, Health and Exercise*,
IFMBE Proceedings 58,
DOI: 10.1007/978-981-10-3737-5_15

B. Respiratory sound pre-processing

Respiratory sound signals are dominated by noises such as heart sound and other artifacts. One of the most significant noises that cannot be eliminated directly is heart sound. Heart sounds arise as a result of the process of opening and closing of the heart valves in the pumping of blood by the heart. Heart sound occupies the frequency range of 20-150 Hz, which overlaps with the low-frequency component of the sound of the lung [13]. Researchers have used various pre-processing methods to filter noise, avoid aliasing and to remove DC offset. [3,14] made up of recordings which are filtered to evacuate heart sounds and artifacts. The respiratory sound signal was high-pass filtered to remove DC offset and low-pass filtered to avoid aliasing. Moving average filter is needed for filtering out unwanted parts of the captured signal to remove large amplitude fluctuations between samples and a much smoother signal is generated [6,10].

C. Features extraction

The earlier researches have used several methods for features extraction to analyze the respiratory sounds. The extraction of features, which is the process of recognize distinctive properties from a signal, plays a significant role in the effectual recognition of the respiratory sounds. The features can be extracted from signals whether in time domain or frequency domain only or maybe can be in time-frequency domain. The feature extraction method can be chosen by two major factors: the domain and the characteristics of the feature vector.

Autoregressive (AR) model, Mel-frequency cepstral coefficient (MFCC), spectral features, and wavelet [3,5,10,14] are the most popular techniques of features extraction that are widely used in computer-based respiratory sound analysis. Respiratory sounds are non-stationary signals, and variety methods such as short-term Fourier transform and discrete wavelet transform can be applied for extracting the transient features of non-linear signals. However, the utilization of time-frequency domain features has been recommended by [4,7,9].

D. AI technique in lung sounds analysis

Over the past few decades, researchers have implemented a numerous AI technique to classify the respiratory sounds between normal and abnormal lung sounds. The most popular AI technique that used by the researchers is K-Nearest Neighbor (k-NN) classifier [4,5,7,14]. Other than that, support vector machine (SVM) classifier is also used as an AI technique in lung sound analysis [3,14] and the classifier of artificial neural networks (ANN) is also used in [10].

II. Methodology

A. Respiratory sounds acquisition

In this studies, CORSA (Computerized Respiratory Sound Analysis) is formulated to protocol that developed by European Respiratory Society (ERS) for breath sound acquisition [14]. The type of recording with 20 sec from the each auscultation point is known as short-term recording as shown in Fig. 1. 15 non-smokers and 15 smokers with 6 years of experience in smoking are recruited as the subjects for data collection. From the three locations; trachea, right lung base and also left lung base, the respiratory sounds are recorded for three times in each location using WISE digital stethoscope. The wireless digital stethoscope acts as air-coupled sensor. The subjects have to be seated with their hands on their thighs as shown in Fig. 2. This is to avoid any possible contact from arm to auxiliary areas [7,9,14,15].

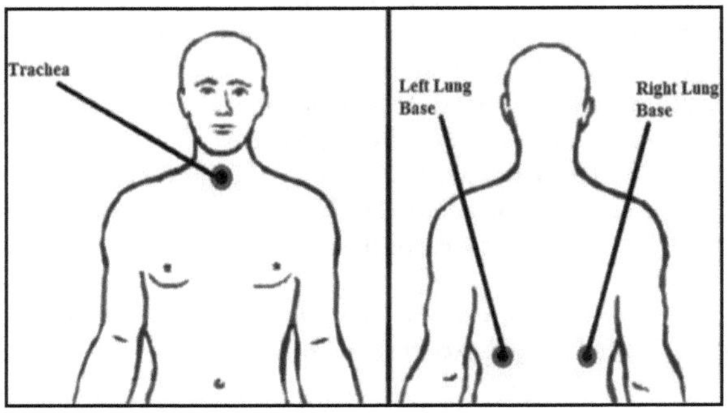

Fig. 1 Auscultation points illustrations [14]

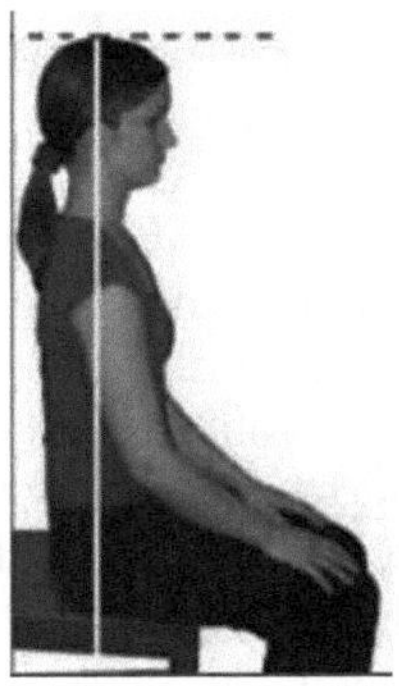

Fig. 2 Sitting position [14]

By using WISE (wireless digital stethoscope), the respiratory sounds are recorded for analysis and classification in this study. WISE is the digital stethoscope that allows patients to diagnose conveniently anytime and anywhere and it gives a better quality of sound recordings and objected diagnosing information that related with heart and lung sounds. WISE is excellent in sound amplification and also good for noise reduction. Fig. 3 below shows the setup of WISE digital stethoscope.

WISE is able to analyze stethoscope sounds using spectrogram and waveform. The lung sound data can be saved in the computer for viewing the sound pattern directly. By using the software VPM3000W, it can function as a visual monitoring, display, recording and analysis system.

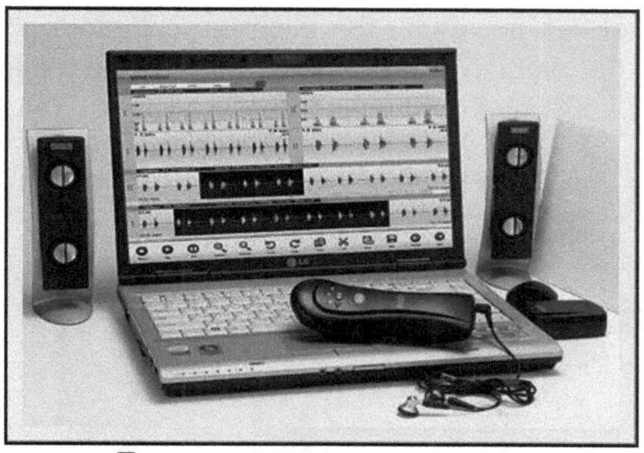

Fig. 3 WISE digital stethoscope [13]

B. Pre-processing

The stage of preprocessing is very important to reduce the noise or unwanted signals. According to CORSA standards, the original sampling rate of the breath sound recordings is 10 kHz, which is then down sampled to 8 kHz for signal pre-processing. The recorded breath sounds will be filtered using high pass filter (70 Hz) with 1st order Butterworth filter and band pass filter (80 Hz to 1600 Hz) with 4th order Butterworth filter. Then the filtered signals will be transformed into spectral features (frequency domain) using Fourier transform.

C. Features extraction

The features can be extracted from signals whether in time domain or frequency domain only or time-frequency domain. In this study, Fourier transform (FT) will be applied for extracting the non-linear signals. FT spectrum of the acoustic signals converts the signal from the time domain to frequency domain, which FT provides good representation of the signal [16]. It is because the respiratory sounds have non-stationary and also non-linear for their characteristics. In this research, the signal will be converted to frequency domain. The features extraction techniques that are widely used in respiratory sounds analysis are energy spectrum, wavelet and entropy. By comparing with the previous study, more researchers used frequency based features, which are energy spectrum and entropy to obtain the respiratory sounds.

D. Classification

The classification to detect the respiratory sounds between smokers and non-smokers is performed by using k-NN classifier that is mostly used as classifier in statistical analysis [7]. The superiority of using k-NN is its simplicity and robustness. It also does recognize normal and abnormal respiratory sounds. Since k-NN classifier gives superior classification accuracy on least redundant features, it is the most suitable classifier to be incorporated with the effective features extraction. Instead of using k-NN classifier, SVM also have a similarity structure but only k-NN that can be considered as a simplicity reason as the extension of SVM to multiple classes is not as natural as k-NN [7].

III. Results and Discussion

A. Experimental lung sound data

Fig. 4 below shows the sample of lung sound data recorded from the 30 subjects on left lung base, right lung base and trachea using WISE digital stethoscope.

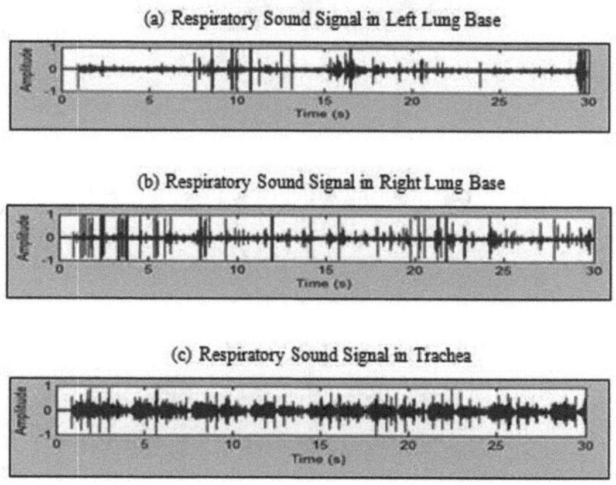

Fig. 4 Respiratory sound signal processing in three locations. (a) Left lung base, (b) right lung base, and (c) trachea.

B. Filtering

The respiratory signal and two types of filtering with high pass filter and band pass filter are shown in the following figures, where 4th order Butterworth band pass filter with cutoff frequency range between 80 Hz to 1.6 kHz and 1st order Butterworth high pass filter with cutoff frequency of 70 Hz are used to filter the noise. The sample signal result is taken from a smoker's subject, which the signal data were collected at the trachea location. For Fig. 5, it shows the original respiratory signal, while in Fig. 6 and Fig. 7 show the filtered respiratory signal using band pass filter and high pass filter simultaneously. From the filtering process, we can identify that band pass filter performs better in removing the unwanted signal (noise) as compared with high pass filter. Other researchers also considered that band pass filter is a more suitable filter method for filtering respiratory sounds signal [13].

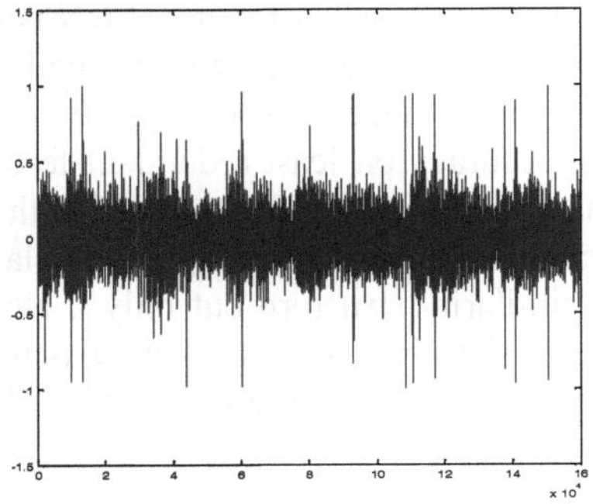

Fig. 5 Original signal

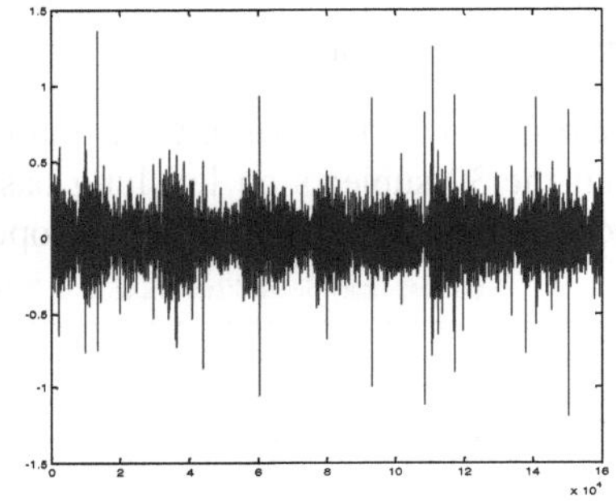

Fig. 6 After filtering using band pass filter 80Hz, 1.6kHz

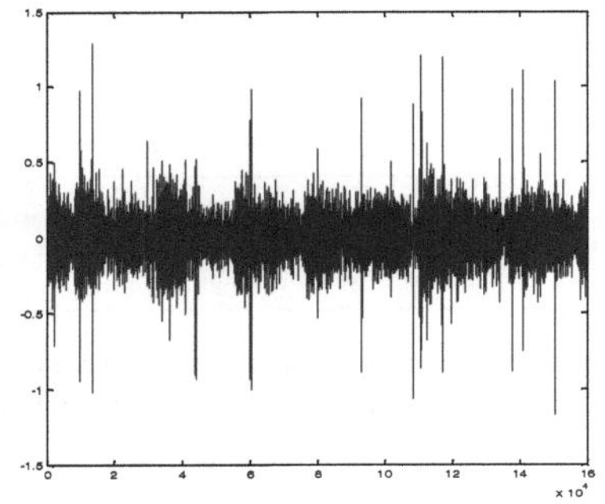

Fig. 7 After filtering using high pass filter 70Hz

C. Features extraction

Two main factors such as domain and characteristic are selected as the method of features extraction [17]. In this study, the respiratory signal must be classified using spectral features. In features extraction, the respiratory signal in time domain was converted into the frequency domain using Fast Fourier Transform (FFT). Other than that, energy spectrum and entropy are extracted as the features of the lung sounds.

D. Classification

The respiratory sounds are collected from three different locations, which are left lung base, right lung base and trachea. From the 5 trials on each location, we can evaluate that the trachea gives a very low accuracy as compared with left lung base and right lung base. Fig. 8 shows the percentage of accuracy for each auscultation point. In trachea, the average accuracy is 54.20%, while for left lung base is 77.50% and right lung base is 83.10%.

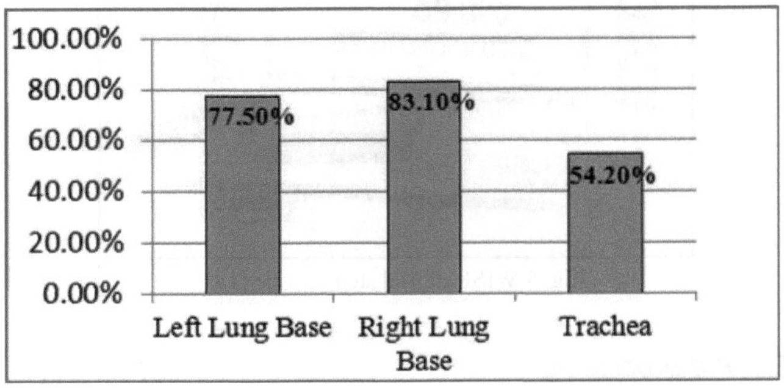

Fig. 8 Accuracy at three auscultation points

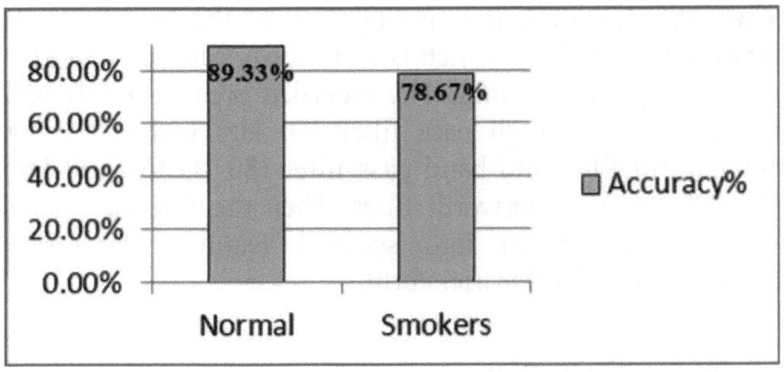

Fig. 9 Accuracy normal and smoker in right lung base

By using the k-value of 9 and entropy as the suitable features, the accuracy between normal respiratory sounds and the smokers has been identified. A normal respiratory sound means that, the subjects involved in this research are not having any disease such as asthma. Aside from that, the airways obstruction pathology that included with wheeze and rhonchi are probability exists in smokers but not all smokers have the airways obstruction. From Fig. 9 above, the accuracy for non-smoker is 89.33%, which is higher than smoker with accuracy of 78.67%.

The most important factor that significance the consequent of the algorithm test was the position of the recording device that substantially on the trachea and chest (left lung and right lung). Most of the recordings were performed on the chest and it has been noted that the chest wall functions as low pass filter that attenuate the high frequencies. So it is to believe that the most accurate result of the respiratory lung sounds can be detected on the chest. Both of the detection and classification between normal and airways obstruction in smokers' respiratory sounds were performed on the k-NN classifier using the entropy as the suitable feature.

IV. Conclusion

In conclusion, a system that captures and processes respiratory sounds has been developed. The respiratory sounds were recorded using WISE digital stethoscope at three locations, left lung base, right lung base and trachea, on the 30 subjects that consists of smokers and non-smokers. The suitable protocol for respiratory sound data acquisition was formulated by following the CORSA standard. The features that are used together with the k-NN classifier can classify the normal respiratory lung sounds and airways obstruction in smokers. k-NN classifier is a more suitable classifier because it gives superior classification accuracy on least redundant features. The accuracy for non-smokers is 89.33%, while for smokers is 78.67.

A problem of environmental background noise is important in the normal clinical setting. Respiratory sounds have more noise while collecting the data. The noise comes from the condition of the room which is not soundproofed. Distinctly, this effect must be diminished as much as possible by protecting the acoustics shielding and the use of suitable sensor. Renewal such as laser microphones, piezo-electric membrane element and the microphone, should in future, provides a more competent method of capturing the sound of the lungs. Besides that, all subjects have an average age between 22 to 34 years and the most of the smokers have only 6 years of experience in smoking. Therefore, for future work, smokers with experience of at least 20 years in smoking will be selected for further analysis.

Acknowledgment

This project is supported by the ScienceFund of the Malaysian Ministry of Science, Technology and Innovation (MOSTI).

Conflict of Interest

The authors declare that they have no conflict of interest.

References

1. Uwaoma C, Mansingh G (2014) Detection and Classification of Abnormal Respiratory Sounds on a Resource-constraint Mobile Device. International Journal of Applied Information System (IJAIS) 7(11):35–40.
2. Sovijärvi ARA, Vanderschoot J, Earis JE (1997) Standardization of Computerized Respiratory Sound Analysis: Current Methods Used For Computerized Respiratory Sound Analysis. Crit Care Med 156:974–987.
3. Palaniappan R, Sundaraj K (2013) Respiratory Sound Classification Using Cepstral Features And Support Vector Machine. In Intelligent Computational Systems (RAICS), 2013 IEEE Recent Advances in, pp 132–136.
4. Charbonneau G, Ademovic E, Cheetham BMG, Malmberg LP, Vanderschoot J, Sovijarvi ARA (2000) Basic Techniques for Respiratory Analysis. European Respiratory Review 10(77):625–635.
5. Oud M, Dooijes EH, Van Der Zee JS (2000) Asthmatic Airways Obstruction Assessment Based on Detailed Analysis of Respiratory Sound Spectra. IEEE Transaction on Biomedical Engineering (47)11:1450–1455.
6. Palaniappan R, Sundaraj K, Sundaraj S, Huliraj N, Revadi SS (2015) A Telemedicine Tool To Detect Pulmonary Pathology Using Computerized Pulmonary Acoustic Signal Analysis. Applied Soft Computing 37:952–959.
7. Yilmaz CA, Kahya YP (2006) Multi-channel Classification of Respiratory Sounds. In Engineering in Medicine and Biology Society, 2006. EMBS'06. 28th Annual International Conference of the IEEE, pp 2864–2867.
8. Sen I, Kahya YP (2005) A Multi-Channel Device for Respiratory Sound Data Acquisition and Transient Detection. In Conf Proc IEEE Eng Med Biol Soc, pp 6658–6661.
9. Palaniappan R, Sundaraj K, Sundaraj S (2014) Artificial Intelligence Techniques Used In Respiratory Sound Analysis – a systematic review. Biomedizinische Technik/Biomedical Engineering 59(1):7–18.
10. Jin F, Krishnan S, Member S, Sattar F (2011) Adventitious Sounds Identification and Extraction Using Temporal – Spectral Dominance-Based Features. IEEE Transactions on Biomedical Engineering, 58(11):3078–3087.
11. Taplidou SA, Hadjileontiadis LJ (2007) Wheeze Detection Based On Time-Frequency Analysis of Breath Sounds. Computers in Biology and Medicine 37(8):1073–1083.
12. Kraman SS, Pasterkamp H, Kompis M, Takase M, Wodicka GR (1998) Effects Of Breathing Pathways On Tracheal Sound Spectral Features. Respiration Physiology 111(3):295–300.
13. Rizal A, Hidayat R, Nugroho HA (2015) Signal Domain in Respiratory Sounds Analysis: Methods, Application and Future Development. Journal of Computer Science 11(10):1005–1016.

14. Forkheim KE, Scuse D, Pasterkamp H (1995) A Comparison of Neural Network Models for Wheeze Detection. In IEEE WESCANEX 95, pp 214–219.
15. Rossi M, Sovijarvi ARA, Piirila P, Vannuccini L, Dalmasso F, Vanderschoot J (2000) Environmental And Subject Conditions And Breathing Manoeuvres Fro Respiratory Sounds Recording. European Respiratory Review 10(77):611–615.
16. Hashemi A, Arabalibiek H, Agin K (2011) Classification of Wheeze Sounds Using Wavelets and Neural Networks. International Conference on Biomed. Eng. and Tech., vol. 11, pp 127–131.
17. Bahoura M, Pelletier C (2004) Respiratory Sounds Classification using Cepstral Analysis and Gaussian Mixture Models. Annual International Conference of the IEEE EMBS, vol. 1, pp 9–12.

Author: Chee Kiang Lan
Institute: School of Mechatronic Engineering, University Malaysia
City: Arau Perlis
Country: Malaysia
Email: lckiang@unimap.edu.my

Design and Development of a Virtual Reality Based Track Cycling Simulator

Z. A. Zakaria, and H. J. Yap

Centre for Product Design and Manufacturing, University of Malaya, Kuala Lumpur, Malaysia

Abstract— **Malaysia Track cycling team has achieved great achievement for the past few years. However, the athlete still facing difficulty of training in this country as we are lack of training facilities. This paper is consist of the design and development of a virtual reality based track cycling simulator to provide a cost saving and portable training facilities for the track cycling athlete and potential athletes. The virtual reality system allow the cyclist to immerse in the virtual environment based on real velodrome track.**

Keywords— **Virtual Reality, Track Cycling, Bicycle Simulator.**

I. Introduction

The development of virtual reality (VR) has begun since 1838 with the creation of illusion using panoramic painting which is then evolved to the creation of stereoscopic photo and viewer, link trainer (the first flight simulator), sensorama until the invention of a ultimate display of virtual reality by Ivan Sutherland in 1965 [1]. This is the starting point on stepping into a virtual world instead of just watching picture in a monitor. The idea presented by Sutherland is "*to make a world in the window look real, sound real, feel real, and respond realistically to the viewer's action*" [2].

Since then, the application of virtual reality (VR) has been adapted to various field including medical [2], automotive [3], aerospace [4], sport [5] and product manufacturing [6]. There are lot of ongoing research to broaden the application of VR in other field. One of the beneficial application of VR that going to be highlighted in this paper is the integration of VR with bicycle trainer.

During the eighteenth century, bicycle is one of the main transport and a lot of collision and accident occurred involving bicycle due to unskillful rider and many other factors. This is beginning of the invention of bicycle trainer. A primary designed has been illustrate in the Figure 1 [7]. It is designed with a home trainer, electric fan and cinematograph so that all the pleasure of a tour can be had in a bedroom [7].
The idea has been adapted to the new technology of bicycle trainer [8-10] that integrate the virtual reality with bicycle trainer.

However, the bicycle trainers in current market are unable to let the user fully immerse in the virtual world as the trainer could not simulate the motion of bicycle in the real environment. Thus, this research is focusing into developing a bicycle simulator with virtual reality environment and a moving 6 degree of freedom (6DOF) platform to let the user totally immerse in the virtual reality world.

Fig. 1. The illustration of bicycle trainer idea at 1886 [7]

The objective of this project is to develop a portable 6DOF motion bicycle simulator with virtual reality. It will imitate the real environment during cycling including all the forces encounter by the cyclist. This research will be beneficial to the track cycling athletes as we are focusing on the environment of cycling in a velodrome.

Track-based cycling is a restricted competition with a lot of regulations to comply with [11]. In Malaysia, track cycling team has a good potential to strive for better achievement. Unfortunately, there are no many training facilities (velodrome) in Malaysia for the track cycling athletes. Alternatively, the athlete will undergo training in the other country that has the training facilities.

The track cyclist need the physical velodrome to train but building one is very costly and take a long time. Moreover, the velodrome is not portable. Thus, many velodrome need to be built around Malaysia for the track bicycle athletes. These facilities also give opportunity for the other young talents to train themselves. However, this will consume a lot of money, space, time as well as time.

The idea of design and developing the track cycling simulator will reduce the cost of building the velodrome as

F. Ibrahim et al. (eds.), *3rd International Conference on Movement, Health and Exercise*,
IFMBE Proceedings 58,
DOI: 10.1007/978-981-10-3737-5_16

well as providing a portable bicycle simulator that is possible to be transfer anywhere.

This report is structured with the explanation of the principal operations of the simulator and describing the parts of the simulator in section II. Next is reviewing the related previous project that can be adapted in this project in section III. Conclusion and future work is later discussed in section IV.

II PRINCIPAL OF OPERATIONS

The principal operations of the bicycle is divided by three parts: hardware, software and display unit as in Figure 2. Each part is connected using communication cable to allow two-way communication. The hardware parts include the 6 degree of freedom (DOF) platform, bicycle, and bicycle trainer with sensor to provide tire resistance on the bicycle.

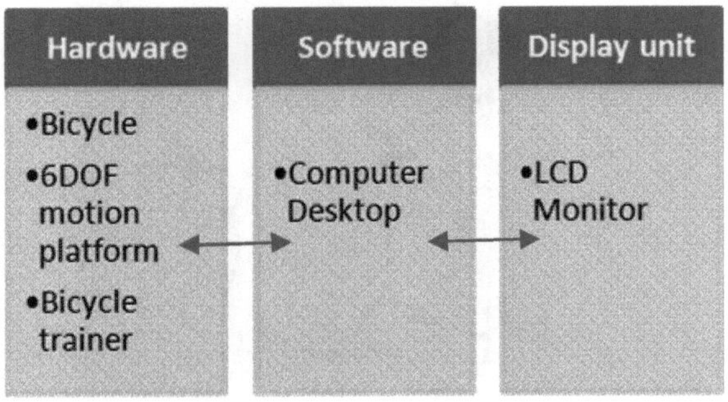

Fig. 2. The principal operations of the bicycle simulator.

The simulator is expected to simulate the real track cycling environment with virtual scene to be displayed in the Head-Mounted Display (HMD). The displayed will also provide information of the cycling activities such as the cycling speed, distance travelled, and inclination position of the bicycle. These information is essential to observe the efficiency and performance of the cyclist. The simulator is also expected to be able to simulate the tire resistance encountered by the cyclist. This scope will be discuss later in this report.

A. Hardware

Hardware is the vital part in this simulator. As mentioned previously it is consist of a 6 DOF motion platform, bicycle, and a bicycle trainer. The 6 DOF motion platform is designed based on the application of the Stewart platform as in Figure 3. It consist of 6 linear actuators that can be extended and allow the platform to experience a certain inclination angle based on the real velodrome track.

The challenge in this design is that by using actuator leg, the platform's height will be more than 2 meters in an actual scale. This is not a safe design considering that it is a moving platform that able to incline to a maximum of 23 degrees. The solution is to replace the actuator with a DC motor as per Figure 4 in order to reduce the height of the platform.

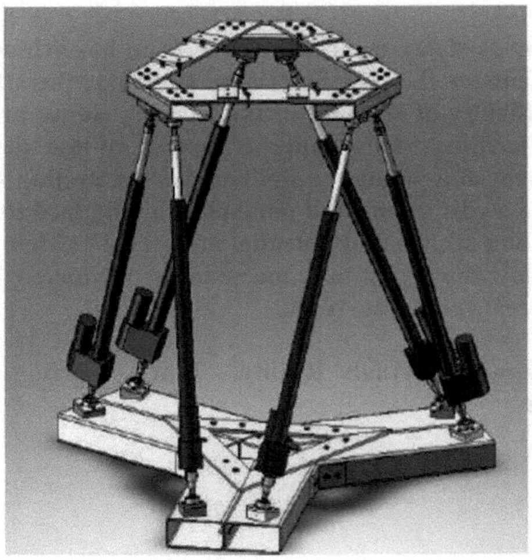

Fig. 3. Example of the 6 DOF Motion platform based on Stewart Platform design. [4]

In this simulator, the maximum inclination is limited to 23 degree by considering the safety factor of the cyclist when operating this simulator. 6 DOF motion platform will be powered by AC-DC power supply.

Fig. 4. Actual scale prototype of 6DOF platform

A real track cycling bicycle will be used in this simulator as simulator wants the cyclist to immerse in the real cycling situation with the real bicycle. It will be mounted at the top of the platform. The bicycle can be detached from the platform to allow the cyclist to use their own bicycle with the simulator. This is essential as we want the cyclist to feel the resistance when using their own bicycle and improved their performance from it. For this research, a bicycle that complied with the Union Cycliste Internationale (UCI) regulations will be used.

The last part is the bicycle trainer. Automatic bicycle trainer will be used as it can change the tire resistance as the cyclist move faster. Manual bicycle trainer is not suitable in this project as the resistance need to be change manually by the cyclist. Nowadays, bicycle trainer in the market came with a cycling software that can integrate the training with computer to provide information such as the speed of the bicycle.

B. Display Unit

In this project LCD monitor is used as the display unit. It will display the virtual reality program and the related information such as the speed of the bicycle, inclination position of the bicycle, tire resistance, total travelled distance, and cycling cadence. This information can be used to evaluate the performance of the cyclist and planning on the aspect to improve their performance.

The virtual reality software used in this project is the Unreal Engine. It will then integrated with the Oculus head mounted display (HMD) and Wii Mote to create two ways communication between the hardware and software.

III PREVIOUS RELATED PROJECTS

Basically there are a few related projects that has been conducted before. There is project conducting design and development of 6 DOF motion platform for driving simulator [3]. The concept used for the 6 DOF motion platform is based on the Stewart platform. The dimension is first calculated by using the Inverse Kinematic using the given position and orientation of the upper platform. The platform is then design and converted into a SimMechanics model generation. The simulator is using closed loop system as shown in figure below.

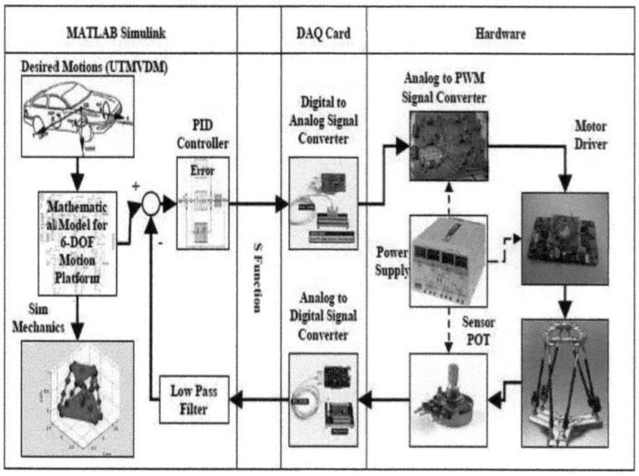

Fig. 5. Motion platform closed loop system. [4]

The maximum angle of inclination for this project's prototype is 20°. However, the success of this prototype is not a warranty that it is applicable into the real simulator considering that the prototype is too high even with a small surface area of moving platform. Safety is a concern here since top of the platform will be occupied by a human. This Stewart platform is also adapted in other projects [12-15]. The concept of moving platform can be adapted to develop the virtual reality bicycle simulator with motion platform.

Another related project is the development of a virtual cycling simulator [16]. This project targeting the road cycling type simulator. Like our project, this project also adapting the virtual reality into the bicycle simulator. The difference is that the bicycle platform is only allow a small magnitude of tilting as it used spring installed at bottom of the platform. This design is not applicable for track cycling as the maximum angle in velodrome is roughly around 45°.

Fig. 6. Arrangement of virtual reality cycling simulator [10]

The virtual reality software used in this paper is Microsoft Visual C++. Dynamic model of bicycle is used to solve the equation of motion for the characteristic of the simulator. This simulator is successful in simulating the dynamic change of the virtual environment including resistance of land profile and also the effect of accelerating as well as decelerating when the bicycle is moving up and down hill.

IV CONCLUSIONS AND FUTURE WORK

The development of virtual reality track cycling simulator will help the cyclist athlete and even potential athlete to train and develop themselves. The Stewart Platform with the immersive of virtual reality is a first step in developing a track cycling simulation.

To allow the cyclist totally immerse in the virtual reality environment, more forces need to be considered in this project such as air drag force and slipping between tire and track. In future, design of the current velodrome track can be program in the virtual reality software so that cyclist can feel like training in a real particular velodrome as preparation for any competition.

Acknowledgemnt

The authors would like to acknowledge for the Ministry of Education of Malaysia for the financial support under Grant No. GA012-2014

Conflict of Interest

The authors declare that we have no conflict of interest in this project.

References

1. Reality, V. (2015, December 25). *History of virtual reality - virtual reality*. Retrieved August 12, 2016, from Virtual Reality, http://www.vrs.org.uk/virtual-reality/history.html
2. Tomasz M. and Michael G. Virtual Reality: History, Applications, Technology and Future.
3. Mohamad K. Design and Development of 6 DOF Motion Platform for Vehicle Driving Simulator.
4. Alejandro R. G. R., Alexandre A. D. R., Marcelo G. G. et al. Automation of Flight Simulator Platform in Immersive Virtual Reality.
5. Yuk-Ming T., Matthew T., Daniel F. et al. The Development of a Virtual Cycling Simulator. Technologies for E-Learning and Digital Entertainment. Berlin : Springer, pp 162-170
6. Qingjin P. Virtual Reality Technology in Product Design and Manufacturing.
7. The Wheel and cycling trade review (1988). (Vol. 26, no. 4 ed.). New York : Wheel and Cycling Trade Review.
8. Home. (2016). *Direct drive indoor turbo trainers*. Retrieved August 12, 2016, from http://turbobiketrainer.com/best-virtual-reality-turbo-trainer-2015-review/
9. Virtual reality. Retrieved August 12, 2016, from https://www.tacx.com/products/software/virtual-reality-advanced
10. CycleOps. (2016). *CycleOps*. Retrieved August 12, 2016, from https://www.cycleops.com/virtualtraining/overview
11. Dan C. Track Cycling – An Introduction.
12. Ajay V. S., Kyle D. B., Ryan R. B. et al. Development of A Virtual Reality Based Power Wheel Chair Simulator.
13. Boian R.F., Bouzit M., Burdea G.C. et al (2004) Dual Stewart Platform Mobility Simulator. IEEE EMBS, pp 4848-4851
14. Dong-Soo K., Gi-Hun Y., Chong-Won L. et al (2001) KAIST Interactive Bicycle Simulator. IEEE International Conference on Robotics & Automation, pp 2313-2318
15. Dong-Soo K., Gi-Hun Y., Yougjin P. et al (2002) KAIST Interactive Bicycle Racing Simulator: The 2nd Version with Advanced Features. IEEE International Conference on Intelligent Robots and System EPPFL, 2961-2966
16. Bo Z. (2005) Design and Implementation of a 6 DOF Parallel
17. Manipulator with Passive Force Control

Author: Yap Hwa Jen
Institute: Department of Mechanical Engineering, Faculty of Engineering, University of Malaya,
Street: Jalan University
City: Kuala Lumpur
Country: Malaysia
Email: hjyap737@um.edu.my

Relationship Between Electromyostimulation and Free Weight Exercises in Multiple Repetition Maximum Strength Test

R.N.J.R. Hussain[1], K.M. Kee[1], R. Razman[2], S.I. Ismail[1], M. Shari[1] and N.M. Ideris[1]

[1] University Technology MARA, Shah Alam, Malaysia
[2] Univeristy Malaya, Kuala Lumpur, Malaysia

Abstract— **Background: Electromyostimulation (EMS) is a training that uses electrical current to stimulate muscle contraction. The progression of EMS training is similar with principles of resistance training which are based on the parameters such as pulse frequency, electrical pulse, duration of rest, duration of contraction, the number of repetitions and the intensity (Hz) used. Rate perceived exertion (RPE) scale was used in monitoring the loading along the duration of EMS training while the percentage of repetition maximum (%RM) is globally use in monitoring loading of resistance training (RT). Since the unit of measurement for loading is different, therefore this study seek to find the relationship between multiple repetition maximum electrical pulses in EMS (miliAmpere-mA) and multiple repetition maximum lift (Kilogram-Kg) in free weight strength test. Methods: A total of 10 recreational athlete (age: 22±0.77, height: 155.6±0.92, weight: 51.6±1.02, %BF: 21.3±1.10) volunteered to participate in strength testing. Participants go through the adaptation of EMS and weight training. Multiple RM (1, 3, 5, 7, 9) was used to measure chest press and squat strength using National Strength Conditioning Association procedure. Data from the tests were analysed using Pearson Correlation. Results: There is significant positive strong relationship $p<.01$ between EMS and load lift for chest press (1RM; $r = .96$, 3RM; $r = .86$, 5RM; $r = .91$, 7RM; $r = .93$; 9RM; $r = .94$) and squat (1RM; $r = .83$, 3RM; $r = .88$, 5RM; $r = .93$, 7RM; $r = .92$, 9RM; $r = .81$). Discussion: The results indicate that both measurements were highly correlated between mA and Kg along the nine repetitions maximum strength test. Furthermore, it showed that both measurements can be applied in training to increase muscular strength.**

Keywords— **Electromyostimulation, Strength, Intensity, Repetition maximum**

I. Introduction

Electromyostimulation (EMS) training is a method of using electrical current to stimulate muscle contraction. This is because muscle contraction in EMS training is different than the traditional training (e.g. weight lifting). Muscle contraction in traditional training is initiated by the central nervous system while muscle contraction in EMS training is controlled by electrical current [1]. Nowadays, use of EMS as a whole body workout is popular and demanding compared to traditional resistance training. EMS training has proven to increase muscular strength in healthy humans [2, 3]. Past study reported increases in maximal voluntary torque during isokinetic solicitations for a wide range of concentric and eccentric angular velocities [4] and during isometric actions for joint angles close to the training angle [5].

Several authors revealed that the improvements from EMS training may originate from neural factors [6-8]. Both muscular activation and electromyographic activity increases after a short duration (<4 weeks) of EMS training [9, 10], whereas significant muscular hypertrophy been showed after longer (>8 weeks) training durations [2, 9]. Though the training method differs with traditional resistance training, both trainings were able to develop and increase athlete's performance. Both trainings also include various principles of training. For example, EMS training are based on the programmed parameters such as the electrical pulse which triggers the excitation of the motor neuron, duration of contraction, duration of rest, number of repetitions and intensity (Hz) used. These parameters were similar to the principles of training that widely used in traditional resistance training which are muscle focus, rest duration, the number of repetitions, duration of training, and intensity (load).

What differs between both trainings is the method of controlling the loading. EMS training used rate perceived exertion (RPE) scale and pain threshold while the percentage of repetition maximum is globally used in monitoring the loading for traditional resistance training. RPE scale been used widely in EMS training might be because of the training involved electrical pulse (milliAmpere) which there is no study showed how to specifically control the loading of miliAmpere (mA) along the training duration.

RPE has been controlled and adjusted by the athletes using their own perceptions of effort when they faced physiological stress on their bodies during undergoing exercises. This method has been demonstrated during steady-state exercise [11] and high-intensity interval cycling [12] where the athletes' reported RPE correlated well to average heart rate [11] and acute changes in heart rate [12]. Furthermore, measuring intensity using RPE is widely used

F. Ibrahim et al. (eds.), *3rd International Conference on Movement, Health and Exercise*,
IFMBE Proceedings 58,
DOI: 10.1007/978-981-10-3737-5_17

in aerobic exercises compare to resistance training [13]. Past studies showed poor correlations between heart rate and RPE responses during short-duration, high-intensity soccer drills [14] and during step dance sessions [15].

In training, progressive overload is an integral part of a successful exercise plan. This is because, proper manipulation of volume, intensity, and recovery phase is vital for optimal results [16]. One of the way to achieve successful exercise plan, a method in measuring loading/intensity must be valid and reliable. RPE and %RM are methods that reliable to be used in monitoring intensity. As been mentioned, high-intensity exercise is difficult to quantify because this type of exercise cannot be objectively evaluated using physiological global measurements such as heart rate [17]. This problem supports the need for a valid and reliable method of monitoring load/intensity in EMS training.

Therefore, the first step is to seek for the relationship between multiple repetition maximum of electrical pulses (mA) and multiple repetition maximum of free weight (Kg) load increment in strength tests.

II. Methods

A. Experimental approach to the problem

Multiple maximum numbers of repetitions (using different percentages of 1RM) was determined for 2 different exercises (chest press and squat). Selected repetition maximum (1RM, 3RM, 5RM, 7RM, 9RM) as it usually been used as training loads in traditional resistance training. Initial 1RM testing was conducted in female recreational athletes (at least 1 - 3 months of consistent resistance training experience). Each participant has 6 testing sessions; each session consisted of 2 different exercises at 5 different repetition maximum, in a balanced and matched order.

B. Subjects

Ten healthy female recreational athletes (age: 22±0.77, height: 155.6±0.92, weight: 51.6±1.02, %BF: 21.3±1.10) volunteered to participate in this study. All ten participants have 1 – 3 months of EMS and RT experience, body fat percentage (%BF) ranged 18.5% to 25%, and free from any musculoskeletal injuries. Participant's characteristics are reported in Table 1. There is no significant difference (p>.05) among participants in body fat percentage and 1RM strength on chest press and squat.

Table 1: Participants characteristics

Characteristics	Mean	SD
Age	22	0.77
Weight	51.6	1.02
Height	155.6	0.92
Body fat percentage (%)	21.3	1.10
1RM chest press (Kg)	26.89	2.85
1RM squat (Kg)	52.44	4.49

C. Experimental design and procedures

All participants have undergone adaptation with EMS and free weight exercises, familiar with proper techniques and testing procedures for the 2 exercises before multiple RM testing. Strength and conditioning specialist verified the participant's technique in 2 exercises. Participant's height, weight, and age were obtained during adaptation. In addition, body composition (i.e., percent body fat) was determined by In-Body test.

D. Testing

Multiple repetition maximum (RM) test of chest press and squat was perform using National Strength Conditioning Association (NSCA) procedure for both electrical pulses and free weight exercises. Briefly, participants performed a warm-up consisting 8 – 10 repetitions using a light weight and low electrical pulses. Multiple RM strength were tested after the warm-up session by increase the load and electrical pulses on subsequent attempts until the participants were unable to complete a full range of motion. Each attempt was separated by 3 – 5 minutes of rest. Both tests were conducted on six different days which three days for normal chest press and squat test in a gymnasium while another three separated days for electrical pulse chest press and squat tests in a studio. Participants were be given 48 hours of rest between all tests. After finish 1 – 9 RM strength test using load, the participants have to do the same thing using electrical stimulation. However, no load (kg) involved during this test but specific muscles were activated (pectoralis major & quadriceps). The milliamp was tuning to reach each participant's 1-9RM test. The milliamp is shown in the figure 1 below. The button will slowly increase 1mA until reach the desire goal (1-9RM).

Fig. 1 miliAmpere (mA) Tuning Button

mA Value for both muscles for each repetition maximum were recorded. Though the unit of mA and Kg cannot be converted into the same unit, however both methods follow the same strength test protocol standardized. After the both value were recorded, then correlation statistic were conducted to find their relationship.

E. Data analysis

This study had utilized a Pearson correlation to determine the relationship between multiple repetition maximum of electrical pulses (mA) and multiple repetition maximum of free weight (Kg) in chest press and squat. IBM Statistic 20 was used to analyse the raw data with the significant level were set at $p<.05$.

III. Results

Strong positive relationship was shown between weight lift and electrical pulses for chest press and squat strength tests. Table 2 and 3 showed that both methods in measuring strength test for multiple repetition maximum were highly correlated with r value above 0.8 at significant level $p<.01$.

Table 2: Relationship between maximum electrical pulses (mA) and repetition maximum (Kg) in chest press

	1MA	3MA	5MA	7MA	9MA
1RM	.83**				
3RM		.88**			
5RM			.93**		
7RM				.92**	
9RM					.81**

**$p<.01$ (2 tailed)

Table 3: Relationship between maximum electrical pulses (mA) and repetition maximum (Kg) in squat

	1MA	3MA	5MA	7MA	9MA
1RM	.96**				
3RM		.86**			
5RM			.91**		
7RM				.93**	
9RM					.94**

**$p<.01$ (2 tailed)

IV. Discussion

The primary finding of this study is both (mA & Kg) multiple repetition maximum test were highly correlated in measuring upper (chest press) and lower (squat) body strength. The significant relationship shown in results indicate that higher load in electrical pulses (mA) correlated with higher load in free weight (Kg) exercises. Load/intensity is important in all training. Individuals train based on specific loads in achieving specific goals of training. Therefore, it is important to know and able to control it during exercise.

In EMS training, there is no specific method in control and monitor the intensity/loading of a training. A study by Kemmler [18] on the whole body EMS asked the participants to appraise the average intensity in each session on a rating scale (RPE) between 1 (very low) and 7 (very high). While another study control the stimulation intensity (range 0 – 100mA) based on his or her pain threshold [2]. These showed that EMS training did not have a specific method in monitoring loading/intensity of training.

Since past studies on EMS used rate perceived exertion (RPE) and pain threshold in training, this study revealed that there is a correlation between EMS and free weight exercise in strength test using NSCA procedure. NSCA procedure is one of the gold standards in determine strength level for upper and lower body. Results from the strength test can be used for further exercise plan as the benchmark of loading/intensity. Therefore, this study provides an idea to use the percentage of maximum electrical current in monitoring the intensity/loading for EMS training.

Manipulation of loading is important in an exercise plan. It describes the amount of weight that participants should lift and become the goals of training for each session [19]. Altering the training load can significantly affect the hormonal, cardiovascular, acute metabolic, and neural responses to training [17, 20]. Loading can become the goals of training for each session depends on individual training status and goals.

Loading and repetition maximum are related in training to increase strength for trained and untrained individuals. In trained individuals, loads greater than 80 - 85% of 1RM were needed to produce further neural adaptations in training [21]. According to Baechle [16], 80 - 85% of 1RM is equal to 6 to 8 repetition maximum. As shown in result from this study, both measurements in EMS and free weight exercises are correlated for 6 to 8 repetition maximum.

Therefore, EMS training can be applied for trained individuals by following the principle of loading (80 – 85% of 1 Maximum electrical current).

Other than that, multiple repetition maximum strength test can be used to predict 1RM load for an individual. There are several equations to predict 1RM such as equation from Brzycki, Baechle, and dos Remedios. This equation need to be select precisely based on the target population and training status. Brzycki and Baechle were usually been used to predict 1RM among trained and untrained individuals. Before an individual is able to predict his or her 1RM, multiple repetition maximum tests should be conducted. Therefore, when a value from multiple repetitions maximum tests is obtained, 1RM can be predict.

Based on the current result, this study also provides additional information on finding the 1 Maximum electrical current for EMS training based on multiple repetition maximum strength test for upper and lower body.

V. Conclusion

Contrary to the known primarily free weight training, EMS training is a new comprising training technology that proven to increase muscular strength as early of 4 weeks. Time saving and low-intensity exercise programs with the impact on fitness are increased rapidly in the fitness industry. Thus, it is important for the exercise and fitness specialist to measure and compare between various existence training to find the most effective and applicable towards achieving specific fitness goals. The first step showed that there is highly correlated between multiple repetition maximum in free weight (Kg) and electrical pulses (mA). This provides a useful information for further measurement in both trainings.

VI. Acknowledgements

We gratefully acknowledge the support of miha bodytec (United Lifestyle) and Faculty of Sports Science gymnasium. None of the authors has any conflict of interest. We were also grateful for the grant support from Research Management Institute (RMI): 600-RMI/MyRA 5/3/LESTARI (32/2016).

VII. References

[1] B. D. Johnston, "ElectroMyoStimulation," *Synergy [Internet]*, 2004.

[2] N. Babault, G. Cometti, M. Bernardin, M. Pousson, and J.-C. Chatard, "Effects of electromyostimulation training on muscle strength and power of elite rugby players," *The Journal of Strength & Conditioning Research*, vol. 21, pp. 431-437, 2007.

[3] T. Hortobagyi, K. Scott, J. Lambert, G. Hamilton, and J. Tracy, "Cross-education of muscle strength is greater with stimulated than voluntary contractions," *Motor Control*, vol. 3, pp. 205-219, 1999.

[4] N. A. Maffiuletti, S. Dugnani, M. Folz, E. Di Pierno, and F. Mauro, "Effect of combined electrostimulation and plyometric training on vertical jump height," *Medicine & Science in Sports & Exercise*, pp. 1638-44, 2002.

[5] N. A. Maffiuletti, G. Cometti, I. Amiridis, A. Martin, M. Pousson, and J. Chatard, "The Effects of Electromyo-stimulation Training and Basketball Practice on Muscle Strength and Jumping Ability," *International journal of sports medicine*, vol. 21, pp. 437-443, 2000.

[6] S. Colson, A. Martin, and J. Van Hoecke, "Re-examination of training effects by electrostimulation in the human elbow musculoskeletal system," *International journal of sports medicine*, vol. 21, p. 281, 2000.

[7] K. Hainaut and J. Duchateau, "Neuromuscular electrical stimulation and voluntary exercise," *Sports medicine (Auckland, NZ)*, vol. 14, p. 100, 1992.

[8] N. A. Maffiuletti, R. Zory, D. Miotti, M. Pellegrino, M. Jubeau, and R. Bottinelli, "Neuromuscular adaptations to electrostimulation resistance training," *American journal of physical medicine & rehabilitation/Association of Academic Physiatrists*, vol. 85, p. 167, 2006.

[9] J. Gondin, M. Guette, Y. Ballay, and A. Martin, "Electromyostimulation training effects on neural drive and muscle architecture," *Medicine and science in sports and exercise*, vol. 37, p. 1291, 2005.

[10] N. A. Maffiuletti, M. Pensini, and A. Martin, "Activation of human plantar flexor muscles increases after electromyostimulation training," *Journal of Applied Physiology*, vol. 92, pp. 1383-1392, 2002.

[11] J. Borresen and M. I. Lambert, "The quantification of training load, the training response and the effect on performance," *Sports Medicine*, vol. 39, pp. 779-795, 2009.

[12] J. M. Green, J. R. McLester, T. R. Crews, P. J. Wickwire, R. C. Pritchett, and R. G. Lomax, "RPE association with lactate and heart rate during high-intensity interval cycling," *Medicine and science in sports and exercise*, vol. 38, p. 167, 2006.

[13] M. L. Day, M. R. McGuigan, G. Brice, and C. Foster, "Monitoring exercise intensity during resistance training using the session RPE scale," *The Journal of Strength & Conditioning Research*, vol. 18, pp. 353-358, 2004.

[14] T. Little and A. G. Williams, "Measures of exercise intensity during soccer training drills with professional soccer players," *The Journal of Strength & Conditioning Research*, vol. 21, pp. 367-371, 2007.

[15] A. Ozkan and A. Kin-Isler, "The reliability and validity of regulating exercise intensity by ratings of perceived exertion in step dance sessions," *The Journal of Strength & Conditioning Research*, vol. 21, pp. 296-300, 2007.

[16] T. Baechle, R. Earle, and T. R. Baechle, *NSCA's essentials of personal training*: Human Kinetics, 2004.

[17] C. Foster, J. A. Florhaug, J. Franklin, L. Gottschall, L. A. Hrovatin, S. Parker, *et al.*, "A new approach to monitoring exercise training," *The Journal of Strength & Conditioning Research*, vol. 15, pp. 109-115, 2001.

[18] W. Kemmler, R. Schliffka, J. L. Mayhew, and S. von Stengel, "Effects of whole-body electromyostimulation on resting metabolic rate, body composition, and maximum strength in postmenopausal women: The training and electrostimulation trial," *The Journal of Strength & Conditioning Research*, vol. 24, pp. 1880-1887, 2010.

[19] W. J. Kraemer, N. Ratamess, A. C. Fry, T. Triplett-McBride, L. P. Koziris, J. A. Bauer, *et al.*, "Influence of resistance training volume and periodization on physiological and performance adaptations in

collegiate women tennis players," *The American Journal of Sports Medicine,* vol. 28, pp. 626-633, 2000.
[20] W. J. Kraemer, N. A. Ratamess, and P. Komi, "Endocrine responses and adaptations to strength and power training," *Strength and power in sport,* pp. 361-386, 2003.
[21] W. J. Kraemer and N. A. Ratamess, "Fundamentals of resistance training: progression and exercise prescription," *Medicine and science in sports and exercise,* vol. 36, pp. 674-688, 2004.

Author: Raja Nurul Jannat Raja Hussain
Institute: Universiti Teknologi Mara
City: Selangor
Country: Malaysia
Email: rajanuruljannat@gmail.com

Analysis and Classification of Forearm Muscles Activities during Gripping using EMG Signals

A.F. Wahab[1], C.K. Lam[1] and K. Sundaraj[2]

[1] School of Mechatronic Engineering, Universiti Malaysia Perlis (UniMAP), Kampus Pauh Putra, Perlis. Malaysia.
[2] Faculty of Electronics and Computer Engineering, Universiti Teknikal Malaysia Melaka (UTeM), Hang Tuah Jaya, Melaka, Malaysia.

Abstract— **Electromyography (EMG) is the recording about the electrical activity of muscle tissue. EMG measures the electrical signal that generated from the muscles during contraction or relaxation action. It has been used in various fields such as medical diagnosis and engineering applications. In this research, EMG signals were recorded from three forearm muscles to analyze and classify forearm muscles activities during gripping. EMG signals were measured from the subjects by applying three different values of grip force. Features such as root mean square, mean absolute value, standard deviation and variances are extracted from the acquired EMG signals. The values of these features were used for the classification in different level of grip strength. The results show that EMG signals' amplitude and frequency from the forearm muscles increased due to the increase of grip strength and different levels of grip force can be classified from the three forearm muscles' activities.**

Keywords— **Electromyography, Hand grip, Feature extraction, Classification**

I. Introduction

The electromyography (EMG) is a diagnostic tool that records the electrical activity of muscle tissue or as an audible signal using electrodes that either attached on the skin or inserted into the muscle. Research in the area of EMG signal has followed several avenues. Early work by Gene Shuman [1] those using forearm electromyograms to classify hand gestures. It describes an experiment in which classification is used to build a model that decides which of a set of hand gestures was made by the subject based on forearm EMGs.

EMG sensor has two types of electrode, surface and intramuscular. Surface EMG has an electrode that placed on the skin surface. The intramuscular EMG is a needle that inserted directly into the muscle. Surface EMG is easier for placement the electrodes compare to intramuscular EMG.

An electromyography (EMG) is studying about the electrical activity of muscle. An EMG is the measurement of electrical current generated by the muscle contraction or relaxation. It is called an electromyogram as a voltage recorded [2]. The action of nerves and muscle is essentially electrical. Information is transmitted along nerves as a series of electrical discharges carrying information in pulse repetition frequency. The EMG sensor consists of a two electrode for electromyography and one ground electrode.

A. Related works

Many authors used different classification techniques and feature set to analyze of hand and forearm muscles activities. Al Omari et al. [4] used the k-NN classifier for analyzed the extracted forearm sEMG signals. The result of accuracy is 94.25% of the average classification and this rate was obtained using the Sample Entropy, RMS, Myopulse Percentage Rate (MYOP) and Difference Absolute Standard Deviation Value (DASDV) set [5]. Haris et al. [5] analyzed the EMG signal based on finger movement recognition for prosthetic hand control and also used this k-NN classifier for classification method. The highest accuracy is 86.11% with value of k is 6 was obtained.

Jiang et al. [6] used the method to import the feature value into an Artificial Neural Network to identity the finger motion. Vigneswari et al. [2] focused on classifying the acquired real time EMG signals in artificial neural networks which produces actions to be performed by the prosthetic hand. Hickman et al. [7] developed artificial neural networks (ANN) as pattern recognition systems to classify sEMG signals into nine select hand motions. The feature is extracted using statistical analysis and pattern classification which were done by linear discriminant analysis (LDA) by [8]. They said that the classification using LDA is simpler compared to others such as support vector machine and neural network. Lee et al. [9] used the LDA method to classify the EMG signal based on grip configuration. Before classification, they used two data reductions of the feature vector which are nonnegative matrix factorization (NMF) and principal component analysis (PCA) that combined with LDA. After classification, NMF-LDA get 98.46% and PCA-LDA get 97% of accuracy.

Shuxiang Guo et al. [10] use this classification method for upper-limb movement. The experiment results show that neural network classifier get the highest recognition accuracy rate about 88.7% during the training process. Support Vector Machine performs better in real-time experiments about 85.9% [10]. Support Vector Machine

F. Ibrahim et al. (eds.), *3rd International Conference on Movement, Health and Exercise*,
IFMBE Proceedings 58,
DOI: 10.1007/978-981-10-3737-5_18

takes less than Neural Network during training process in term of time consumption and more time in term of real-time computation.

II. METHODOLOGY

A. Subjects selection

The subjects have been chosen from male only, where a total number of 10 subjects are selected. All subjects are normal and healthy human who do not have any serious injury history at the forearm, hand, and also the fingers. This is to ensure they can perform ordinary hand grip during the data collection so EMG signals can be recorded.

B. Procedures and tasks

Participants have been informed of the experimental procedures and they agreed with the informed consent. They have to exert handgrip forces using their right hand. They were seated at a standardized position which is sitting straight up with the forearm flexed in 90 degrees horizontally. The forearm is supported with a cushion to give the subjects a comfortable situation to rest their forearm. The hand gripper tool is used to measure the hand grip force. The hand grippers have a fixed value for each force respectively. The force of gripping is set by three different measurements, 50N, 100N and 150N. Participants must hold the grip for 5 seconds and the EMG signal is recorded within these 5 seconds. Each grip force will be repeated for 3 times with an interval of 5 to 10 minutes of rest. This step is continued for 5 trials and repeated the same by the other two forces, 100N and 150N. This is the same procedure with [11].

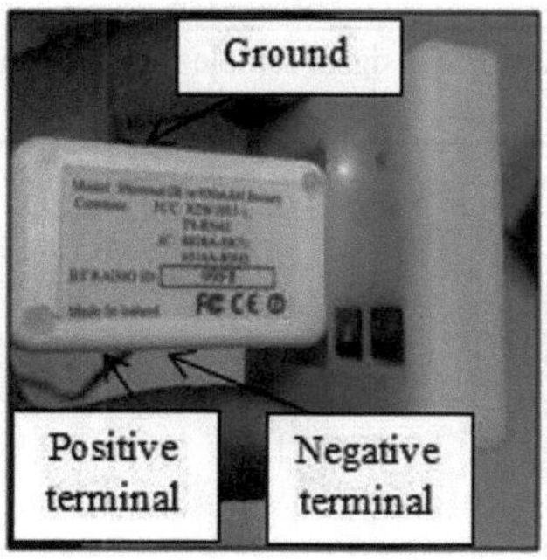

Fig. 1 Shimmer EMG Kit

C. EMG acquisition

The EMG signals are acquired using the Shimmer EMG Development Kit. The Shimmer EMG kit as shown in Fig. 1 includes two surface EMG electrodes with a grounding electrode. EMG signal can be measured by this device and the data is saved into a digital database for further process.

D. Placement of sensor

In this project, the muscle that attached with the EMG sensor is flexor digitorum superficialis (FDS), which has maximum fatigue when perform the gripping action [2]. Other two muscles are flexor digitorum profundus (FDP) and extensor digitorum (ED). According to these three muscles, three pairs of EMG sensors have been used and attached on the muscles. According to Lee et al. [9], these three muscles are mostly related to the movement of the thumb, forefinger, middle finger and ring finger that produce EMG signals. Before attaching the sensor, the skin on the forearm will be cleaned. Fig. 2 and Fig. 3 below show the location of the sensor and the type of muscles that involve during the gripping action.

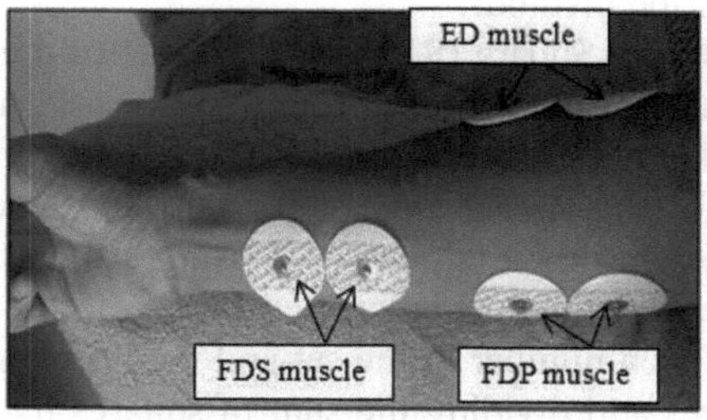

Fig. 2 Placement of Sensors

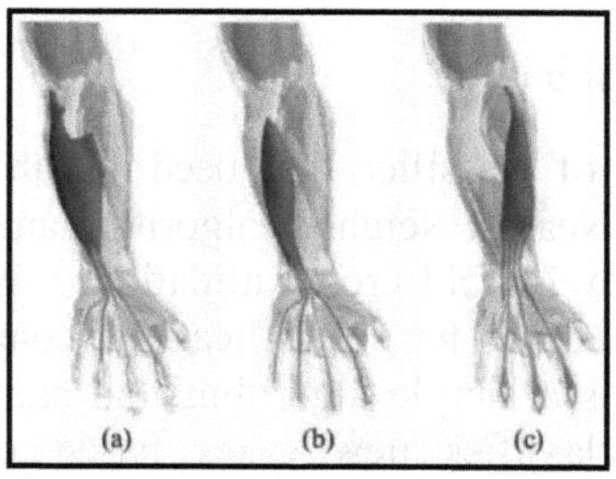

Fig. 3 Three types of muscle for grip motion. (a) FDS, (b) FDP and (c) ED [9]

E. EMG signal pre-processing

After the sample data is recorded using EMG sensors, it has to be processed before it can be used for the next steps, which are features extraction and classification. The signals require some filtering process to remove the noise that occurred during the data acquisition. The signal is converted from time domain to frequency domain using Fast Fourier Transform (FFT) before applying the filter.

In this project, the type of filter that is chosen is band-stop filter and band-pass filter. The cutoff frequency of band filter is in range between 20-500 Hz. A full wave rectifier is used to convert the raw EMG signal to a single, negative polarity into positive polarity of digital signal processing and smoothing process for last steps of signal pre-processing.

F. Feature extraction

Four features has been used in feature extraction to extract the information from the EMG signal, which are Mean Absolute Value (MAV), Root Mean Square (RMS), Standard Deviation (STD) and Variances (VAR). These four features are mostly used in the features extraction by many researchers [1, 2, 4, 8, 10], especially RMS. RMS is a method that has capabilities of calculating the value of finite sampled of signal sampled at uniform sampling rate. It is used for calculating the optimal value of the maximum potential of outputs. All this features are extracted from the filtered signal and after that it can classify the signal during classification methods.

Root Mean Square (RMS) is the value of the continuous voltage that results the same power disposal as the time-averaged power disposal of the voltage. The definition is similar for the RMS value of the current. Several studied used RMS as their signal processing technique [1, 12]. Variance is defined as calculated by taking the differences between each number in the set and the mean, squaring the differences and dividing the sum of squares by the number of values in the set.

G. Classification

The type of classifier that used for this classification method is k-Nearest Neighbor algorithm and also as known as k-NN with 10 fold cross validation. k-NN is a simpler classifier method for classification compare to other methods. It is a simple algorithm that stores all available cases and classifies new cases based on a similarity measure. According to Al Omari et al. [4], their research used this classifier and the accuracy was around 77% to 94%. k-NN is measured by a distance function, in this case, the correlation is used as the distance function. According Haris et al. [5], they used this distance function with a value of 4 and their result of accuracy is about 82.5%.

III. Results and Discussion

A. Results of signal pre-processing

The EMG signals from the 10 subjects were recorded by the Shimmer device where three sensors consist of two electrodes for positive and negative terminal and one electrode for ground terminal were used. The data was obtained by gathering EMG signals around 45 to 50 seconds for each trial and the sampling rate of the signal is 1 kHz. Therefore, each EMG channel has around 45000 to 50000 sample data per 45 to 50 seconds. The raw EMG signal sample for each force from the muscles is shown in Fig. 4, Fig. 5 and Fig. 6.

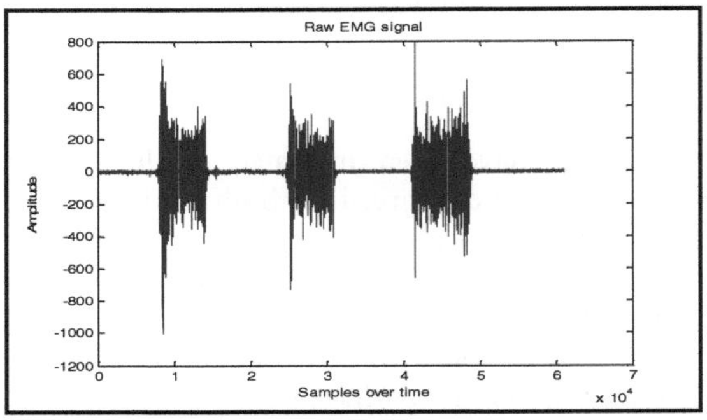

Fig. 4 Raw EMG signal sample for 50N grip force

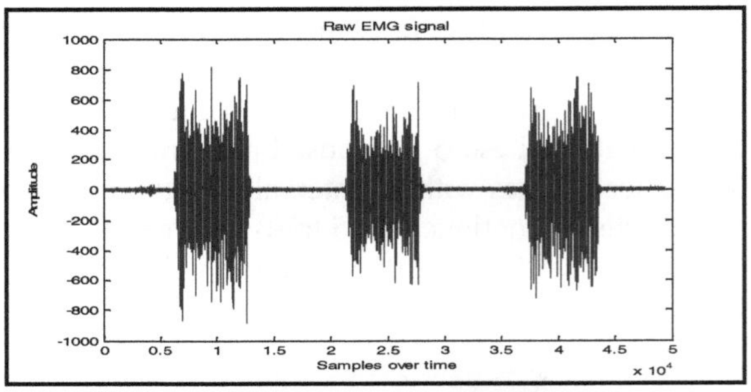

Fig. 5 Raw EMG signal sample for 100N grip force

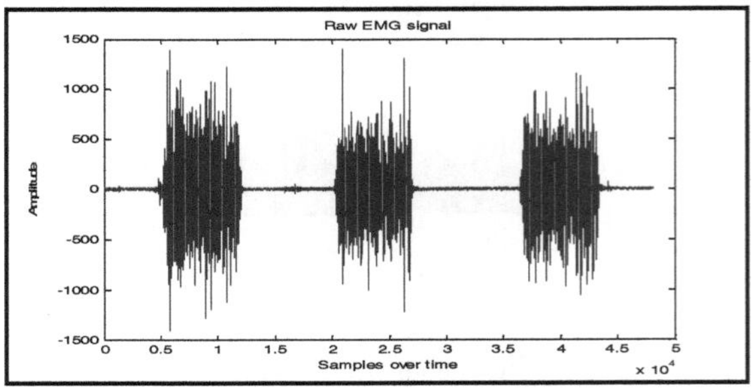

Fig. 6 Raw EMG signal sample for 150N grip force

From the raw EMG, it was converted to frequency domain using FFT. After that, the filter was applied to the signal to remove the noise. The signal was filtered twice, firstly by Butterworth band-stop filter then followed by band-pass filter. After that, the full-wave rectified was used to convert signal from negative portions into all positive ones and lastly, the smoothing process has been done and the signal data will be used for feature extraction. Smoothing was used to reduce the noise and curve fitting of the signal pattern to make smoother shape of signal. It used fifth-order Butterworth low pass filter. Fig. 7 shows the EMG signal pattern after the filtering process was applied and Fig. 8 shows the signal after applying the full-wave rectifier along with smoothing process.

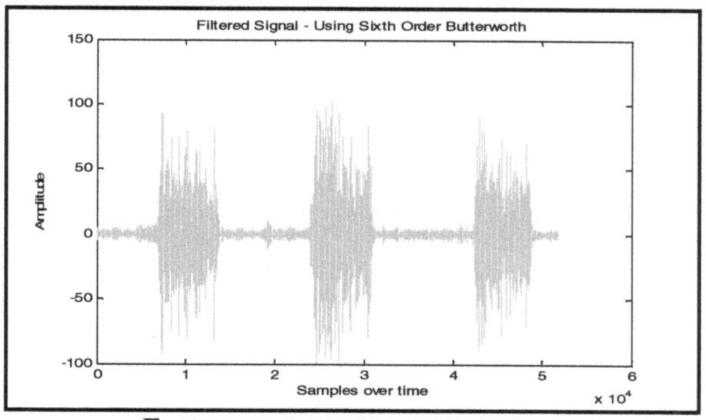

Fig. 7 EMG signal after all noise filtered

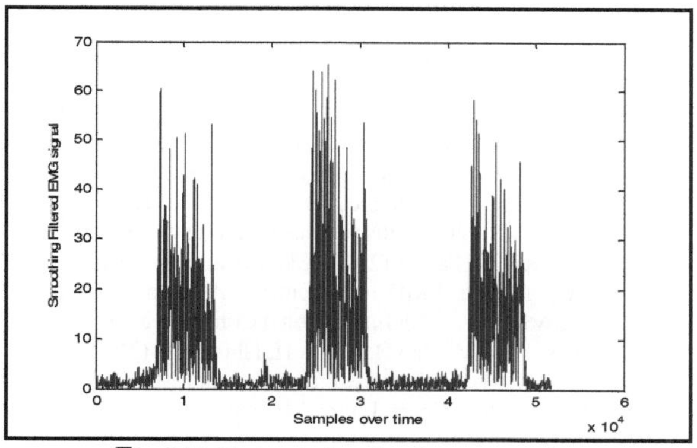

Fig. 8 EMG signal after signal pre-processing

B. *Results of features extraction*

A total number of four features have been extracted from the signal, which are MAV, RMS, STD and VAR. Feature values from 10 subjects for each muscle and each applied force have been recorded. The value of feature is extracted from whole of the signal data. Table 1 shows the sample value of each feature for gripping force of 50N at FDS muscle.

Table 1 Features value for 50N force on FDS muscle

Subject	(MAV)	(RMS)	(STD)	(VAR)
1	9.65	16.39	13.25	175.7807
2	12.175	21.29074	17.46626	305.07022
3	9.50163	15.76631	12.58168	158.29867
4	10.4086	19.30736	16.26162	264.44036
5	10.031	18.31997	15.32988	235.00534
6	15.9131	29.14525	24.41786	596.2321
7	6.76809	9.903862	7.230529	52.280549
8	9.57453	15.53726	12.23674	149.73785
9	8.4507	15.24181	12.68469	160.90138
10	10.2119	17.10212	13.71873	188.20368

The values of feature for each force at each muscle have been extracted and recorded. All features value has been used for classification in order to determine the accuracy. Value of each features depend on the signal that has went through pre-processing process.

C. *Results of* classification

The type of classifier that is used for this classification is k-NN with 10 fold for cross validation and the distance function is used as the correlation. 20 samples were used for training process and 5 samples were used for testing process to identify the accuracy. Classifications were divided into three sections, which classify for each force that applied during gripping process with 50N, 100N and 150N. In this cases, the value of the four features were used together to classify each force. So, the accuracy of classification was calculated from the four features which are RMS, MAV, SD and VAR. Table 2 shows the accuracy for each force that were classified using the four features and also the average of accuracy for six trials.

Table 2 Percentage of accuracy for each force

No. of Trial	Force 50N (%)	Force 100N (%)	Force 150N (%)
1	86.67	66.67	60.00
2	73.33	60.00	73.33
3	80.00	60.00	66.67
4	86.67	53.33	66.67
5	86.67	66.67	66.67
6	80.00	60.00	73.33
Average	82.22%	61.11%	67.78%

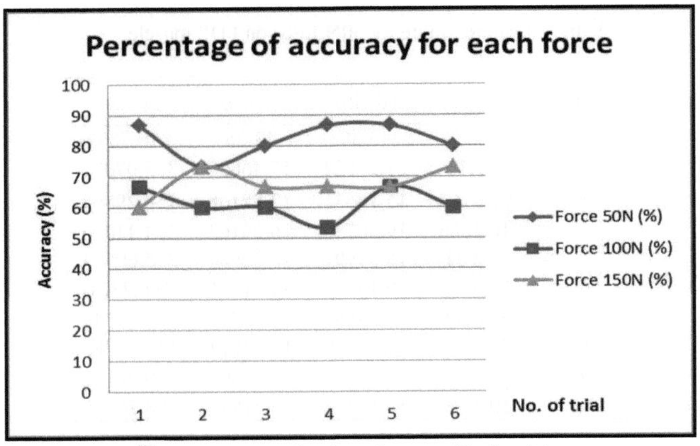

Fig. 9 Accuracy for three forces with six trials

From Table 2 and Fig. 9, the gripping force is recorded the highest of percentage accuracy with 86.67% is for 50N force. Second force is 100N with the accuracy value is about 66.67% and percentage accuracy is 73.33% for gripping force of 150N. So, from the result, it can be determined that the accuracy range for 50N is between 70% to 87% accuracy. The accuracy range for 100N is between 50% to 67% accuracy and for 150N force the range is between 60% to 74% accuracy. This is due to the reason that some of the subjects' forearm vibrates uncontrollably when they are exerting the grip force of 100N and 150N during data collection.

IV. Conclusion

In this research, analysis of the signal processing is applied to the raw EMG signal data that to improve the signal with can classify them into three different forces. When the hand was performing gripping action in certain strength, the forearm muscles will produce the movement of muscles which can be detected by using EMG sensors.

The data has been analyzed with the signal processing method and after that, classified it into three different forces. For future research, EMG signals can be applied in any applications field such as medical field, where it can use for clinical diagnostic tool to identify neuromuscular diseases. Other recommended for future work is analyzing the EMG signals recorded from handicap persons. Comparison between handicap person and normal person can be analyzed. Various classifiers can be tested for the classification of signal, so that the difference in accuracy can be determined to identify which classifier has higher accuracy.

Acknowledgment

This research is supported by the Fundamental Research Grant Scheme (FRGS) from the Ministry of Higher Education (MoHE) Malaysia.

Conflict of Interest

The authors declare that they have no conflict of interest.

References

1. Shuman G (2009) Using forearm electromyograms to classify hand gestures. In Bioinformatics and Biomedicine, 2009. BIBM'09. IEEE International Conference on, pp 261–264.
2. Vigneswari SV, Savithri D, Mahendran VS (2015) Gripping Force Measurement and EMG Classification For Hand Functions. International Journal of Applied Engineering Research 10(7):17315–17326.
3. Al Omari F, Gunhai L (2014) Analysis of extracted forearm sEMG signal using LDA, QDA, K-NN classification algorithms. Open Automation and Control Systems Journal 6:108–116.
4. Haris M, Chakraborty P, Rao BV (2015) EMG signal based finger movement recognition for prosthetic hand control. In 2015 Communication, Control and Intelligent Systems (CCIS), pp 194–198.
5. Jiang MW, Wang RC, Wang JZ, Jin DW (2006) A method of recognizing finger motion using wavelet transform of surface EMG signal. In Engineering in Medicine and Biology Society, 2005. IEEE-EMBS 2005. 27th Annual International Conference of the, pp 2672–2674.
6. Hickman S, Mirzakhani AS, Pabon J, Alba-Flores R (2015) A case study on tuning artificial neural networks to recognize signal patterns of hand motions. In Southeast Con 2015, pp 1–4.
7. Raurale SA (2014) Acquisition and processing real-time EMG signals for prosthesis active hand movements. In Green Computing Communication and Electrical Engineering (ICGCCEE), 2014 International Conference on, pp 1–6.
8. Lee J, Kim J, Park J (2013) Classification of grip configuration using surface EMG. In Control, Automation and Systems (ICCAS), 2013 13th International Conference on, pp 173–176.
9. Guo S, Pang M, Gao B, Hirata H, Ishihara H (2015) Comparison of sEMG-Based Feature Extraction and Motion Classification Methods for Upper-Limb Movement. Sensors 15(4):9022–9038.
10. Mohideen AJH, Sidek SN (2011) Development of emg circuit to study the relationship between flexor digitorum superficialis muscle activity and hand grip strength. In Mechatronics (ICOM), 2011 4th International Conference On, pp 1–7.
11. Tarata MT (2003) Mechanomyography versus electromyography, in monitoring the muscular fatigue. Biomed Eng Online 2(3):1–10.

Author: Chee Kiang Lam
Institute: Universiti Malaysia Perlis
Street: School of Mechatronic Engineering, Pauh Putra Campus
City: Arau
Country: Malaysia

Body Weight Distribution and Centre of Pressure Displacement Velocity on The Shooting Score of Malaysian Rifle and Pistol Athletes- A Pilot Study

Y. Selva and S. Joseph

National Sports Institute of Malaysia, Kuala Lumpur, Malaysia

Abstract— **Posture stability and consistency is one of the important factors in shooting. The study aims to investigate the correlation between the stability parameters and score for air rifle and air pistol shooters. It also explores and compare the dominant inter variable correlation among pistol and rifle shooters. 11 Malaysian shooting athletes (6 air rifle and 5 air pistol) participated in the study. The data were collected (Male athletes fired 60 competition shots in 75 minutes while female athletes fired 40 competition shots within 50 minutes) during training. Scoring value was obtained from electronic scoring system and Scatt system. Simultaneously, Tekscan pressure insoles were used to collect data from all shooters. Dependent variables such as body weight distribution percentage, anterior-posterior displacement velocity and medial-lateral displacement velocity were extracted for analysis from all shots. Bivariate correlation analyses were applied to obtain correlation strength from analyzed parameters. Overall, there were no strong correlations between investigated variables with score for both air pistol and rifle athletes. The highest correlation value for air pistol shooters was found between anterior-posterior displacement velocity of right feet and medial- lateral displacement velocity of left feet (r = 0.948). Meanwhile, air rifle shooters exhibit the highest correlation value between the medial-lateral displacement velocity and anterior-posterior displacement velocity of left feet (r = 0.991). Based on intra-individual analysis, moderate correlation between score and percentage of weight placed on both feet existed for one female air pistol shooter.The scores obtained by shooter are not highly influenced by the body weight distribution percentage and centre of pressure displacement velocity. The air pistol shooters display a higher number of strongly correlated variables than air rifle shooters. Inter-relationship between the variables existed for pistol shooters where displacement velocity in right feet influenced the displacement velocity in left feet and vice versa. Based on the intra-individual analysis that was conducted, 2 pistol athletes and 1 rifle athlete was found to have correlation between investigated variables and score.**

Keywords— **Shooting, Body weight, Pressure, Rifle, Pistol**

I. Introduction

A globally static posture with forearm and hand movements with small amplitude and rate is a common term used to characterize shooting.[1] Precise body control, strong mental focus and accurate decision making are also other requirement of the sport. Ability of athlete to reproduce this elements and trigger to hit the centre of a tiny target which is about 1mm diameter circle in the air rifle event determines their scores.[2] For pistol shooters, a steady posture in the upper limb is important for shot accuracy as most coaches believe that elite shooters must develop an optimal "fixation strategy" and eliminate unwanted movements of the body at the same time.[3] Ability to maintain equilibrium while remaining in a static position is known as static balance. Meanwhile, ability to maintain equilibrium while moving from one position to another is known as dynamic equilibrium.[4]

Pistol shooting for example is a static sport that requires a strict control of body in order to be able to align the rear sight aperture to foresight using proprioceptive feedback and gaze fixation either on the target directly or on the target and the weapon to produce precise shots[5]. For pistol athletes, changes in the position of centre of pressure on the X axis are mainly related to body movements while changes in the Y axis are related to the shoulder and wrist movement control. They were also found to influence the vertical and lateral movements of the gun respectively.[6]

The point of application of the resultant where the external forces are applied is known as COPor centre of pressure [7]. COP is the common variable used by researchers to analyze body movement and stability of stance. The correlation analysis between body sway and performance could provide the index of relatedness between body sway and performance. The strength of relationships can also be assessed using this analysis [8].

Among novice rifle shooters, postural balance and rifle stability were accounted for the 26% of the variance in the shooting accuracy [9]. Low rifle and body sway are also identified as the main performance determining factors and is the determinant factor between an elite and a novice athlete [10].

II. Methodology

11 National level shooters comprising of 5 air pistol (Male: 2, Female: 3) and 6 air rifle athletes (Male: 4, Female: 2) were involved in this study. All athletes are right handed. Participating athletes were representatives of Malaysia in the 2015 SEA games held in Singapore. For air rifle shooters, they were permitted to wear special shoes,

F. Ibrahim et al. (eds.), *3rd International Conference on Movement, Health and Exercise*,
IFMBE Proceedings 58,
DOI: 10.1007/978-981-10-3737-5_19

jacket and glasses. Meanwhile, air pistol shooters wore special shoes and glasses. Air pistol athletes do not wear any special jacket.

Figure 1: Posture adapted by athlete during 10m air pistol (left) and air rifle (right) respectively.

Body weights of athletes are first measured using a weighing scale. Data collection was conducted using Tekscan pressure measurement insoles. Venue of data collection is the shooting range in Majlis Sukan Negara Bukit Jalil and Pahang respectively. The insoles were cut according to the size of athlete's feet and inserted into their shoes. The insoles are then calibrated according to athlete's body weight. Calibration procedures are in accordance with the Tekscan manual for calibration. Athletes were then permitted to train and fire sighting shots or also known as trial shots. The number of shots is in accordance to athlete's preferences within a time limit of 15 minutes. During the actual shots, male athletes are required to fire 60 shots within a time limit of 75 minutes while female athletes are required to fire 40 shots within a time limit of 50 minutes. The timing provided to the athletes to complete shooting all the shots is in accordance to the standard set by the International Shooting Sport Federation. Data is recorded once the shooter is aiming to shoot at target and stopped once the shot is fired by the athlete. Values of shooting score are obtained from the Meyton electronic scoring system and Scatt system respectively.

Data obtained from the insoles were analyzed using Microsoft excel data. Body weight distribution, standard deviation of anterior-posterior displacement velocity and medial-lateral displacement velocity for each shot were obtained from the raw data. Body weight distribution is the calculation of the percentage from the overall body weight that is placed on each foot by the athlete. The displacement velocity data are all calculated in the unit cm/s.

The average value of each variable for each athlete is then compiled and analyzed using the SPSS16.The data is also analyzed intra-individually.

III. Results

Nomenclatures used for all variables in the research are as enlisted below:

Parameter	**Definition**
Left_Feet	Percentage of weight placed on left feet
Right_Feet	Percentage of weight placed on right feet
AP_L	Anterior-posterior displacement velocity of left feet
AP_R	Anterior-posterior displacement velocity of right feet
ML_L	Medial-lateral displacement velocity of left feet
ML_R	Medial-lateral displacement velocity of right feet

Table 1: Correlation for 10 m air pistol

Variables	Left_Feet	Right_Feet	AP_L	AP_R	ML_L	ML_R
Left_Feet	1.000					
Right_Feet	-1.000	1.000				
AP_L	-0.885*	0.885*	1.000			
AP_R	-0.761	0.761	0.904*	1.000		
ML_L	-0.614	0.614	0.784	0.948*	1.000	
ML_R	-0.539	0.539	0.723	0.939*	0.903*	1.000
Score	0.209	-0.209	-0.221	-0.395	-0.196	-0.589

* Correlation is significant at the 0.05 level (2 tailed)

Table 2: Correlation for 10m air rifle

Variables	Left_Feet	Right_Feet	AP_L	AP_R	ML_L	ML_R
Left_Feet	1.000					
Right_Feet	-1.000	1.000				
AP_L	-0.655	0.655	1.000			
AP_R	-0.047	0.047	0.752	1.000		
ML_L	-0.661	0.661	0.991*	0.738	1.000	
ML_R	0.143	-0.143	0.520	0.926*	0.500	1.000
Score	-0.311	0.311	0.618	0.414	0.689	0.090

* Correlation is significant at the 0.05 level (2 tailed)

From the Table 1 where the correlations for air pistol shooters is analyzed, it was found that there is strong correlation between percentage of weight placed on left and right feet with anterior-posterior displacement velocity. For the left feet correlation value is r = -0.885 while for right feet the correlation value is r = 0.885. Strong correlations (r = 0.904) existed between anterior-posterior displacement velocity of left feet and right feet; between anterior-posterior displacement velocity of right feet and medial-lateral displacement velocity of left feet (r = 0.948); between anterior-posterior and medial-lateral displacement velocity of right feet r = 0.939); and between the medial-lateral displacement velocity of left feet and right feet (r = 0.903).

Based on findings in Table 2 where correlation for air rifle shooters is analyzed, strong correlation (r = 0.991) was found between the anterior-posterior and medial-lateral displacement velocity of left feet as well as right feet (r = 0.926). Only moderate correlation was found between score and AP_L (r = 0.618) as well as between score and ML_L (r = 0.689) for the air rifle shooters.

Correlation analysis was also conducted intra-individually between athletes. Based on the results, 1 female pistol shooters have significance correlation between the investigated parameter and score. The female athlete was reported to have moderate correlation between the percentages of weight placed on both feet (r = 0.509, left feet), r = -0.509, right feet) and score. The female athlete also has a weak correlation (r = 0.344) between medial-lateral displacement velocity and score. For the rifle shooters, based on individual analysis, only one female shooter has correlation between anterior-posterior displacement velocity and score (r = -0.349).

Based on the inter-individual analysis, no correlation was found between the variables and score for the air pistol event (Table 1) but there were correlations found in the intra-individual analysis. The finding from the data is similar to findings by Ball et al (2003) where the correlation analysis of inter-individual elite pistol shooters reveals that the relationship between body sway and performance is not correlated. However, significance was found by the author when analysis was conducted intra-individually [8]. Furthermore, in a study investigating pistol and running target shooting, the elite athletes were able to manage body sway and gun movement independently[10]. From the findings of the study, not all athletes have shown correlations between investigated variables and score. Ability of athletes to control body sway and its influence on score is dependent on individual ability and level of expertise.

In comparison, pistol group showed higher number of correlated variables than air rifle group. For the air rifle athletes, their posture enables them to move as a block compared to pistol shooters who have an open stance that facilitates more body sway.

For air rifle displacement velocity no strong correlation was found between variables of unilateral limb. No strong inter limb relationship existed. This is unlike pistol shooting. It can be attributed to the fact that in both principles athletes adapt a different standing stance.

IV. Conclusions

Percentage of body weight distribution and centre of pressure displacement velocity is not found to influence the score of air pistol athletes based on the correlation analysis. Air pistol athletes exhibited significant correlations in more number of variables compared to air rifle shooters. Moderate correlation was found between the score and variables for air rifle athletes and none for pistol athletes.

Inter-individual relationship between the variables was dominant for pistol shooters where displacement velocity in right feet influenced the displacement velocity in left feet and vice versa. Based on intra-individual analysis, 1 pistol athlete and one rifle athlete has shown correlation between the investigated variable and score.

Conflict of Interest

The authors declare that they have no conflict of interest.

Statement of Informed Consent

The coaches and athletes were informed prior to data collection regarding the process and procedure of the data collection. Data collection was conducted only with the consent of both parties.

Statement of Human and Animal Rights

Neither humans nor animals were harmed in the data collection.

Acknowledgment

Special acknowledgment to the Malaysian National Shooting team coaches as well as athlete for their faith in me and their cooperation throughout data collection. Other than that, I would like to thank Mr Edin Kardin Sugarwanda for his assistance in data analysis and interpretation. I would also like to thank Mr Azwari Ahmad for his contributions in making the data collection a success.

References

1. [M. Bertollo, C. Robazza, W. N. Falasca, M. Stocchi, C. Babiloni, C. Del Percio, N. Marzano, M. Iacoboni, F. Infarinato, F. Vecchio, C. Limatola, and S. Comani, "Temporal pattern of pre-shooting psycho-physiological states in elite athletes: A probabilistic approach," Psychol. Sport Exerc., vol. 13, no. 2, pp. 91–98, 2012.
2. D. R. Mullineaux, S. M. Underwood, R. Shapiro, and J. W. Hall, "Real-time biomechanical biofeedback effects on top-level rifle shooters," Appl. Ergon., vol. 43, no. 1, pp. 109–114, 2012.
3. W.-T. Tang, W.-Y. Zhang, C.-C. Huang, M.-S. Young, and I.-S. Hwang, "Postural tremor and control of the upper limb in air pistol shooters.," J. Sports Sci., vol. 26, no. December, pp. 1579–1587, 2008.
4. H. Negahban, N. Aryan, M. Mazaheri, A. A. Norasteh, and M. A. Sanjari, "Effect of expertise in shooting and Taekwondo on bipedal and unipedal postural control isolated or concurrent with a reaction-time task," Gait Posture, vol. 38, no. 2, pp. 226–230, 2013.
5. G. Herpin, G. C. Gauchard, A. Lion, P. Collet, D. Keller, and P. P. Perrin, "Sensorimotor specificities in balance control of expert fencers and pistol shooters," J. Electromyogr. Kinesiol., vol. 20, no. 1, pp. 162–169, 2010.
6. D. Mon, M. S. Zakynthinaki, C. A. Cordente, A. J. M. Antón, B. R. Rodríguez, and D. L. Jiménez, "Finger Flexor Force Influences Performance in Senior Male Air Pistol Olympic Shooting," PLoS One, vol. 10, no. 6, p. e0129862, 2015.
7. K. Gianikellis, J. J. Pantrigo, and A. Vara, "Stabillometry applied on the analysis of individual technique in the air-rifle shooting ," Kostas Gianikellis, Juan J. Pantrigo, and Antonino Vara Faculty of Sport Sciences, University of Extremadura, Spain," Biomechanics, pp. 1999–2002, 2001.
8. K. A. Ball, R. J. Best, and T. V. Wrigley, "Inter- and intra-individual analysis in elite sport: Pistol shooting," J. Appl. Biomech., vol. 19, no. 1, pp. 28–38, 2003.
9. K. Mononen, N. Konttinen, J. Viitasalo, and P. Era, "Relationships between postural balance, rifle stability and shooting accuracy among novice rifle shooters," Scand. J. Med. Sci. Sport., vol. 17, no. 1990, pp. 180–185, 2007.
10. G. Sattlecker, M. Buchecker, E. Müller, and S. J. Lindinger, "Postural Balance and Rifle Stability During Standing Shooting on an Indoor Gun Range Without Physical Stress in Different Groups of Biathletes," Int. J. Sport. Sci. Coach., vol. 9, no. 1, pp. 171–184, 2014.

Author: Yallini Selva
Institute: National Sports Institute of Malaysia
Street: Jalan Barat
City: Bukit Jalil, Kuala Lumpur
Country: Malaysia
Email: yalliniselva89@gmail.com

Effect of *Eurycoma Longifolia* Jack Extract on Lipolysis in Collegiate Athletes: Pilot study

P.J. Lim[1], C.S. Gan[2], T.K. Khong[1], S.H. Hamzah[1] and A. Yusof[1].

[1] Department of Exercise Science, Sports Centre, University of Malaya, Malaysia.
[2] Department of Molecular Medicine, Faculty of Medicine, University of Malaya, Malaysia.

Abstract— **Studies have shown that usage of EL could promote lipid lowering and improvement in muscle strength. In view of that, lipid mobilization and utilizations would be the main mechanisms involved. The purpose of the investigation was to determine whether acute consumption of *Eurycoma longifolia* Jack extract would influence lipolysis among athletes. Ten male athletes were randomly assigned either to *Eurycoma longifolia* Jack group (EL, n = 5) or placebo group (PG, n = 5). Athletes from both groups ingested 1.7 mg/kg of body weight of either EL or PG for three days prior to the treadmill exercise test (65% of VO_{2max}) for an hour. Plasma free fatty acid (FFA), glycerol and triglycerides (TG) levels were measured at pre-exercise and post-exercise on days 0 and 3. Plasma FFA in EL group was significantly lower than that in PG after 3 days ($p < 0.05$), and plasma glycerol level was significantly increased in EL compared to PG ($p < 0.05$). The results of this study suggest that EL supplementation could promote fat lipolysis, resulting towards better energy yield. Therefore, EL can be considered as an ergogenic aid to improve performances and to boost energy production.**

Keywords— **Athletes, *Eurycoma longifolia* Jack, Exercise, Fat, Lipolysis**

I. Introduction

Eurycoma longifolia Jack (EL) or commonly known as 'Tongkat Ali' has been well recognized by Malaysian to improve strength in sexual activities, ailments to many disease and enhance one's health [1, 2].

To date, numerous studies have shown that usage of EL could promote lipid lowering and improvement in muscle strength. Studies on EL supplementation have been reported to enhance exercise performances by stimulating the effect of lipid mobilization and utilizations [3-6]. Due to its potential as ergogenic aids, many athletes have attempted to use EL as a supplement to improve exercise performances.

Lipid mobilization and utilization have been closely linked with endurance performances as lipid could give a better yield of energy [7, 8]. The increase rate of lipolysis are shown when the release of glycerol are augmented, resulting in release of free fatty acids (FFA) from triglycerides (TG) in adipose tissue as source of fuel [9, 10]. Although numerous studies have been done on humans, the evidence is still lacking and scarce. In addition, studies done to investigate the effect of EL supplementation on aerobic performances is limited. As most positive results on lipid lowering are shown in strength training, and physiologically the underlying mechanisms involved are similar, this study aimed to investigate whether EL supplementation can affect lipid lowering capacity in endurance athletes. In other words, the purpose of this study was to investigate the effect of acute EL supplementation on lipolysis biomarkers in endurance athletes.

II. Methodology

A. Participants

Ten male athletes were recruited and assigned either to EL (n = 5) or placebo, PG (n = 5) groups using a randomized, double-blinded placebo-controlled study design. The participants were healthy and did not consumed any medication or supplements that may interfere with the effect of EL or lipolysis (such as caffeine). All participants were fully informed of the experimental trials and risk associated before given a written informed consent and participant information sheet for the study participation. All procedures and protocols were approved by University Malaya Research Ethics Committee (Reference Number: UM.TNC2/RC/H&E/UMREC-42) at the University of Malaya.

B. Supplementation

Each EL capsules used in the study contained 1.7 mg/kg of body weight aqueous EL extract powder (Physta™ Standardize Tongkat Ali Extract, Biotropics Malaysia) [11] whereas, the PG capsules contained an equal amount of corn starch. Both EL and PG capsules were made of gelatin and are identical in size and color. All athletes were instructed to consume one capsule in the morning everyday for 3 days.

C. Preliminary trials

The speed and VO_{2max} was determined during an incremental exercise test on a calibrated treadmill (h/p/cosmos-quasar, Germany) before the 3 days EL supplementation. VO_{2max} was plotted against speed and speed was determined at the point whereVO_{2max} reaches 65%.

F. Ibrahim et al. (eds.), *3rd International Conference on Movement, Health and Exercise*,
IFMBE Proceedings 58,
DOI: 10.1007/978-981-10-3737-5_20

Table 1: Demographic information of the participants

Characteristics	EL (n = 5)	PG (n = 5)	p-value
Age (year)	25.0 ± 3.0	25.0 ± 3.8	0.859
Body weight (kg)	63.8 ± 11.9	60.0 ± 5.7	0.534
Height (cm)	172.2 ± 6.0	171.2 ± 2.9	0.747
Body mass index (kg/m^2)	21.4 ± 2.4	20.5 ± 1.7	0.514
Body fat mass (kg)	9.8 ± 4.3	8.6 ± 4.2	0.659
Percentage body fat (%)	15.0 ± 5.3	13.7 ± 5.4	0.710

D. *Exercise trials*

Participants were required to arrive at the Physiology Lab two hours prior to the exercise. Body measurement using an electronic body composition analyzer (Inbody, Korea) and fasting blood was measured prior to the exercise trial. Prior an hour to the exercise trial, participants were given a standardized breakfast consist of a piece of bread (Gardenia) and a glass (250 ml) of water. During the experimental trial, the participants ran at 65% of their VO_{2max} for an hour and the exercise heart rate was recorded every 15 mins. Blood samples were collected from the antecubital area after the trials for plasma FFA, glycerol and TG analyses. Water was given to participants after the run to prevent dehydration.

E. *Blood Analyses*

Plasma blood levels for FFA, glycerol and TG were determined before and after 3 days of EL supplementation. Blood collected were centrifuged at 3000rpm for 15 minutes at 4 °C (Heraus™ Multifuge™ X1 Centrifuge Series, USA). Aliquots of samples were analyzed using spectrophotometer (Epoch BioTek, USA) at wavelength according to the assay kit manufacturer's protocol.

F. *Statistical Analyses*

Data was presented with mean ± standard deviation (SD). A two-way ANOVA was performed to detect the mean difference of plasma FFA, glycerol and TG during 3 days between two groups. Difference between supplementation and duration were determined by independent *t*-test. Statistical significant was set and accepted at $p < 0.05$. Aside from that, Cohen's d effect was also calculated [12].

III. Results

A. *Participant characteristics*

The characteristics of the participants were listed in Table 1. There were no significant difference between EL and PG groups on age, height, body mass index, body fat mass and percentage body fat. This shows that the compliance of both groups were similar.

B. *Body Composition Analyses*

Although the body weight, body fat mass and percentage body fat have a small decrement in the EL group, it was not statistically different to placebo group. The EL group, on average, have decrement in body weight (d = 0.4, medium), body fat mass (d = 0.0, small) and percentage body fat (d = 0.1, small) compared to PG group.

C. *Blood Analyses*

Blood plasma levels of FFA, glycerol and TG were carried out to determine whether there would be effect of EL on lipolysis. There was a significant general interaction between the effect of treatments and duration on FFA ($p < 0.05$) (Figure 1). There were also significant differences between supplement ($p < 0.01$) and between duration on EL group ($p < 0.05$). As for glycerol, there was significant changes observed in duration in EL group ($p < 0.05$) (Figure 2). However, no significant changes was observed in TG levels (Figure 3).

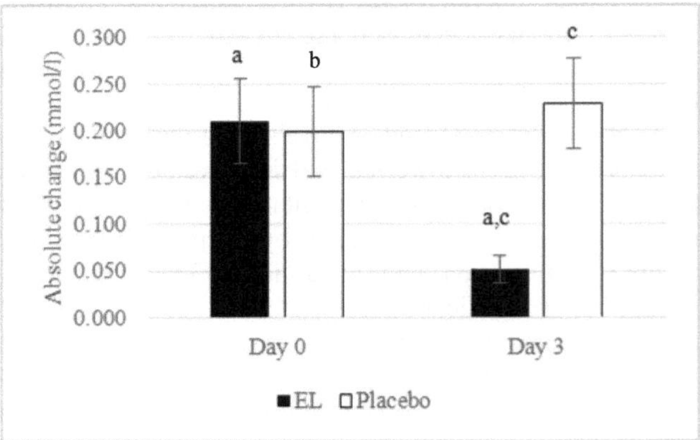

Figure 2: Free fatty acid concentrations during days 0 and 3 in EL supplementation and placebo. Values are mean, error bar represent SEM and significant difference ($p < 0.05$) between groups are represented by the same letter above the bar.

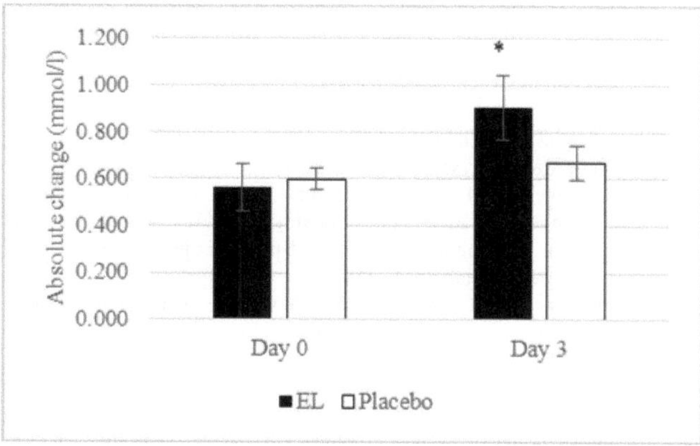

Figure 3: Glycerol concentrations during days 0 and 3 in EL supplementation and placebo. Values are mean, error bar represent SEM and significant difference ($p < 0.05$) to EL day 0 are indicated by an asterisk.

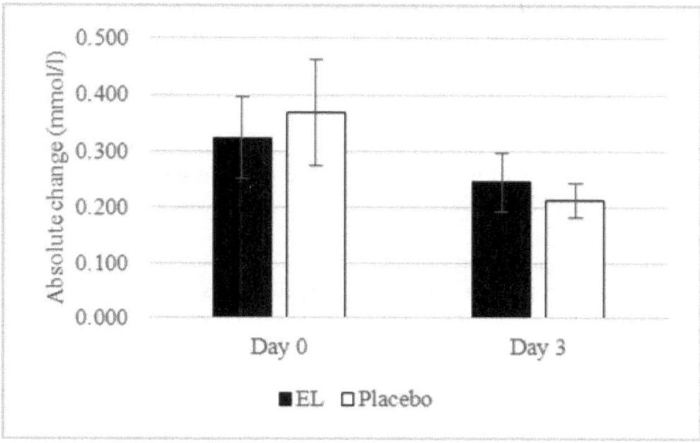

Figure 4: TG concentrations during days 0 and 3 in EL supplementation and placebo. Values are mean and error bar represent SEM.

IV. Discussion

The present study showed that three days of EL supplementation showed significant changes in FFA and glycerol. This demonstrates that consumption of EL supplementation prior to exercise is effective in regulating lipid constituents, FFA and glycerol, during endurance exercise.

Our findings are similar to that of Hamzah and Yusof [3] who reported that consuming 100 g/day of EL supplement increases free fat mass in healthy adults compared to placebo. The improvement in this study can be best explained by the alteration in lipolysis as reduction of FFA and concomitant increase in glycerol as shown in Figures 1 and 2.

The mechanism underlying changes in lipolysis with aerobic exercise training have been explored by many researchers [7, 8, 13, 14]. Aerobic exercise has an impact on substrate metabolisms by increasing fatty acid oxidation to synthesize ATP [15]. This could explained the lowering FFA levels in this study as after exercise, most of the FFA have been converted to energy to sustain the athlete's training test.

The study also explored the glycerol levels during exercise in each group. An increase was observed in day 3 in EL group suggests that EL supplementation promotes fat utilizations. According to Romijn and team, glycerol in blood are indication of lipolysis and as glycerol increases, total FFA from lipolysis made available for oxidation at the rate of two-threefold greater than the rate of oxidation [16]. Thus, EL supplementation clearly promotes lipolysis, resulting improvement in exercise performances in humans.

Logically, lipolysis accounts for a decrease in TG level as it is broken down to form FFA and glycerol. Although TG was not significantly different between EL and PG groups, a decrease was observed in Figure 3. This could be due to fat replacing the role of glucose as the primary energy source as exercise continues [17, 18]. During a prolonged exercise, lipolysis promotes the release of FFA and glycerol from TG (fat) into the blood circulation, therefore, leading to a decrease in TG concentration in blood plasma following exercise [16, 19].

In conclusion, EL supplementation at 1.7 mg/kg of body weight promotes lipid lowering following an endurance exercise. The amount and duration is safe as larger dosage and longer duration have been prescribed in other studies [3, 5, 6, 11, 18]. Thus, the study demonstrated support the ergogenic property of EL supplementation in terms of promoting lipolysis to provide better energy yield during endurance exercise in this subset of participants. Therefore, EL supplementation can be considered as an ergogenic aid among athletes to improve aerobic performances and to boost energy production.

Acknowledgement

Authors wished to acknowledge the Institute of Research Management & Monitoring of University of Malaya for the funding support (PPP- PG022-2014A) in this research and the help and support from coaches and athletes who participated in this study.

Conflict of Interest

The authors declared that they have no competing interests.

References

1. Ang H H, Cheang H S, and Yusof A P M (2000) Effects of Eurycoma longifolia Jack (Tongkat Ali) on the initiation of sexual performance of inexperienced castrated male rats. Experimental Animals 49:35-38
2. Indubala J and Ng L (2000) Herbs: The green pharmacy of Malaysia. Kuala Lumpur: Vinpress Sdn. Bhd, 2002, pp 76-77
3. Hamzah S and Yusof A (2003) The ergogenic effects of Eurycoma longifolia Jack: a pilot study. Br J Sports Med 37:464-470
4. Low B S, Choi S B, Abdul Wahad H et al. (2013) Eurycomanone, the major quassinoid in Eurycoma longifolia root extract increases spermatogenesis by inhibiting the activity of phosphodiesterase and aromatase in steroidogenesis. J Ethnopharmacol 149:201-207
5. Henkel R R, Wang R, Bassett S H et al. (2014) Tongkat Ali as a potential herbal supplement for physically active male and female seniors—a pilot study. Phytother Res 28(4):544-550.
6. Yusof S M, Zakaria Z, Karim A A H et al. (2016) Enhancement Effects of Tongkat Ali (Eurycoma longifolia) Supplementation on Performance Functions Following Strength Training in Middle-Aged Women. ICoSSEET Proc vol.2, International Colloquium on Sports Science, Exercise, Engineering and Technology, Singapore, 2016, pp 3-13
7. Huijsman E, van de Par C, Economou C et al. (2009) Adipose triacylglycerol lipase deletion alters whole body energy metabolism and impairs exercise performance in mice. Am J Physiol Endocrinol Metab 297(2):E505-E513.
8. Murase T, Haramizu S, Shimotoyodome A et al. (2006) Green tea extract improves running endurance in mice by stimulating lipid utilization during exercise. Am J Physiol Regul Integr Comp Physiol 290(6):R1550-R1556.
9. Carlson L and Oro L (1963) Studies on the relationship between the concentration of plasma free fatty acids and glycerol in vivo. Metabolism 12: p. 132-142
10. Vaughan M and Steinberg D (1963) Effect of hormones on lipolysis and esterification of free fatty acids during incubation of adipose tissue in vitro. J Lipid Res 4(2): p. 193-199
11. Li C H, Liao J W, Liao P L et al. (2013) Evaluation of acute 13-week subchronic toxicity and genotoxicity of the powdered root of Tongkat Ali (Eurycoma longifolia Jack). Evid Based Complement Alternat Med 2013:11
12. Cohen J (1988) Statistical power analysis for the behavioral sciences. 2nd Lawrence Erlbaum Associates. Hillsdale, NJ
13. Krssak M, Petersen K F, Bergeron R et al. (2000) Intramuscular Glycogen and Intramyocellular Lipid Utilization during Prolonged Exercise and Recovery in Man: A 13C and 1H Nuclear Magnetic Resonance Spectroscopy Study. J Clin Endocrinol Metab 85(2):748-754.
14. Durstine J L, Grandjean P W, Davis P G et al. (2001) Blood lipid and lipoprotein adaptations to exercise. Sports Med 31(15):1033-1062.
15. Kulaputana O, Thanakomsirichot S, Anomasiri W (2007) Ginseng supplementation does not change lactate threshold and physical performances in physically active Thai men. J Med Assoc Thai, 90(6): 1172-9.
16. Romijn J, Coyle E F, Sidossis L S et al. (1993) Regulation of endogenous fat and carbohydrate metabolism in relation to exercise intensity and duration. Am J Physiol 265(3):E380-E391.
17. Bursztyn P (1990) Physiology for sportspeople: a serious user's guide to the body. 1990: Manchester University Press.
18. Muhamad A S, Chen C K, Ooi F K et al. (2010) Effects of Eurycoma longifolia Jack supplementation on recreational athletes' endurance running capacity and physiological responses in the heat. Int J Appl Sports Sci 22(2):1-19.
19. Arner P, Kriegholm E, Engfeldt P et al. (1990) Adrenergic regulation of lipolysis in situ at rest and during exercise. J Clin Invest 85(3):893.

Author: Ashril Yusof
Institute: Department of Exercise Science, Sports Centre, University of Malaya.
Street: Jalan University
City: Kuala Lumpur
Country: Malaysia
Email: ashril@um.edu.my

Anthropometric Profiles of Malaysian Elite Swimmers

L.H. Loo[1], N.C. Wilson[2] and W.J. Chai[3]

[1] National Sports Institute of Malaysia/Penang Satellite Center, Penang, Malaysia
[2] National Sports Institute of Malaysia/Sport, Health and Activity Research Malaysia (SHARM), Bukit Jalil, Malaysia
[3] National Sports Institute of Malaysia/Research and Education Division, Bukit Jalil, Malaysia

Abstract— **This study was aim to compare the anthropometric profiles of Malaysian state and national swimmers, compare the basic physical characteristics with world class swimmers, compare the world class data of 1991 with more recent international swimmer data. Sixty-three state swimmers (25 females, 38 males) and 18 national swimmers (10 females, 8 males) were assessed. The national swimmers were compared with 401 world class swimmers from the 1991 Perth World Championships and other recent international data. Anthropometric profiles for Malaysian swimmers were collected according to the standardized ISAK protocol. Somatotype was derived by anthropometic variables according to Heath-Carter somatotype scale. State swimmers were younger, lighter, shorter and smaller on girth measurements compared to national swimmers. World class swimmers were older, heavier, taller ($p<0.01$) and possessed greater humerus breadth ($p<0.05$) compared to Malaysian national swimmers. Malaysian swimmers were consistently shorter and lighter compared to other international swimmers. Malaysian male swimmers had a similar ecto-mesomorphic somatotype (2-5-3) when compared to world class athletes; however, female national swimmers had higher endomorphy ratings ($p<0.05$) than world class swimmers. This study identified key anthropometric characteristics (greater weight, height, and limb girth) which may be beneficial to swimming performance and should be emphasized in talent identification programs.**

Keywords— **Swimmer, Anthropometry, Somatotype, Talent Identification**

I. Introduction

Body composition is important to sports performance, especially in swimming [1]. In addition, anthropometry is a recognized factor that distinguishes between swimmers competing at different levels [2], and studies shown direct relationship between anthropometric measurements and swimming potential [3, 4]. For example, height associated with swimming performance [5], long limbs are beneficial in propulsive forces development [6].

Somatotype is used to assess the relative body shape and composition, but it may change as a result of growth, aging, exercise and nutrition [7]. Typically there is little variation in somatotypes of elite swimmers competing in specific events [8]. Studies [3] reported somatotype of elite female swimmers shown higher endomorphy and lower mesomorphy compared to male swimmers [4, 8].

To date there is no comparative data on anthropometric characteristics of Malaysian and world class swimmers. Thus, the purposes of this study were to compare the anthropometry profile of state and national Malaysian swimmers and compare their basic physical characteristics with world class swimmers. This study also aims to determine the changes of world class swimmer physique since 1991. This study may identify key anthropometric characteristics which are associated with successful swimming performance in Malaysian swimmers, which may also be useful for talent identification in Malaysia.

II. Methods

A. Subjects

Data of 63 state swimmers (38 male, 25 female) was included in this study. Mean age for male and female state swimmers were 15.3±1.6 and 14.9±2.1 years old, with body mass 59.2±6.7 kg and body stature 168.2±6.2 cm compared to 49.1±7.9 kg and 157.6±7.0 cm respectively. Eighteen national swimmers (8 male, 10 female) were recruited in this study. Mean age for male and female national swimmers were 18.3±2.3 years old and 17.1±2.5 years old, with body mass 70.4±3.7 kg and body stature 177.9±4.1 cm compared to 57.5±6.2 kg and body stature 163.5±4.0 cm respectively. The world class data consisted of 401 swimmers (231 male, 170 female) were used in comparison with this study. Mean age 21.3±2.7 years old, 78.4±7.1 kg, and 183.8±7.1 cm for male compared to 19.6±2.9 years old, 63.1±5.9 kg and 171.5±7.0 cm respectively.

B. Measures

F. Ibrahim et al. (eds.), *3rd International Conference on Movement, Health and Exercise*,
IFMBE Proceedings 58,
DOI: 10.1007/978-981-10-3737-5_21

Table 1 Physical characteristic of different groups of swimmers (mean + standard deviation)

Variables	Male			Female		
	State (n=38)	National (n=8)	World Class (n=231)	State (n=25)	National (n=10)	World Class (n=170)
Age (years)	15.3±1.6**[a]	18.3±2.3	21.3±2.7**[b]	14.9±2.1*[a]	17.1±2.5	19.6±2.9**[b]
Weight (kg)	59.2±6.7***[a]	70.4±3.7	78.4±7.1**[b]	49.1±7.9**[a]	57.5±6.2	63.1±5.9**[b]
Height (cm)	168.2±6.2***[a]	177.9±4.1	183.8±7.1*[b]	157.6±7.0*[a]	163.5±4.0	171.5±7.0**[b]
Sum of 6 SF (mm)	49.9±10.7	50.3±12.3	45.8±9.5	77.8±19.2	79.0±23.3	72.6±19.6
Arm relaxed G (cm)	28.3±2.5**[a]	31.5±1.7	32.0±1.7	26.1±2.9*[a]	29.1±2.5	28.8±1.7
Arm flexed G (cm)	29.8±2.2***[a]	33.7±1.9	34.5±1.8	26.3±2.9**[a]	29.9±2.2	30.1±1.7
Waist G (cm)	70.7±4.3***[a]	77.5±2.0	79.4±3.5	65.2±4.3**[a]	70.0±3.1	69.4±3.7
Hips G (cm)	87.3±5.1**[a]	93.7±2.6	94.8±3.6	86.0±6.9**[a]	92.9±4.6	93.1±3.8
Calf relaxed G (cm)	35.1±1.9**[a]	37.2±1.0	37.6±1.8	31.9±2.4**[a]	33.9±1.6	34.9±1.7
Humerus B (cm)	6.9±0.4	7.0±0.2	7.4±0.4*[b]	5.7±0.4**[a]	6.1±0.3	6.4±0.4*[b]
Femur B (cm)	9.6±0.3*[a]	9.8±0.2	9.9±0.5	8.6±0.5	8.8±0.3	9.1±0.5
Somatotype	2.2-5.1-3.1	2.3-5.2-3.0	1.9-5.0-2.9	3.7-3.5*[a]-3.1	3.6 -4.3-2.5	2.9*[b]-3.9 -3.0

SF= Skinfold, G= Girth, B=Breadth

* Significant difference at $p < 0.05$ — [a] Significant difference between State and National groups

** Significant difference at $p < 0.01$ — [b] Significant difference between National and World class groups

*** Significant difference at $p < 0.001$

Anthropometric measurements for Malaysian swimmers were obtained by International Society of The Advancement of Anthropometry (ISAK) certified anthropometrists using standard protocols [9], included body mass, stature, seven skinfold sites (triceps, subscapular, biceps, supraspinale, abdominal, front thigh, medial calf), five girths (arm relaxed and flexed, waist, hips, calf) and two bone breadths (biepicondylar humerus, femur). A total of six skinfold sites (excluding biceps) was used to compare the physical characteristic between the different groups. Somatotype was derived according to Heath-Carter [7].

C. Procedures

State swimmers data was obtained from the 2006 and 2008 Malaysian Games Anthropometry Projects (MGAP 06 & MGAP 08) conducted during Malaysian Games (SUKMA XI Kedah 2006 & SUKMA XII Terengganu 2008) respectively. The MGAP projects were approved by the National Sports Council of Malaysia (NSC) and National Sports Institute of Malaysia (ISN). A written consent form was completed by all subjects. National swimmers was recruited among those competed in the SEA Games Laos 2009. Nine overlapped subjects being removed from state data but retain in national data. World class swimmers' data was retrieved from published data of the World Championship Anthropometry Project in Perth, 1991 [3], which represents the largest scale and most complete international data set available. As the world class data is relative old, a simple comparison (age, body height and weight) of the world class swimmers data [3] was made with the current international swimmers data [4, 10, 11] to determine any significant change of physique of international swimmers since 1991.

D. Statistical Analysis

Statistical analyses were done using SPSS version 17.0 to compare state and national data (Mann-Whitney test), national and world class data, world class data with recent international data [4, 10, 11] (t-test). Effect size (ES) was used to determine the magnitude of differences calculated by the Cohen d, modified by Hopkins [12].

III. Results

The male and female world class swimmers were older (ES: 1.12 & 0.87, moderate) and taller (ES: 0.84 & 1.17, moderate) than national swimmers (Table 1). Both male and female state swimmers were significantly younger (ES: 1.77 large, 1.02 moderate) than national swimmers. Thus, not surprisingly shorter (ES: 1.68 large, 0.96 moderate), lighter (ES: 1.81 large, 1.16 moderate), and had smaller girths (ES: 0.93 - 1.85). These differences were considered large for males except for calf girth (ES: 1.20, moderate) and femur breadth (ES: 0.71, moderate). The differences for females were moderate except for arm flexed girth (ES: 1.36, large) and humerus breadth (ES: 1.10, moderate). There were no significant difference in sum of skinfolds between the state and national swimmers.

Although national swimmers were significantly lighter than world class swimmers (ES: 1.14 (male), 0.95 (female) moderate), there were no significant differences in sum of skinfolds or girth measurements. However, world class swimmers had a greater humerus breadth compared to national swimmers (ES: 0.76 moderate).

Malaysian male swimmers were ecto-mesomorphy (2-5-3), no significant difference between state and national groups (Figure 1). However, state female swimmers shown significantly lower mesomorphy compared to national counterparts (p=0.028, ES: 0.87, moderate). State female swimmers were central somatotypes (3.7-3.5-3.1) while national female swimmers were endo-mesomorphic (3.6-4.3-2.5), showed less muscularity and relatively higher fatness compared to national female swimmers. Male swimmers for both national (2.3-5.2-3.0) and world class (1.9-5.0-2.9) had similar somatotypes. However, the physique of national female swimmers (3.6-4.3-2.5) differed from the world class swimmers (2.9-3.9-3.0), with higher relative fatness (ES: 0.78 moderate) and less linearity.

IV. Discussion

The world class swimmers were older and taller than national swimmers, which are in accordance with Carter et al [3], reported the best male swimmers were significant older and taller than the rest. Carter et al [3] also stated age could be important factor to distinguish better swimmers, due to physiological and structural maturity. As the world class data is relatively old, comparison of the data was made with current international swimmers data (Table 2). There were no significant differences in stature [4, 10, 11], whereas body weight was significantly heavier in AIS swimmers [13], but lighter in the South African swimmers [4], although this may have been because of their younger age.

Table 2: Comparison data between national, 1991 world class swimmers and recent data

Gender	Level of Swimmers	Sample Size (n)	Age (years)	Height (cm)	Weight (kg)	Reference
Male	AIS[1]	24	22±2**[a]	185±0.05**[a]	80±6**[a]	Anderson et al, 2008
	AIS[1]	46	21	-	82.1±7.9**[ab]	Pyne et al. 2006
	Top ten for South African Swimming Championship	16	18.7±3.3**[b]	182.6±5.8**[a]	74.7±10.1**[a]*[b]	Coetzee, 2007
	World class	231	21.3±2.7	183.8±7.1	78.4±7.1	Carter et al, 1994
	National	8	18.3±2.3	177.9±4.1	70.4±3.7	Present data
Female	AIS[1]	16	20±3**[a]	174±0.07**[a]	64±6**[a]	Anderson et al, 2008
	AIS[1]	31	20	-	64.9±9**[a]	Pyne et al, 2006
	NCAA[2] Division 1 (USA)	18	19.5	170.1±6.7**[a]	63.3±5.4**[a]	Peterson et al, 2006
	Top ten for South African Swimming Championship	10	17.0±2.4**[b]	170.5±4.5**[a]	61.6±5.4	Coetzee, 2007
	World class	170	19.6±2.9	171.5±7.0	63.1±5.9	Carter et al, 1994
	National	10	17.1±2.5	163.5±4.0	57.5±6.2	Present data

[1] Australian Institute of Sports

[2] National Collegiate athletic Association

* Significant difference with World class group ($p<0.05$)

** Significant difference with World class group ($p<0.01$)

[a] Significant difference between National data and other recent international data

[b] Significant difference between World class data (1991) and other recent international data

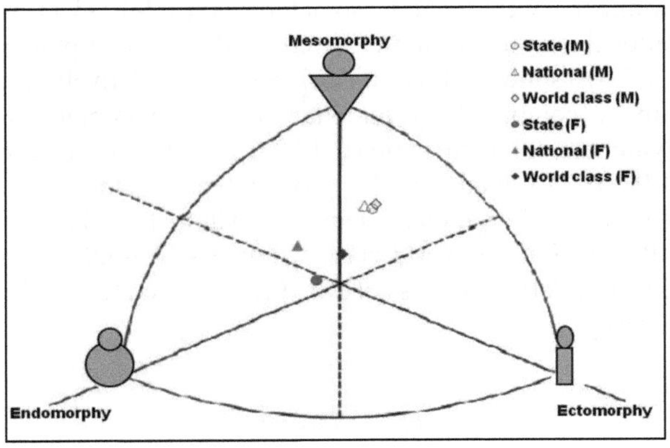

Figure 1 Somatoplots of swimmers by gender and groups

Overall the height of international swimmers has not changed significantly since 1991. Thus, irrespective of whether the national swimmers are compared with the 1991 [8] or with the more recent studies [4, 10, 11, 13] the national swimmers are consistently shorter and lighter. This puts Malaysian swimmers at disadvantage as studies [6, 14] have stated greater height and longer limbs are associated with better performance. Long limbs produce larger stroke lengths; require less power when swimming [15]. Body length is important for sprinter during starting, kick-off in turning and stretching to the finish [16]. Long pull path supported by long body segments will provide sufficient power to produce greater propulsive force especially for freestyle and backstroke event [8]. Higher body weight (lean mass) is beneficial for swimmers to improve the effective propulsion and performance [16, 17]. World class swimmers' humerus breadth was greater than the national swimmers, similar to study [3] stated top finishing male swimmers in freestyle and breaststroke sprint events had greater humerus breadth than the rest.

Maglischo [17] stated higher body fatness may increase body surface area, which positively correlate to higher body drag forces and swim time. All international and male Malaysian swimmers had the advantage of highest relative muscularity and lowest relative fatness in somatotyping. However, Malaysian female swimmers had the disadvantage of higher relative fatness. Burke [18] stated female swimmers experience a converse change (substantial deposition of body fat and smaller gain in muscle mass) in puberty, which can be detrimental to swimming performance.

Swimmer's physiques will change by age and training [8], but identify athletes at early age with the anthropometric characteristics most suited to swimming would be advantageous. With respect to world class swimmers, tall individuals with long levers and high fat-free mass would be in advantage [6, 14, 16]. Heavier and taller swimmers are potentially with larger arm span and surface areas; the longer stroke length can produces higher velocity, stroke index and critical velocity, therefore better performance [19].

Since Malaysian swimmer's anthropometry profile has been identified, training programs should emphasize in lean mass gain and decrease body fat mass [10, 12]. However, swimming performance not solely relies on body composition but the integration of multiple disciplines of sports science, coaching and technology.

V. Conclusions

Malaysian national swimmers were relatively smaller in body limb and size compared to world class swimmers. Malaysian male swimmers had similar somatotypes to male world class swimmers, while Malaysian female swimmers indicated higher fatness and less linearity compared to their world class counterparts. Therefore it could be recommended that female Malaysian swimmers focus on gaining lean body mass and limit fat gain.

In summary, anthropometric profiles for young Malaysian swimmers have been identified. Greater body weight, height, arm circumferences (relaxed and flexed) and lower values in all skinfolds are advantageous for performance. Optimum physique contributes to a several parts of successful participation, and interacts with other sport science disciplines as a foundation for swimming performance. Further investigation is recommended using ISAK full profiles, which may identify more anthropometric characteristics that could optimize swimming potential. Such approach may lead to the development of anthropometric profile for swimmer talent identification program, in addition to performance based selection.

Acknowledgment

The authors would like to thank NSC and ISN for funding MGAP in 2006 and 2008, appreciation to the athletes as subjects, support and cooperation by the secretariat of SUKMA XI Kedah and SUKMA XII Terengganu, state sports councils, MGAP working committees, coaches and managers. Thanks to Sports Nutrition Centre staffs in data collection and feedback given on this paper from Cheah Boon Chong, Albert Tan, Dr Barry Wilson and Dr Martyn Beaven of ISN.

Conflict of Interest

The authors declare that they have no conflict of interest.

References

1. Zampagni ML, Casino D, Benelli P, Visani A, Marcacci M, De Vito G (2008) Anthropometric and strength variables to predict freestyle performance times in elite master swimmers. J Strength Cond Res 22(4):1298-1307
2. Cicchella A, Jidong L, Jürimäe T, Zini M, Passariello C, Rizzo L, Stefanelli C (2009) Anthropometric comparison between young Estonian and Chinese swimmers. J. Hum. Sport Exerc 4(2):154-160
3. Carter JEL, Ackland TR, Mazza JC, Ross WD (1994) Kinanthropometry in Aquatic Sports: A Study of World Class Athletes. Champaign IL: Human Kinetics
4. Coetzee B (2007) Proportionality and sexual dimorphism in elite South African crawl stroke swimmers. In Marfell-Jones M, Olds T (Eds) Kinanthropometry X. London:Routledge pp 243-262
5. Geladas ND, Nassis GP, Pavlicevic S (2005) Somatic and physical traits affecting sprint swimming performance in young swimmers. Int J Sports Med 26:139-144
6. Grimstom SK, Hay JG (1986) Relationships among anthropometric and stroking characteristics of college swimmers. Med Sci Sports Exerc 18:60-68
7. Carter JEL, Heath BH (1990) Somatotyping – Development and Applications. New York: Cambridge University Press.
8. Carter JEL, Marfell-Jones M (1994) Somatotypes. In Carter JEL, Aukland T, Mazza J, Ross W (Eds) Kinanthropometry in Aquatic Sports. Champaign IL: Human Kinetics, pp 55-82
9. Marfell-Jones M, Olds T, Stewart A. & Carter L (2006) International standards for anthropometric assessment. ISAK: Potchefstroom, South Africa
10. Anderson ME, Hopkins WG, Robert A, & Pyne DB (2008) Ability of test to predict competitive performance in elite swimmers. J Sports Sciences, 26(2):123-130
11. Peterson HL, Peterson CT, Reddy MB, Hanson KB, Swain JH, Sharp RL, Alekel DL (2006) Body composition, dietary intake, and iron status of female collegiate swimmers and divers. Int J Sport Nutr Exerc Metabolism 16:281-295
12. Hopkins WG (2004) How to interpret changes in an athletic performance test. Sportsci.org 8:1-7
13. Pyne DB, Anderson ME, Hopkins WG (2006) Monitoring changes in lean mass of elite male and female swimmers. Int J. Sports Phys and Performance 1:14-26
14. Geladas ND, Nassis GP, Pavlicevic S (2005) Somatic and physical traits affecting sprint swimming performance in young swimmers. Int J Sports Med 26:139-144
15. Reilly T (1990) Swimming. In Reilly T, Secher N, Snell P, Williams C (Eds) Physiology of Sports. London: E & F.N Spon, pp 217-257
16. Norton K, Olds T, Olive S, Craig N (2002) Chapter 11: Anthropometry and sports performance. In Norton K, Olds T (Eds) Anthropometrica: A textbook of body measurement for sports and health courses. Sydney: University of New South Wales, pp 287-352
17. Maglischo EW (2003) Swimming Fastest. Champaign, III: Human Kinetics.
18. Burke L (2007) Swimming and Rowing. In Burke L. Practical Sports Nutrition. Champaign IL: Human Kinetics, pp 141-168
19. Vitor FM, Bohme MT (2010) Performance of young male swimmers in the 100 meters front crawl. Pediatr Exerc Sci 22: 278-287

Author: Loo Lean Hiong
Institute: National Sports Institute of Malaysia
Street: Penang States Sports Council, Off Jalan Sultan Azlan Shah
City: Batu Uban, Pulau Pinang
Country: Malaysia
Email: lhloo@isn.gov.my / lhloo84@gmail.com

Effects of Toe-out and Toe-in Gait with Varying Walking Speeds on Knee Adduction Moment and Mechanical Work Done-A Pilot Study

S.S. Khan[1], S. J. Khan [1] and J. Usman[1, 2]

[1] Department of Biomedical Engineering, University of Malaya, Kuala Lumpur, Malaysia
[2] Centre of Applied Biomechanics, University of Malaya, Malaysia

Abstract— **Knee joint is subjected to loads during activities of daily living. Higher loads can cause deterioration of the joint and malalignment. Toe-in and toe-out gait are among the techniques that modify the posture in order to minimize these loads. Several studies have reported their effects in reducing knee joint load. No effects of these techniques, however, have been reported with varying walking speeds on knee adduction moment (KAM) and mechanical work done at lower limb joints. The aim of this study was to investigate the effects of self-selected toeing-in and toeing-out with self-selected normal, slow and fast walking speeds on first and second peaks of KAM, individual lower limb joints mechanical work and total lower limb mechanical work done during level walking. A pilot study was conducted using cinematographic gait analysis of 5 healthy young adults (age: 28 years; weight: 58.3 kg, height: 1.6 m) walking at self-selected normal, slow and fast walking speeds for each of the three foot positions: straight (natural), toe-out and toe-in. Repeated measures ANOVA ($p < 0.05$) was applied with pairwise comparison to find the differences between groups. The results showed that there are significant effects of changing foot progression angle on knee joint loads and on positive and negative muscle work done. Also, the analyses showed that walking speed has a prominent influence on the relationship of foot progression with knee joint load and with mechanical work. Therefore, it is suggested that walking speed should also be considered while prescribing toe-out and toe-in gait. Further researches with a broader spectrum of walking speeds may identify the optimal speed for each foot position.**

Keywords— **Knee Adduction Moment, Work Done, Toe-out, Toe-in, Walking Speed.**

I. Introduction

The Activities of Daily Living (ADLs) put load on the weight bearing lower limb joints particularly on the knee joint. An excessive load can cause cartilage degeneration leading to increased risk of disorders like knee osteoarthritis [1]. In order to maintain healthy cartilage and medio-lateral alignment of the limbs, the load on knee joint is to be kept within normal ranges. A prime surrogate measure of knee joint load is knee adduction moment (KAM) [2]. An increased KAM represents elevated medial knee compartment loads, leading to one of the key contributors of Global Burden of Disease, the medial knee osteoarthritis [3].

In order to reduce KAM, there are majorly two types of techniques currently in existence: orthoses (such as knee braces and wedged insoles) and gait retraining methods. The latter includes varying walking speed, changing foot progression angle (toe-in and toe-out gait) and trunk lean.

Gait modifications have been found to have considerable effect on KAM and Ground Reaction Force (GRF) in both early and late stance phase. Effects are varied according to sample type (knee osteoarthritic or healthy) and degree of implementation among studies. Trunk lean was found to have considerable effects in reducing peak KAM values, but is aesthetically unpleasant. Participants opted out of trunk lean modification in a multi-parameter gait retraining program because of discomfort, difficulty to maintain and imbalance [4]. Increasing walking speed has been reported to increase KAM [5]. Although decreasing walking speed decreases the peak KAM values but it also increases the duration of load exposure to the knee joint [5, 6]. Toe-in and toe-out gait tend to reduce early and late stance KAM values respectively by shifting the body's Centre of Pressure (CoP). However, their undesirable effects in increasing late and early stance KAM values respectively have also been reported [5, 7].

So far studies have reported foot progression angle change with self-selected/ controlled constant walking speed and the effects of varying speed in combination with toe-in/ toe-out gait modification have not been investigated. It is a matter of investigation since it is difficult to maintain constant walking speed out of the laboratory setting. Furthermore, there is no reported investigation found on the effects of toe-in/ toe-out on mechanical work done.

Mechanical work done is a measure of the amount of power generated and absorbed at particular joint during gait cycle [8]. Ideally at constant speed and level walking energy is generated and absorbed by the muscular system but the net work done is zero [9]. However, there is no evidence that the net work done will remain zero if we produce postural change during gait (toe-out and toe-in gait).

The purpose of this study was to study the effects of toe-in and toe-out gait modifications with varying self-selected walking speeds on (1) KAM. (2) individual lower limb

F. Ibrahim et al. (eds.), *3rd International Conference on Movement, Health and Exercise*,
IFMBE Proceedings 58,
DOI: 10.1007/978-981-10-3737-5_22

joints mechanical work and total lower limb mechanical work done during level walking.

II. MATERIALS AND METHODS

A. Sample Information

Five healthy participants (2 males, 3 females) were recruited for this pilot study through community advertisement and the word of mouth. The mean and standard deviation (SD) values (Mean (SD)) of the sample taken were: age: (28 (4.3)) years, weight: (58.3 (11)) kg, height: (1.60 (0.2) m.

B. Inclusion Criteria

The inclusion criteria for the subjects included healthy young adults, aged 20 to 35 years, having a BMI of less than 25 kg/ m^2 (non-overweight and non-obese [10]) and having no physical disability. The participants were excluded on the basis of any neurological or musculoskeletal disorder or inability to adopt novel gait pattern.

C. Ethical Approval

Approval was obtained from University of Malaya Research Ethics Committee (UMREC). All participants provided written consent for the study.

D. Gait Modifications

The study deals with the effects of two gait modification techniques that deal with foot progression angle. The participants walked with straight foot, self-selected toe-out angle (making V shape with their feet) and toe-in angle (making A shape with their feet). The toe-out and toe-in angles were calculated by the degrees by which the foot vector (directed from the ankle joint centre to the second meta-tarsal marker) deviates from the progression axis of the walkway [11].

E. Data Collection

Data collection was performed via VICON Motion Capture System (100 Hz; Vicon, Oxford Metrics, Oxford, UK), consisting of five infrared-sensitive cameras. PlugIn gait model was used to attach infrared markers to the participant's skin or on the surface of the shoe that was directly above the bony prominence with the help of adhesive double-sided tape. Sixteen bony prominences defined by this model are anterior and posterior superior iliac spines, lateral thigh, lateral femoral epicondyle, lateral shank, calcaneus, lateral malleolus and second metatarsal head. Two embedded force plates (1000Hz; AMTI, USA) in the walkway were used for kinetics determination. The driver software for the system was VICON Nexus while VICON Polygon was used to generate reports.

The participants were asked to walk with provided shoes with three self-selected walking speeds i.e. normal (**N**), fast (**F**) and slow (**S**) for each of the three foot positions i.e. straight (*ST*), toe-out (*TO*) and toe-in (*TI*). Hence, each participant was required to walk using nine combinations.

F. Variables of Interest

The parameters of interest were first and second peak KAM (fKAM and sKAM) and individual joint and total limb mechanical work. Peak KAM values were normalized by dividing them by weight times height and taking them as percentage.

Positive mechanical work W_+ (concentric muscle work) at a joint was obtained by integrating positive power values. While negative work W_- (eccentric muscle work) was obtained by integrating the negative power values. Net work W_{net} is the sum of positive and negative work values. Total limb work W_T is the summation of individual net work values of hip, knee and ankle joints. All work values were calculated for each of the nine test conditions.

G. Statistics

Shapiro-Wilk test was applied to the data to assess normality. Differences in first and second peak KAM values (fKAM and sKAM) and work done for all the test conditions were evaluated using repeated measures ANOVA ($p < 0.05$) with pairwise comparison (LSD method) using IBM SPSS version 20 (SPSS Inc., USA).

III. RESULTS

A. Walking Speed

Walking speed for **N** (1.18 m/s) was 27.9% faster than **S** (0.85 m/s) and 17.4% slower than **F** (1.43 m/s).

Foot Progression Angle

The mean self-selected foot progression angles for *ST*, *TO* and *TI* were 10.8°, 29.4° and -3.1° respectively.

B. First Peak Knee Adduction Moment (fKAM)

Significant increase was observed at **S** for *TO* (13.6%) while no change for *TI* when compared with *ST* (1.7 N-m/ %BW*Ht) as shown in Figure 1. At **N** *TO* showed higher value of fKAM (21%), while *TI* showed lower values (7.3%) when compared with *ST* (1.9 N-m/ %BW*Ht). Similarly, at **F**, *TO* increased fKAM (11.2%), while *TI* decreased fKAM (21.6%) when compared with *ST* (2.6 N-m/ %BW*Ht).

C. Second Peak Knee Adduction Moment (sKAM)

TO showed reduced sKAM at all three speeds **S** (2.8%), **N** (20.4%) and **F** (16.7%) as compared to *ST* (S: 1.39, N: 1.37and F: 1.49N-m/ %BW*Ht) as shown in Figure 1. *TI* showed increased sKAM at all three speeds **S** (15.1%), **N** (21.8%) and **F** (10%) as compared to *ST* (**S**: 1.39 N-m/ %BW*Ht, **N**: 1.37 N-m/ %BW*Ht and **F**: 1.49N-m/ %BW*Ht).

D. Mechanical Work at Hip Joint

No significant changes were observed in W_+ for both *TO* (19.8J) and *TI* (17J) as compared to *ST* (18J) at **N** as depicted in Figure 2. *TO* increased W_+ at **S** (43.5%) and **F** (13.3%) when compared with ST (**S**: 10.1J and **F**: 19.5J respectively). While *TI* increased W_+ at S (43.5%) and decreased at **F** (18.8%) when compared with *ST*.

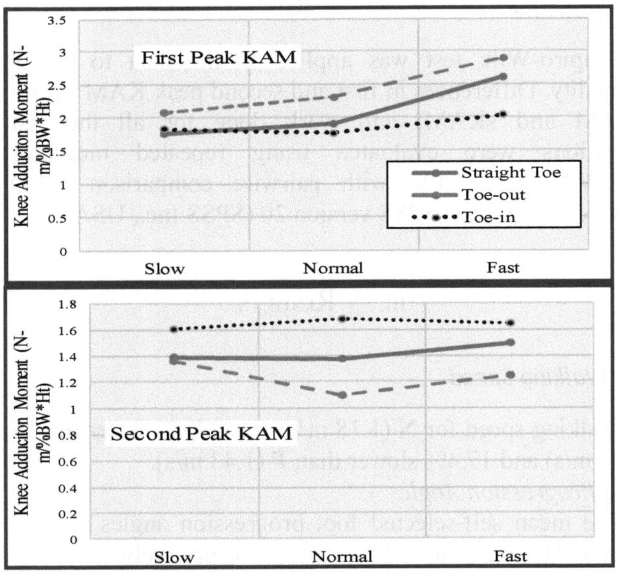

Figure 1: The mean knee adduction moment (KAM) values for the self-selected normal, slow and fast walking speeds with straight-toe, toe-out and toe-in gait conditions normalized to 100% of the stance phase.

Significant decrease in W_- was observed at **N** for both *TO* (39.8%) and *TI* (37.7%) as compared to *ST* (-5.3J). At **F**, *TI* decreased (54.5%) W_-, while *TO* increased (27.7%) it as compared to *ST* (-4.2J). Conversely at **S**, *TI* increased (92%) W_-, while *TO* decreased (36%) it as compared to *ST* (-1.5).

W_{net} for Hip joint showed the following trend:
At **S**: *TO* (13.5J) > *TI* (11.7J) > *ST* (8.6J)
At **N**: *TO* (16.6J) > *TI* (13.7J) > *ST* (12.7J)
At **F**: *TO* (16.8J) > *ST* (15.3J) > *TI* (14J)

E. Mechanical Work at Knee Joint

No significant changes were observed for W_+ at **S** and **N** for *TO* and *TI* respectively as compared to *ST* (**S**: 5.4J; **N**: 7.5J) as shown in Figure 2.. Significant reductions were observed for W_+ at **S** and **N** for *TI* (41.6%) and *TO* (9.3%) respectively as compared to *ST* (**S**: 5.4J; **N**: 7.5J). At **F**, W_+ was decreased by *TO* (10%) while increased by *TI* (21.9%) from *ST* (11J).

At **S**, significant reduction in W_- was observed for both *TO* (22.2%) and *TI* (20.9%) as compared to *ST* (-23.4J). At both **N** and **F**, significant increase in W_- was observed for *TO* (**N**: 9.7%; **F**: 21%) and *TI* (**N**: 38.5%; **F**: 41.4%) as compared to *ST* (**N**: 26.7J; **F**: 30.4J).

W_{net} for Knee joint showed the following trend:
At **S**: *ST* (-18J) > *TI* (-15.3J) > *TO* (-12.4J)
At **N**: *TI* (-29.9J) > *TO* (-22.4J) > *ST* (-19.1J)
At **F**: *TI* (-29.6J) > *TO* (-26.8J) > *ST* (-19.3J)

F. Mechanical Work at Ankle Joint

W_+ was decreased for *TO* (61%) and *TI* (59%) from *ST* (27.2J) at **N**. At **S**, *TO* decreased W_+ by 20%, while *TI* had no significant change from *ST* (22.34J). Insignificant changes were observed for *TO* and *TI* at **F**, as compared to *ST* (30.8J).

W_- was increased for *TO* (**N**: 36.1%; **F**: 20.4%) and *TI* (**N**: 55.5%; **F**: 9.6%) from *ST* (**N**: -25.2J; **F**: -25J) at **N** and **F**. At **S**, *TO* increased W_- immensely (178%) while *TI* reduced it insignificantly (7.6%) from *ST* (-24.9J).

W_{net} for ankle joint showed the following trend:
At **S**: *TO* (-51.6J) > *ST* (-2.7J) > *TI* (-0.2J)
At **N**: *TI* (-28.1J) > *TO* (-23.7J) > *ST* (2J)
At **F**: *TO* (12.4J) > *TI* (5.5J) > *ST* (5.3J)

G. Total Limb Mechanical Work

W_T was found to be negative in all the conditions except two: *ST* at **F** (1.3J) and *TO* at **F** (2.4J) as depicted in Figure 2. Maximum W_T was observed for *TO* at **S** (-50.5J) followed by *TI* at N (-44.3J) and *TO* at **N** (-29.5J).

IV. Discussion

The objective of the study was to analyze the effects of toe-out and toe-in gait modifications with varying walking speeds on knee joint load and total limb mechanical work. It was observed that these gait modifications had considerable influence on peak KAM values and work done at lower limb joints. The faster walking speed of 1.43 m/s was closer to the fast walking speeds of previous studies but the normal walking speed was considerably slower than previously reported normal range (1.33 m/s) [12]. The toe-out and toe-angles are in accordance with the previous works [13, 14].

Toe-out generally increased the first peak KAM and reduced the second peak of KAM, while toe-in gait reduced the first peak KAM and increased second peak KAM. This behaviour is in accordance with the previous works that show opposing effects of toe-out and toe-in gait on KAM [5, 14]. However, our study found that these behaviours are also considerably influenced by walking speed. Walking faster increased the first peak KAM for all three foot positions as compared to natural (straight foot). It has been reported that GRF increases at faster walking speeds at both early and late stance, resulting in higher KAM values [15]. Lesser sensitivity in first peak KAM was observed when the participants slowed down their speeds. This suggests that there might be medio-lateral shifting of the GRF vector by toeing-in but the magnitude of vector is closer to the magnitude of body weight at slower speed and not much can be done to reduce the KAM. Hence no further benefits can be reaped out of toe-in gait in reducing first peak KAM when one slows down the walking speed.

Changes with walking speed for peak KAM at late stance were more pronounced with toe-out gait with 23.8% and 13.7% increase in second peak KAM at slow and fast walking speeds, respectively. Compared with the first peak KAM, the second peak KAM was more sensitive to increasing the walking speed than decreasing it. It may be prescribed to walk at normal walking speed with toe-in gait when intending to reduce early stance knee load.

The total limb mechanical work was found to be negative (eccentric) at slow and normal walking speeds in all foot positions. The total positive work increased with increasing walking speed in straight toe condition. This relation was more pronounces with toe-in and toe-out conditions. Toe-out appeared to be more energy absorbing (higher eccentric work) at slow and normal walking speeds than straight or toe-in gait. The largest contribution to this energy dissipation was from the hip and ankle joints.

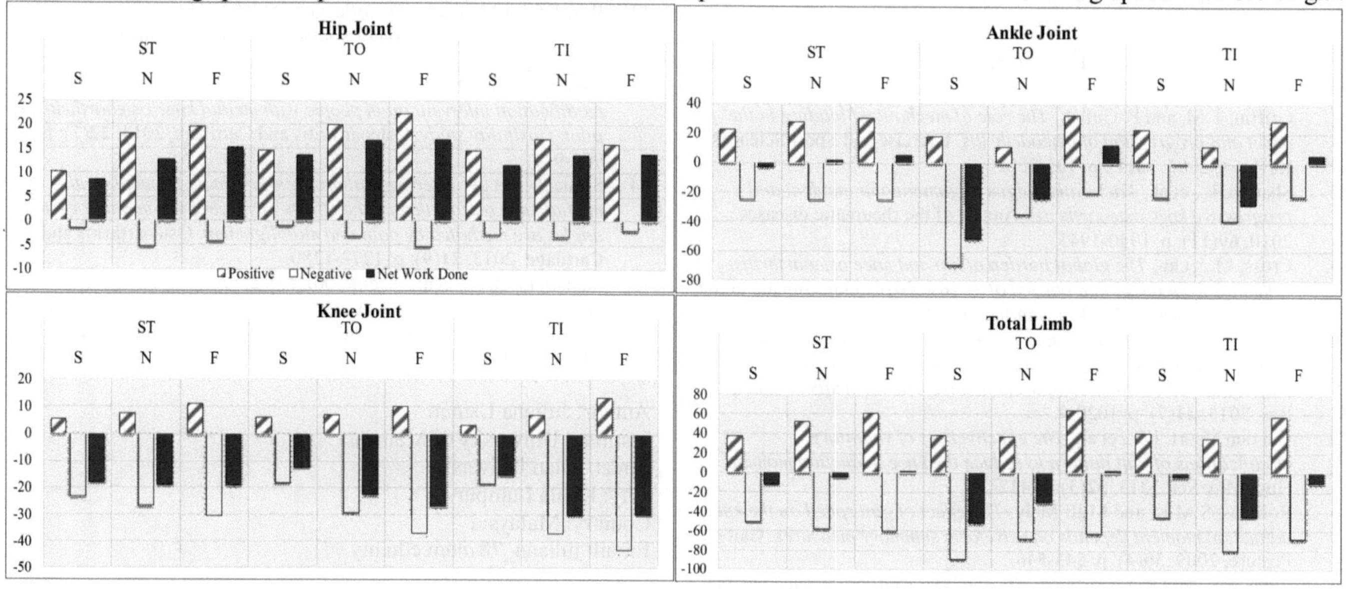

Figure 5 Mean joint mechanical work values for all test conditions. The vertical axis depicts mechanical work done in Joules. ST is straight (natural gait), TO is toe-out gait, TI is toe-in gait, S is slow (0.85 m/sec) speed, N is normal speed (1.18 m/s)

V. Conclusions

The peak KAM values were influenced by toe-out and toe-in gait modifications at all walking speeds. Negative work at knee joint was mostly affected by toe-out and toe-in at fast walking speed, and at ankle joint by toe-out at slow speed. Changing the walking speed altered the peak KAM and work done values and the relation between toe-out or toe-in gait to normal foot position.

Acknowledgment

The study was funded by University of Malaya Research Grant, Project Number RP-020C-13AET.

Conflict of Interest

The authors declare having no conflict of interest.

References

1. Griffin, T.M. and F. Guilak, *The role of mechanical loading in the onset and progression of osteoarthritis.* Exercise and Sport Sciences Reviews, 2005. **33**(4): p. 195-200.
2. Sharma, L., et al., *Varus and valgus alignment and incident and progressive knee osteoarthritis.* Annals of the rheumatic diseases, 2010. **69**(11): p. 1940-1945.
3. Cross, M., et al., *The global burden of hip and knee osteoarthritis: estimates from the global burden of disease 2010 study.* Annals of the rheumatic diseases, 2014: p. annrheumdis-2013-204763.
4. Shull, P.B., et al., *Six-week gait retraining program reduces knee adduction moment, reduces pain, and improves function for individuals with medial compartment knee osteoarthritis.* J Orthop Res, 2013. **31**(7): p. 1020-5.
5. van den Noort, J.C., et al., *The effectiveness of voluntary modifications of gait pattern to reduce the knee adduction moment.* Hum Mov Sci, 2013. **32**(3): p. 412-24.
6. Robbins, S.M.K. and M.R. Maly, *The effect of gait speed on the knee adduction moment depends on waveform summary measures.* Gait & Posture, 2009. **30**(4): p. 543-546.
7. Lynn, S.K. and P.A. Costigan, *Effect of foot rotation on knee kinetics and hamstring activation in older adults with and without signs of knee osteoarthritis.* Clinical Biomechanics, 2008. **23**(6): p. 779-786.
8. Chen, I., K. Kuo, and T. Andriacchi, *The influence of walking speed on mechanical joint power during gait.* Gait & Posture, 1997. **6**(3): p. 171-176.
9. DeVita, P., J. Helseth, and T. Hortobagyi, *Muscles do more positive than negative work in human locomotion.* J Exp Biol, 2007. **210**(Pt 19): p. 3361-73.
10. Organization, W.H., *Obesity: preventing and managing the global epidemic.* 2000: World Health Organization.
11. Rutherford, D., et al., *Foot progression angle and the knee adduction moment: a cross-sectional investigation in knee osteoarthritis.* Osteoarthritis and Cartilage, 2008. **16**(8): p. 883-889.
12. Burnfield, J.M., et al., *The influence of walking speed and footwear on plantar pressures in older adults.* Clinical Biomechanics, 2004. **19**(1): p. 78-84.
13. Hunt, M.A. and J. Takacs, *Effects of a 10-week toe-out gait modification intervention in people with medial knee osteoarthritis: a pilot, feasibility study.* Osteoarthritis and Cartilage, 2014. **22**(7): p. 904-911.
14. Simic, M., et al., *Altering foot progression angle in people with medial knee osteoarthritis: the effects of varying toe-in and toe-out angles are mediated by pain and malalignment.* Osteoarthritis and Cartilage, 2013. **21**(9): p. 1272-1280.
15. Stansfield, B.W., et al., *Sagittal joint kinematics, moments, and powers are predominantly characterized by speed of progression, not age, in normal children.* Journal of Pediatric Orthopaedics, 2001. **21**(3): p. 403-411.

Author: Juliana Usman
Institute: University of Malaya
Street: Jalan University
City: Kuala Lumpur
Country: Malaysia
Email: juliana_78@um.edu.my

The Effects of Varying Level of Glucose and Fructose on Brain Activation During Mouth Rinse

A.H. Abu Bakar[1], S. Mat Isa[1], M.H. Mohd Rasli[1], I.L. Shuaib[1], R. Singh[2], S. Derbyshire[3], H.K. Kamaruddin[1,4] and A.M. Che Muhamed[1]

[1] Advanced Medical and Dental Institute, Universiti Sains Malaysia
[2] Physiology Unit Faculty of Medicine, AIMST University
[3] Faculty of Arts and Social Sciences, National University of Singapore
[4] Faculty of Sports Science and Recreation, Universiti Teknologi MARA Perlis

Abstract*— Mouth rinsing a carbohydrate (CHO) solution has been shown to activate receptors within the oral cavity that is related to reward and pleasure centres of the brain. This brain activation is linked to enhance endurance exercise performance. However, it remains unclear, whether the caloric content or the level of sweetness of a solution may influence the level of brain activation during mouth rinse. Therefore, the aim of this study was to examine the effects of varying level of caloric content made up from glucose and fructose on brain activation during mouth rinse. Eleven healthy male participants (Age: 21 ± 2 years; stature: 168 ± 7 cm; body mass: 61.4 ± 6.4 kg; and peak rate of oxygen consumption (VO2max): 51.3 ± 2.2 ml.kg.min-1) volunteered to participate in this study. Participants were randomly assigned to perform 6 trials of mouth rinsing. During each trial, participants were asked to rinse a CHO solution for 10 seconds while fMRI was performed. Three sets of caloric-content solutions were prepared. The first set of caloric-content was at 19 kcal/g consisting 6% of glucose and 5.3% of fructose. The second set was 59 kcal/g consists of 18% of glucose and 15.9% of fructose and the last set was 79 kcal/g consists of 24% of glucose and 21.2% of fructose. The neuroimaging results showed that there was no clear trend on brain activation when rinsing with high caloric content of the CHO solution. However, when rinsing a sweeter solution of fructose, a larger magnitude of the insula/frontal operculum region of the brain was activated. The current observation suggests that the level of sweetness and not the caloric-contents potentially be the main determinant of brain activation.

Keywords— **Mouth rinsing, Caloric-Contents, Brain Activation**

I. INTRODUCTION

Carbohydrate mouth rinse was first shown to improve an hour of cycle time trial in a study by Carter et al. (2004). This improvement in endurance performance was speculated to be due to stimulation of receptors in oral cavity which in turn activates central afferent effects [1]. Since then, there have been growing evidences on the positive ergogenic effects of carbohydrate mouth rinse on endurance exercise performance [2-10].

To date, the only study that had examined brain activation with carbohydrate mouth rinse and its consequences on exercise performance was carried out by Chambers et al. (2009). They concluded that rinsing a glucose solution would activate the anterior cingulate cortex and striatum which is involved in motor control and reward to enhance exercise performance.

Carbohydrate mouth rinse have been shown to be more effective in enhancing endurance exercise performance when in a fasted state compared to fed state [3, 11]. However, it remains unclear whether the effectiveness of carbohydrate mouth rinse is influenced by the level of caloric content or the level of sweetness.

The purpose of this study was therefore to examine if the caloric content and/or the sweetness level of a carbohydrate solution has a direct influence on the brain activation when performing mouth rinsing.

II. MATERIALS AND METHODS

A. Subjects

Eleven healthy and active subjects (mean ± SD; age: 21 ± 2 years; stature: 168 ± 7 cm; body mass: 61.4 ± 6.4 kg; and peak rate of oxygen consumption (VO_2max): 51.3 ± 2.2 ml.kg.min^{-1}) volunteered to participate in this study. All the subjects were male and active with their endurance sports. All subjects were briefed on all measurement and functional magnetic resonance imaging (fMRI) scan examination procedures. Before obtaining a written consent forms, the individuals who expressed an interest in the study were verbally informed about the objective, procedures, the measurements involve and the demands that study would place upon them as well as informed on the possible risks. All procedures had prior approval from the Universiti Sains Malaysia Human Research Ethics Committee.

F. Ibrahim et al. (eds.), *3rd International Conference on Movement, Health and Exercise*,
IFMBE Proceedings 58,
DOI: 10.1007/978-981-10-3737-5_23

B. *Experimental Design*

The experimental design used in magnetic resonance imaging (MRI) examination was an event-related design protocol. The 6 mouth rinse solutions were blinded and randomized between the subjects. Each subject visited the MRI chamber on 7 separate occasions and once at the laboratory for VO_2max test. The first visit at MRI chamber was for familiarization phase. The familiarization protocol was same with the actual test protocol. This trial was conducted to ensure the subjects do not experience claustrophobia or panic attack during the MRI scanning tests. Subjects need to undergo 6 MRI scan examinations with different solutions of mouth rinse in each trial.

All MRI scanning examinations commenced at the evening period, two hours after having similar lunch meal. The purpose to have a meal before undergoes MRI examination is to avoid any brain activation caused by the hungry state. Subjects also were needed to drink 3 L of water one day prior to the MRI examination to avoid dehydration. All subjects were asked to provide a urine sample to test the hydration status before the MRI examinations. A urine sample was measured using a refractometer (Pal-10s, Atago, Japan).

Gastrointestinal (GI) comfort procedure, adopted from Rollo et al. (2010) was use to determine subjects' gastrointestinal comfort level after rinsed and ingested the testing solutions. The GI comfort was rated by using a 12 point scale with anchors provided at 0 = neutral, 4 = uncomfortable, 8 = very uncomfortable and 12 = painful. The GI comfort was collected from the subjects before and after MRI scanning procedure.

Blood glucose concentration was collected and determined by a finger prick samples (Roche Diagnostics GmbH, Accu-chek Performa, Mannheim, Germany). The blood glucose sample was taken every pre and post of the MRI scan examinations.

C. *Preliminary: VO_2max test*

The first visit at the laboratory consisted of measurement of stature and body mass for each participant. After a standardized warm-up, all subjects undergo a maximal graded exercise test on a motorised treadmill (HP Cosmos, Nussdorf, Germany). The graded exercise test consisted of four levels of submaximal running speeds of 7, 9, 11 and 13 km/hr with four minutes duration of interval time for each running speed. After the completion of submaximal running test, subjects were engaged with 5 minutes of active rest on the treadmill at a speed 3.5 km/hr. Following the completion of active rest, the workload was set at 12 km/h with 2% increment of treadmill gradient for every two minutes until exhaustion. Respiratory gas exchange was measured throughout the test using a calibrated gas system (TrueOne2400, Parvomedics, Utah, USA). The maximal graded tests were conducted in a climatic chamber set to a thermoneutral environment (20°C of ambient temperature and 40% of relative humidity).

D. *fMRI Procedure*

On the arrival at MRI chamber, subjects need to rinse with warm water to clear the oral. Then, subjects were asked to cover their eyes with eye mask and lie down on the MRI bed. Two delivery tubes were placed on the left side of their mouth. The first delivery tube is for CHO solution and the second delivery tube is for distilled water. Each tube was connected to an automatic injector via syringe. Then, subjects were given to grasp a signal controller on both sides, which they only have to press the right-side button when every solution reaches their tongue to notify that the solution already inside their mouth. In case of emergency, subjects were asked to press the button repeatedly.

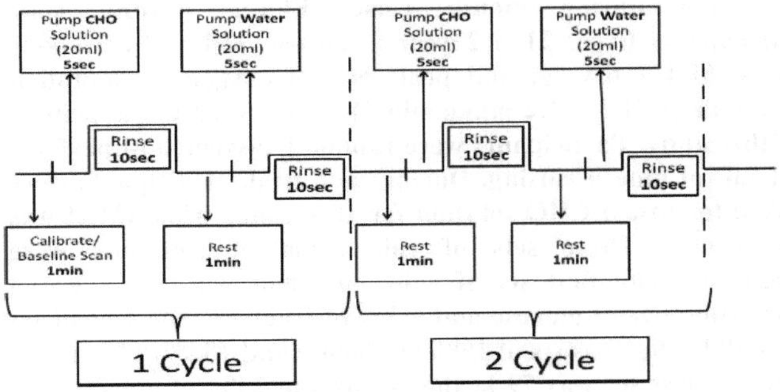

Figure 1 fMRI protocol for 2 cycles

At the beginning of each trial, (Fig. 1) there was a 60 s calibrate and baseline scan period following which 20 mL of test solution was automatically delivered into the mouth for 5 s period via the first delivery tube. Subjects were instructed to press the signal button given earlier and make one tongue movement without move the head and jaw to distribute the solution in the oral cavity. Then, 10 s after the delivery of the stimulus, subjects were cued by a touch stimulus to swallow the solution. After swallowing the solution, subjects rest at 60 s periods. Following the rest period, 20 mL of distilled water was automatically delivered through the second delivery tube for 5 s period and again subjects were instructed to press the signal button at the controller and rinse by one tongue movement. After 10 s rinsed, subjects were cued again to swallow the distilled water. One cycle was counted after subjects completed rinse with both stimulus solution and distilled water. Each cycle lasted for 150 s. After the completion of a cycle, the next cycle will start again

Table 1. Coordinates of activation cluster peaks and the T-score values

Solutions	Brain Regions	Coordinates	T-score
6% of Glucose	Operculum/ Insula	-42, 10, 14	6.18
	DLPC	44, 0, 54	8.11
18% of Glucose	Operculum/ Insula	40, 2, -8	6.96
	DLPC	-36, 34, 18	4.76
24% of Glucose	Operculum/ Insula	36, -6, 10	5.49
	DLPC	44, -4, 52	5.23
5.3% of Fructose	Operculum/ Insula	36, -4, 10	5.50
	DLPC	18, 66, 2	7.37
15.9% of Fructose	Operculum/ Insula	-56, -4, 12	5.76
	DLPC	-34, 20, 52	5.46
21.2% of Fructose	Operculum/ Insula	-56, -4, 12	9.49
	DLPC	44, -2, 52	6.15

immediately with the same procedure. During each visit, the subject was asked to do 8 cycles of the mouth rinse.
It is estimated that each MRI session will take between 25-30 minutes (including calibration phase and scanning phase).

E. Experimental solutions and mouth-rinsing procedure

The solution of mouth rinse was prepared by a nutritionist of Lifestyle Science Cluster of Advanced Medical and Dental Institute. Each subject was given either a 20 mL bolus of glucose (Sim Company, Penang, Malaysia) or fructose (Sim Company, Penang, Malaysia) solutions to rinse during the MRI scan examinations for 10 s period. The subjects were asked to complete 6 MRI examinations by rinsing with 6 different types of solutions and concentrations: 6%, 18% and 24% of glucose and 5.3%, 15.9% and 21.2% of fructose. The solutions were stored at room temperature (20°C). Every 1 gram of glucose and fructose were made up in 100 mL of distilled water to have 1% of concentration.

To test the amount of caloric-contents for each CHO in 1 gram, a few sample of glucose and fructose were tested on bomb calorimeter (6200 Isoperibol Calorimeter, Moline, Illinois, USA) tests. The calorie for 1 gram of glucose was 3.279 kcal and 1 gram for fructose was 3.712 kcal. To match the calories with 6% of glucose (19 kcal/g), fructose was prepared in 5.3% of concentrations. For 18% of glucose (59 kcal/g), fructose was prepared in 15.9% of concentration to match their calories. To have equal calories of 24% of glucose (79 kcal/g), fructose should be 21.2% of concentration.

F. fMRI data acquisition

The experiments were conducted at the Clinical Trial Complex, Advanced Medical and Dental Institute by using Signa HDxt 1.5 T., where T2* weighted EPI slices were acquired every 2 s (TR = 2) with echo time = 50 ms. The matrix size was 64 × 64 with the field of view (FOV) = 24 cm and slice thickness = 4 mm with spacing = 1.0.

Acquisition was carried out during the task performance yielding 600 volumes in total for each session.

G. fMRI data analysis

The imaging data was analyzed using Statistical Parametric Mapping version 12 (SPM 12) (University College London, London, England) & Matlab 7.8.0 (R2010b) (Natick, Massachusetts, USA) to require numbers of activated voxel (represent area of activation) and T-score (represent intensity of activation). Reported *p* values based on this group analysis for particular regions of interest were corrected for the number of comparisons made within each region using the small volume correction (SVC) procedure [12].

H. Statistical Analysis

All statistical analysis and evaluations were performed by using IBM SPSS statistical version 22 (SPSS Inc.,Chicago, Ill., USA). All values are presented as mean ± SD. Wilcoxon Signed-Rank test was used to detect the significant differences between pre and post blood glucose concentration and paired sample *t* test was used to measure the significant differences in pre and post GI comfort. One sample *t* test was used to indicate the difference of hydration state of subjects before the the MRI scan examinations.

III. Results

The mouth rinsing task produced activation across the expected sensory motor network as been reported by Chambers et al. (2009). Table 1 identifies coordinates of activation cluster peaks and the T-score values as the indicator for activation intensity.

A. Brain response to different concentration of glucose

The brain activations were revealed during mouth rinse with all the three concentrations of glucose. Activation was found in the insula/ operculum which was believed to be the human primary taste cortex [10, 13]. The 18% of glucose solution showed a slightly higher on the activation intensity compare with 6% and 24% of glucose.

Activation also was found in dorsolateral prefrontal cortex (DLPC). DLPC was suggested to have a role in the preparation and selection of cognitive response [14]. The 6% of glucose was illustrated to have highest value of T-score followed by 24% and 18% of glucose solutions.

B. Brain response to different concentration of fructose

Fructose solutions also demonstrated to have activation on both brain regions (Insula/Operculum and DLPC). The highest concentration of fructose (21.2%) showed the greater value of activations of insula/operculum followed by 15.9% and 5.3% of fructose with slightly different in both.

The activation of DLPC during fructose mouth rinsing showed to have high amount of the intensity at 5.3% followed by 21.2% and 15.9%.

C. Responses to the low calories of glucose and fructose solutions (19 kcal/g)

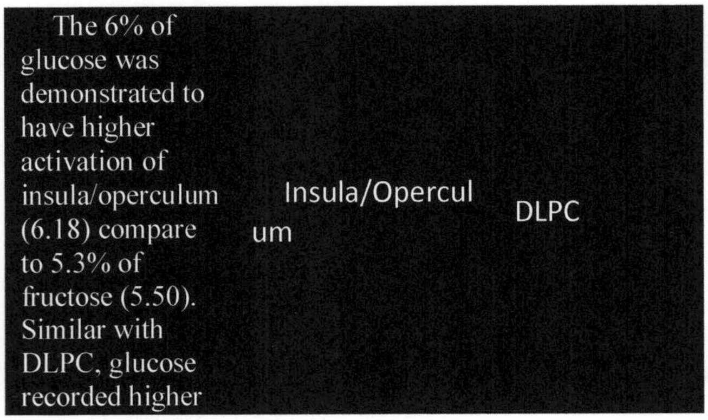

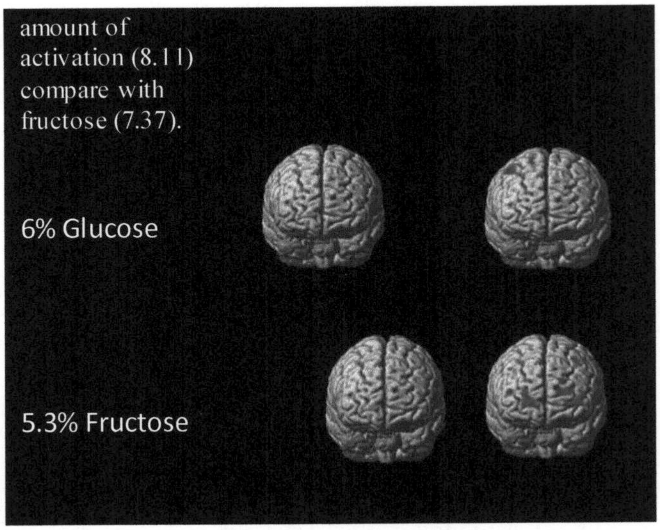

Fig 2. Activations in the insula/operculum and DLPC for 6% glucose and 5.3% fructose

D. Responses to the medium calories of glucose and fructose solutions (59 kcal/g)

The cortical response for medium calories revealed the 18% of glucose to have slightly higher activation in insula/operculum (6.96) compare with 15.9% of fructose (5.76) but the fructose have slightly higher in DLPC activation (5.46) compare with glucose solution (4.76).

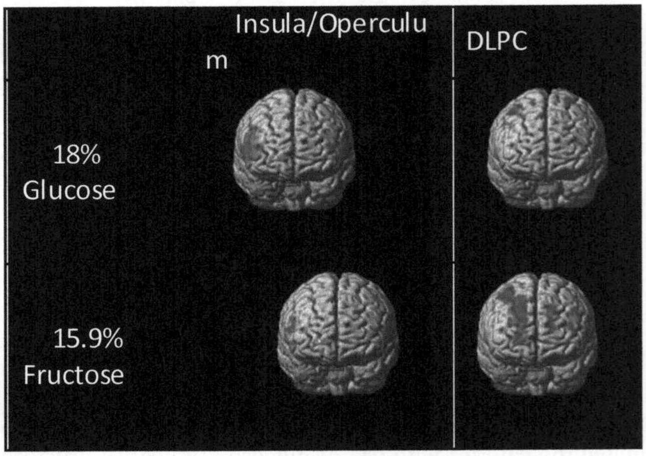

Fig. 3 Activations in the insula/operculum and DLPC for 18% glucose and 15.9% fructose

E. Responses to the high calories of glucose and fructose solutions (79 kcal/g)

In high calorie-content, 21.2% of fructose stated greater amount of activation in insula/operculum (9.49) compare with 24% of glucose (5.49). Equivalent to the activation of

DLPC, fructose stated the higher amount in activation (6.15) compare with glucose (5.23).

Table 2. Blood glucose concentration and GI comfort level (N = 11)

	Blood glucose concentration (mmol.L^{-1})		GI conmfort level	
	Pre	**Post**	**Pre**	**Post**
6 % Glucose	4.6 ± 0.7	4.7 ± 0.6	0 ± 1	0 ± 1
18% Glucose	4.7 ± 0.5	4.8 ± 0.5	0 ± 0	0 ± 0
24% Glucose	4.9 ± 0.8	5.0 ± 0.6	0 ± 0	1 ± 1
5.3% Fructose	4.5 ± 0.9	4.8 ± 0.7	0 ± 0	0 ± 0
15.9% Fructose	4.4 ± 0.8	4.3 ± 0.6	0 ± 0	1 ± 1
21.2% Fructose	4.8 ± 0.5	4.9 ± 0.4	0 ± 1	1 ± 1

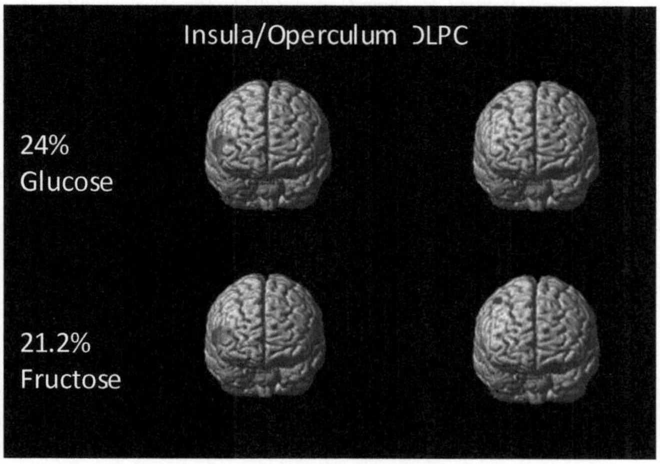

Fig. 4 Activations in the insula/operculum and DLPC for 24% glucose and 21.2% fructose

F. Blood Glucose Concentration

There was no significant difference in blood glucose concentration on pre and post trial for each solution (6% glucose: p = 0.531; 18% glucose p = 0.061; 24% glucose p = 0.552; 5.3% fructose p = 0.054; 15.9% fructose p = 0.833; 21.2% fructose p = 0.359)

G. Gastrointestinal Comfort

GI comfort was measured to determine subjects' gastrointestinal comfort level after rinsing and ingesting the testing solutions. There was no significant difference on GI comfort level before and ingest the solutions (6% glucose p = 0.341; 18% glucose p = 0.082; 24% glucose p = 0.053; 5.3% fructose p = 0.167; 15.9% fructose p = 0.221; 21.2% fructose p = 0.676)

H. Urine Specific Gravity (U_{sg})

Urine specific gravity (U_{sg}) was measured prior to each trail. The results showed that there were significant differences for all solutions which indicate that all the subjects were in euhydrated condition before start the MRI scan examinations (6% glucose: 1.016 ± 0.004, p = 0.002; 18% glucose: 1.014 ± 0.004, p = 0.001; 24% glucose: 1.016 ± 0.005, p = 0.012; 5.3% fructose: 1.016 ± 0.004, p = 0.005; 15.9% fructose: 1.014 ± 0.003, p = 0.001; 21.2% fructose: 1.017 ± 0.004, p = 0.020).

IV. Discussion

The aim of the present study was to examine the influence of caloric content and sweetness level of CHO solutions, made up of varying levels of glucose and fructose on brain activation during mouth rinsing. The presence of CHO in the mouth come into views to activate a novel energy signalling pathway capable of improving human performance [15]. The imaging result showed that brain response to different concentration of glucose (6%, 18% and 24%). The current assumption is that the brain responses to glucose in the oral cavity may intervene emotional and behavioural responses associated with rewarding stimuli [10]. The present study found that glucose activated theinsula/frontal operculum in all three concentrations. Insula/frontal operculum was believed to be the human primary taste cortex [10, 13]. Another brain region, DLPC, was believed have a role in the preparation and selection of cognitive response [14] also activated in all three concentration. 18% of glucose was resulted in higher response in insula/frontal operculum and 6% of glucose stated the high value of activation in DLPC. In fructose, the imaging data also showed to have activation in insula/frontal operculum and DLPC for all three concentrations (5.3%, 15.9% and 21.2%). 21.2% of fructose stated the highest activation in insula/frontal operculum and 5.3% of fructose showed to have greater activation in DLPC.

Comparing between types of CHO, for the low concentration, 6% glucose showed higher activation on both brain regions (insula/frontal operculum and DLPC) compare to 5.3% fructose. In contrast, high concentration of fructose (21.2%) was generated a greater activation on insula/frontal operculum compare with 24% of glucose. Evidently, rinsing fructose which is sweeter than glucose [16] and contains a higher caloric content, resulted in a larger magnitude in insula/frontal operculum.

From the imaging data, 6% and 18% of glucose had activated other regions of brain (anterior cingulate cortex, orbitofrontal cortex and striatum) but did not appear during rinse with other solutions. This limitation was due to the lack of image sensitivity when using a 1.5 tesla MRI system. Earlier studies that had reported more comprehensive areas of brain activation with CHO mouth rinsing had a used 3.0 tesla system, which has a greater image sensitivity [10, 17, 18].

Based on the earlier findings of Chambers et al. (2009) and Turner et al. (2014), using a higher concentration of mouth of mouth rinse solution may enhance the saturation of the oral receptors leading to a greater stimulation of reward pathways but based on the present results, further investigation on the mechanism is needed.

V. Conclusion

In conclusion, when rinsing a higher caloric-content with sweeter solution potentially will trigger and lead to a larger magnitude of insula/frontal operculum and DLPC. The current observation suggests that the level of sweetness together with the high caloric-contents potentially be the main determinant of brain activation. This study thus suggests athletes competing in endurance exercise can potentially enhance their performance by mouth rinsing a high caloric-content and sweeter solution at regular intervals.

Acknowledgment

The authors would like to thank the subjects for their time and effort. Thank you for those people who assisted in the study.

Conflict of Interest

The authors declare that they have no conflict of interest.

References

1. Carter J, Jeukendrup A, Mann C et al. (2004a) The effect of glucose infusion on glucose kinetics during a 1-h time trial. Med Sci Sports Exerc, 36: p. 1543 - 1550
2. Rollo, Cole M, Miller R et al. (2010) Influence of mouth rinsing a carbohydrate solution on 1-h running performance. Med Sci Sports Exerc. 42(4): p. 798-804
3. Fares E J, Kayser B (2011) Carbohydrate mouth rinse effects on exercise capacity in pre- and postprandial States. J Nutr Metab. 2011: p. 385962
4. Lane S C et al. (2013) Effect of a carbohydrate mouth rinse on simulated cycling time-trial performance commenced in a fed or fasted state. Appl Physiol Nutr Metab. 38(2): p. 134-9
5. Wright B F, Davison G (2013) Carbohydrate Mouth Rinse Improves 1.5 h Run Performance: Is There A Dose-Effect? International Journal of Exercise Science. 6(4): p. 8
6. Sinclair J, Bottoms L, Flynn C et al. (2014) The effect of different durations of carbohydrate mouth rinse on cycling performance. Eur J Sport Sci. 14(3): p. 259-64
7. Pottier A, Bouckaert J, Gilis W et al. (2010) Mouth rinse but not ingestion of a carbohydrate solution improves 1-h cycle time trial performance. Scand J Med Sci Sports. 20: p. 105 - 111
8. Fraga C, Velasques B, Koch A J et al. (2015) Carbohydrate mouth rinse enhances time to exhaustion during treadmill exercise. Clin Physiol Funct Imaging
9. Carter A, Jeukendrup A, Jones D (2004b) The effect of carbohydrate mouth rinse on 1-h cycle time trial performance. Med Sci Sports Exerc. 36: p. 2107 - 11
10. Chambers E, Bridge M, Jones D (2009) Carbohydrate sensing in the human mouth: effects on exercise performance and brain activity. J Physiol. 587: p. 1779 - 1794.
11. Che Muhamed A M, Mohamed N G, Ismail N et al. (2014) Mouth rinsing improves cycling endurance performance during Ramadan fasting in a hot humid environment. Appl Physiol Nutr Metab. 39(4): p. 458-64
12. Worsley K J, Marrett S, Neelin P et al. (1996) A unified statistical approach for determining significant signals in images of cerebral activation. Hum Brain Mapp. 4(1): p. 58-73
13. O'Doherty J, Rolls E T. Francis S et al. (2001) Representation of pleasant and aversive taste in the human brain. J Neurophysiol. 85(3): p. 1315-21
14. Rowe J B, Toni I, Josephs O et al. (2000) The prefrontal cortex: response selection or maintenance within working memory? Science. 288(5471): p. 1656-60
15. Turner C E, Byblow W D, Stinear C M et al. (2014) Carbohydrate in the mouth enhances activation of brain circuitry involved in motor performance and sensory perception. Appetite. 80(0): p. 212-219
16. Lowette K, Roosen L, Tack J et al. (2015) Effects of high-fructose diets on central appetite signaling and cognitive function. Front Nutr. 2: p. 5
17. Page K A, Chan O, Arora J et al. (2013) Effects of fructose vs glucose on regional cerebral blood flow in brain regions involved with appetite and reward pathways. Jama. 309(1): p. 63-70
18. Haase L, Cerf-Ducastel B, Murphy C (2009) Cortical activation in response to pure taste stimuli during the physiological states of hunger and satiety. Neuroimage. 44(3): p. 1008-21

Author: Ahmad Munir Che Muhamed
Institute: Lifestyle Science Cluster, Advanced Medical and Dental Institute, Universiti Sains Malaysia
City: Kepala Batas, Penang
Country: Malaysia
Email: ahmadmunir@usm.my

Entropy Based PAPR Reduction for STTC System Utilized for Classification of Epilepsy from EEG Signals Using PSD and SVM

S. K. Prabhakar and H. Rajaguru

Department of ECE, Bannari Amman Institute of Technology, Sathyamangalam, India

Abstract— **Epilepsy is one of the major disorders of the brain that affect the nervous system and is characterized by the recurrent seizures. The day to day life of the patient is severely disturbed because of the abrupt and unpredictable nature of the epileptic seizures. An investigative technique which provides comprehensive information about the classification, analysis and diagnosis of brain conditions is Electroencephalography (EEG). The useful information about the different diseases affecting the brain especially epilepsy are given by the frequency and energy content of this signal. As the recordings made from the EEG are quite large and difficult to process, Power Spectral Density (PSD) is employed here to reduce the dimensions of the entire data. Then the dimensionally reduced EEG data is transmitted through the Space Time Trellis Coded Multiple Input Multiple Output Orthogonal Frequency Division Multiplexing (STTC MIMO OFDM) system. As the system suffers a high Peak to Average Power Ratio (PAPR), entropy based Partial Transmit Scheme (E-PTS) is proposed to reduce the PAPR and Bit Error Rate (BER) is analyzed in the receiver side. Also at the receiver side, Radial Basis Function Kernel Based Support Vector Machine (SVM) is employed to classify the epilepsy from EEG signals. The performance metrics analyzed here are Specificity, Sensitivity, Time Delay, Quality Value, Accuracy, Performance Index, PAPR and BER.**

Keywords— **PSD, Epilepsy, EEG, PAPR, E-PTS**

I. INTRODUCTION

Epilepsy is one of the serious disorders of the brain and has to be treated well [1]. About 1-2 % of the general population is affected by this common neurological disorder named epilepsy. It is best characterized by means of recurrent and unprovoked seizures. The epileptic seizures are nothing but neurological dysfunctions that are observed with the help of abnormal electrical activity of the brain. Sometimes convulsions are associated with seizures because of its behavioral correlations. Seizures too sometimes may not have any clear external manifestations. By monitoring the brain activity, the non-convulsive seizures can be easily detected. To analyze, interpret and investigate the electrical activities of the brain, the scalp potential resulting from the activity of the brain should be recorded [2]. The recorded signal represents the potential difference between its two respective positions and is called electroencephalogram. A lot of data is generated with these long term EEG measurements and to receive it manually is a pretty exhaustive task and therefore automatic seizure detection methods came into existence. The structured organization of the paper is as follows: In section 2, the materials and methods are discussed along with the usage of PSD as the dimensionality reduction technique. In section 3, the STTC MIMO-OFDM System design is discussed along with the PAPR Reduction scheme using Entropy based Partial Transmit Scheme. In section 4, the usage of RBF Based SVM is employed at the receiver side to classify the epilepsy risk levels from EEG Signals. Section 5 deals with the results and conclusion.

II. MATERIALS AND METHODS

A. *Data Acquisition*

For this research study, the EEG data for totally 20 different epileptic patients were taken from Sri Ramakrishna Hospital, Coimbatore, India. The recordings were done for nearly half an hour each. The obtained recorded signals were continuous in nature and each and every recorded signal was easily divided into 2 second epoch duration. However, for each and every patient, the total number of channels is 16 and it is measured over 3 different epochs. The entire recordings are preferred in European Data Format (EDF). Since 50 Hz is the maximum frequency, the sampling frequency is about 200 Hz. Under each epoch for a single channel in a patient, there are totally 400 values obtained and for all the 20 patients, the data to be processed is very huge and therefore PSD is employed here as the dimensionality reduction technique [3].

B. *PSD as the dimensionality reduction technique:*

The power spectrum $S_{xx}(f)$ of a particular time series $x(t)$ explains the distribution of power into its respective frequency components composing that particular signal. Based on the Fourier analysis, any physical signal can be easily decomposed into a number of discrete frequencies over a continuous range. The statistical average of a certain

F. Ibrahim et al. (eds.), *3rd International Conference on Movement, Health and Exercise*,
IFMBE Proceedings 58,
DOI: 10.1007/978-981-10-3737-5_24

signal which includes noise is analyzed easily in terms of its frequency content and is called spectrum [3]. The PSD refers to the spectral energy distribution that would be traced per unit time, since the total energy of such a signal over all time is generally infinite in nature. The spectrum of a particular physical process $x(t)$ contains all the essential and useful information about the nature of X. The major reasons of using PSD as a dimensionality reduction technique are because of its property. The spectrum of a real valued process is real and an even function of frequency and is represented as

$$S_{xx}(-w) = S_{xx}(w)$$

The auto covariance function can be easily reconstructed with the help of Inverse Fourier Transform if the process is purely indeterministic and continuous. Because of these two properties it is used here as a dimensionality reduction technique.

III. STTC MIMO-OFDM SYSTEM

A. System Design:

The dimensionally reduced values are then transmitted through the STTC MIMO-OFDM System. MIMO-OFDM technology is used because of its higher spectral efficiency and channel capacity of the system. To achieve the benefits of MIMO-OFDM, an effective method of Space-Time Coding is used [4]. The inner relationship of signals between space domain and time domain is established in MIMO and so higher diversity gain and coding gain are obtained. The system can achieve high speed and high capacity communication transmission capacity. It can also improve the performance of the system by mitigating noise and multipath fading. STTC is based on trellis coded modulation. STTC provides the highest and maximum possible code gain and diversity gain without increasing the transmitting power and bandwidth. STTC can effectively resist fading, noise and Intersymbol Interference (ISI). When STTC is combined with the MIMO-OFDM the system can easily obtain the transmitting and receiving diversity gain.

B. PAPR Reduction Using Entropy Based PTS Technique

The multicarrier signal is generally the sum of many independent signals modulated onto the sub channels which have equal bandwidth. The input data block as a vector is given as $X = [X_0, X_1, \ldots, X_{N-1}]^T$. The PAPR is given by $PAPR = \frac{P_{peak}}{P_{average}} = 10\log_{10}\frac{\max|x_k|^2}{E\left[|x_k|^2\right]}(dB)$. If the correlation of the transmitting signal sequences reduces, the PAPR of the system also decreases efficiently [5]. The most commonly used performance measure for PAPR reduction technique is CCDF. The probability is denoted as PAPR of a data block which exceeds a given threshold. The performance of PAPR based on this is given as

$$CCDF = P_r(PAPR > PAPR_0)$$

$$CCDF = \left(1 - (1 - e^{-PAPR_0})^N\right)^U$$

where $PAPR_0$ = threshold value of PAPR, N = subcarrier number and U = number of phase sequences vector. The simulation parameters are clearly listed in Table 1.

Table 1 Simulation Parameters

Modulation used	QPSK
MIMO System analyzed	2 x 2 MIMO-OFDM
Number of Parallel Channels to transmit/No of carriers	512
FFT length	1024
Maximum symbols loaded	1e5
Symbol rate	250000
No of time slots	2
Length of Guard Intervals (points)	128
Window function	Blackman-Haris
HPA Model	SSPA
No of frames	10
No of OFDM symbols/ frame	4
Bandwidth	5 MHz
Oversampling factor	4

Steps to implement E-PTS algorithm:

a) The input data X is partitioned into smaller 'M' disjoint sub-blocks.
b) The partitioned sub-blocks are converted to time domain from frequency domain using the N-point IFFT.
c) The time domain sequences are combined with the complex phase factor in order to minimize the PAPR. The PAPR can be reduced by the weighted combination of M sub blocks.
d) The Exhaustive Entropy [6] is applied to the complex phase factors.

e) The percentage of the ith exhaustive component approximately equals to its probability P_i, then using the formula

$$H(s) = -\sum_i P_i . \log P_i$$

the estimation of the random property of a particular sequence can be done. $H(s)$ is the exhaustive entropy property of the sequence S. If the possibilities of all the components equal approximately then the exhaustive entropy of a sequence will be larger.

f) The main aim of PTS scheme is to design an optimal phase factor for the respective sub-block set that reduces the PAPR.

g) To find the best phase factor is a difficult problem and so entropy based PTS technique is used to find the optimal phase factor.

C. *Support vector machine (Radial Basis Function kernel)*

For a curve fitting problem, RBF networks can be used to find set weights. SVM's are a particular set of training samples which are extracted by the algorithm and it is used to trace the optimal plane [7]. The margin of separation is then found out. The respective distance between the closest data point and the separating hyperplane is called as the margin of separation. The margin of separation should be maximized and SVM's goal is to find the particular hyperplane which enables this job. Also, SVM tends to find the separating hyperplane to be optimal. Optimality is defined here as follows

$d_i.(w^T x_i + b) \geq 1$ for $i = 1,....., N$ training samples; where $d_i = \pm 1$. Also, the weight vector 'w' should minimize the cost function as $\theta(w) = \frac{1}{2} w^T w$.

The input data is transmitted from a dimension (m_0) which has a lower dimension to higher dimension feature space (m_1). The weights w_j, nothing but the separating hyperplane is estimated in the feature space. To construct the optimal hyperplane, the inner-product kernel is used. Inner product kernel is nothing but a function which replaces an activation function of a single neuron network. Inner product kernel is a function and is represented as $K(x, x_i) = \phi^T(x)\phi(x_i)$, where ϕ= set of the transformation function $\phi_j(.)$, $j = 1,...., m_1$. Here m_1 is the dimensionality of higher order space and x_i is the ith training sample. Depending on the definition of the kernel, different learning algorithms are built such as polynomial learning machine, RBF networks, and 2-layer perceptrons.

IV. Results and Conclusion

In this part, the PAPR Results, the detailed BER Results, and the Classification Results are explained well.

A. *PAPR Results*

The PAPR results and the BER Analysis results are clearly illustrated here. It is evident from the careful analysis of figure 1, it is apparent that when QPSK modulation is utilized and for different collection values of phase/weighing factors (W), the graph is plotted and it provided a less PAPR reduction of about 1.5 dB when the weighing factor is 4 at CCDF=10^{-3} respectively. The Bit Error Rate is also analyzed as it provided in Table 2.

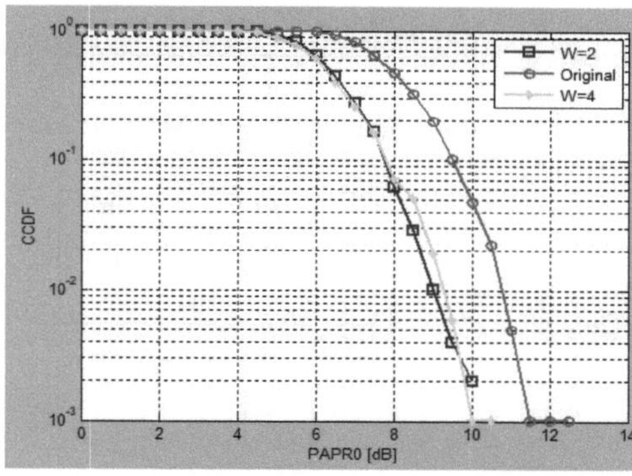

Figure 1 PAPR Reduction Using E-PTS Technique

Table 2 BER Analysis for the system

SNR	BER
0	0.0819
2	0.0629
4	0.0279
6	0.0149
8	0.0084
10	0.0030
12	0.0006

B. Classification Results

For PSD as dimensionality reduction techniques and RBF-SVM as a Post Classifier, based on the Performance Index, Quality values, Time Delay and Accuracy the results are computed in Table 3. The formulae for the Performance Index (PI), Sensitivity, Specificity, and Accuracy are given as follows

$$PI = \frac{PC - MC - FA}{PC} \times 100$$

where PC – Perfect Classification, MC – Missed Classification, FA – False Alarm,

The Sensitivity, Specificity, and Accuracy measures are stated by the following

$$Specificit\ y = \frac{PC}{PC + MC} \times 100$$

$$Sensitivit\ y = \frac{PC}{PC + FA} \times 100$$

$$Accuracy = \frac{Sensitivit\ y + Specificit\ y}{2} \times 100$$

The Time Delay and the Quality Value Measures are given by the following

$$Time\ \ Delay = \left[2 * \frac{PC}{100} + 6 * \frac{MC}{100}\right]$$

$$Quality\ \ Values = \frac{10}{\left[\frac{FA}{100} + 0.2\right] * Time\ \ Delay}$$

Table 3 Average Performance Analysis Values of 20 Patients

Parameters	Obtained Values
PC (%)	98.44
MC (%)	1.56
FA (%)	0
PI (%)	98.41
Sensitivity (%)	100
Specificity (%)	98.44
Time Delay (sec)	2.0624
Quality Values	24.2435
Accuracy (%)	99.22

V. Conclusion

Thus in this paper, when PSD is used as a dimensionality reduction technique and when it is transmitted through STTC MIMO-OFDM System, and when it is classified through Radial Basis Kernel Function Based Support Vector Machine, a perfect classification rate of about 98.44% is obtained, Performance Index of about 98.41% is obtained, Sensitivity of 100%, specificity of about 98.44%, time delay of about 2.062 seconds, quality value of 24.2435 and overall accuracy of about 99.22 is obtained. When exhaustive entropy based Partial Transmit Sequence (E-PTS) is employed to reduce the PAPR of the system, the PAPR is reduced by 1.5 dB. Future works plan to implement various other types of post classifiers and system design for the perfect classification of epilepsy from EEG Signals.

References

1. Tzallas AT, Tsipouras MG & Fotiadis DI (2009), 'Epileptic seizure detection in EEGs using time-frequency analysis', IEEE Trans. on Inf. Tech. in Biomed., vol. 13, pp. 703-710.
2. R.Harikumar, P.Sunil Kumar, "Frequency behaviors of electroencephalography signals in epileptic patients from a wavelet Thresholding perspective", Applied Mathematical Sciences, Vol. 9, 2015, no.50, 2451-2457
3. F.Vernotte, "Estimation of the Power Spectral Density of phase: comparison of three methods, "Proceedings of the 1999 Joint Meeting of the European Frequency and Time Forum, vol.2, pg:1109-1112, 1999
4. Harikumar Rajaguru, Sunil Kumar Prabhakar, " Clinical Health Care for Long Distance using Matrix Factorization and Mahalanobis Based Sparse Representation Measures for Epilepsy Classification from EEG Signals", International Journal of Pharmaceutical Sciences Review and Research, Vol.38(1), May-June 2016, Article 24, Pages:144-148
5. Sunil Kumar Prabhakar, Harikumar Rajaguru, "Factorization and Particle Swarm Based Sparse Representation Classifier for Epilepsy Classification Implemented for Wireless Telemedicine Applications", IFBME Proceedings (Springer), 6th International Conference on the Development of Biomedical Engineering, pp: 474-478
6. Lei Ning et al, "An exhaustive entropy based SLM method for PAPR reduction of OFDM Systems", 6th International ICST Conference on Communication and Networking in China (China com), pg: 118-122, 2011.
7. Harikumar Rajaguru, Sunil Kumar Prabhakar, "LDA, GA and SVM's for Classification of Epilepsy From EEG Signals", Research Journal of Pharmaceutical, Biological and Chemical Sciences, Vol.7(3), pp:2044-2049, May 2016

Author: Sunil Kumar Prabhakar
Institute: Department of Electronics and Communication Engineering, Bannari Amman Institute of Technology
City: Sathyamangalam
Country: India
Email: Sunilprabhakar22@gmail.com

Carbohydrate Mouth Rinse Enhances Time to Exhaustion of Running Performance Among Dehydrated Subjects

H.K. Kamaruddin[1,2], C.H. Ooi[2] A.H. Abu Bakar[1] and A.M. Che Muhamed[2]

[1] Lifestyle Science Cluster, Advance Medical and Dental Institute, Universiti Sains Malaysia, Kepala Batas, Penang, Malaysia
[2] Faculty of Sports Science and Recreation, Universiti Teknologi MARA Perlis, Arau, Perlis, Malaysia

Abstract— **Mouth rinsing with carbohydrate (CHO) solution can effectively enhance endurance performance by stimulating areas of the brain that are related to arousal and motivation. It remains unclear whether the effectiveness of CHO mouth rinsing is influenced by the level of hydration. Hence, this study was conducted to examine the effect of mouth rinsing a CHO solution on the running performance among dehydrated subjects. Twelve well-trained subjects (age: 21 ± 1 years; stature: 170.8 ± 3.5 cm; weight: 58.5 ± 3.5 kg; VO_2max: 58.3 ± 3.3 $mL.kg^{-1}.min^{-1}$) completed two sub-maximal running exercises at an intensity 70% VO_2 max until exhaustion while in a dehydrated state (2% body weight deficit). During the run, participants had mouth rinsed 25 mL of either a 6% CHO solution or placebo (PLA) at 15 min intervals. Plasma glucose (P_{glu}), plasma lactate (P_{lac}) and psychological measures were sampled intermittently during the exercises. Gas exchange, perceived exertion (RER), heart rate (HR), mean skin temperature (T_{sk}) and rectal temperature (T_r) were recorded during trials. The running time to exhaustion (TTE) was significantly (p = 0.000) shorter when mouth rinsing a PLA (76.8 ± 3.9 min) solution as compared to that when rinsing with a CHO solution (81.2 ± 4.1 min). There was no significant effect on the metabolic, psychological, thermoregulatory and cardiovascular variables between the two trials. This study demonstrated that participants in a dehydrated state were able to enhance their performance when they rinsed their mouth using a CHO solution as against a PLA.**

Keywords— **Oral sensing, Endurance capacity, Dehydration, Running, Carbohydrate.**

I. Introduction

Mouth rinsing a carbohydrate (CHO) solution has been shown to improve endurance performance [1-6]. Carter et al. [1] were the first to demonstrate an improvement in endurance cycling performance with CHO mouth rinsing when compared to rinsing a placebo (PLA) solution. The improvement in endurance exercise performance with CHO mouth rinse was initially postulated to be linked to the enhancement of the central effect through afferent signals from CHO receptors in the mouth [1-6]. This proposition was later confirmed by Chambers et al. [22] who reported that mouth rinsing a CHO solution for 10 s had activated certain areas of the brain, such as the anterior cingulate cortex and ventral striatum. These areas are known to facilitate the behavioral responses observed during exercise [7]. Moreover, an energy-signalling pathway that controls and promotes feeding behavior through transmission of information related to energetic, useful nutrients was stimulated in the case of CHO present in the mouth [7].

Previous studies have demonstrated that the effectiveness of CHO mouth rinse on endurance ~~mode~~ exercise is related to pre-prandial state [4, 8], rinsing duration [9] and environmental condition [2, 10, 11]. However, to date, it remains unclear whether the dehydration status of an individual engaging in endurance exercise will directly influence the effectiveness of CHO mouth rinse. To our knowledge, Arnaoutis et al. [12] was the first to examine the effect of mouth rinsing on dehydrated subjects during exercise. In their study, the effect of mouth rinsing of water was compared to ingestion of water on cycling time-trial performance among dehydrated subjects. The results showed no improvement in the cycling time to exhaustion when mouth rinsing with water as compared to ingestion. The study by Arnaoutis et al [12], had used water instead of CHO solution which was used in earlier studies that reported an improvement in endurance exercise performance [1-6]. Thus, this study had extended the work by Arnaoutis et al [12] by using CHO solution for the mouth rinsing intervention among dehydrated subjects.

Limited studies have been carried out to examine the effect of CHO mouth rinse on running performance [5, 6, 14, 15]. The finding remains inconclusive with some studies reported an improved running performance [6, 13, 14], while others reported no ergogenic benefit [5, 15].

The purpose of this study, therefore, was to determine the effect of CHO mouth rinse on time to exhaustion running performance on dehydrated runners.

II. Materials and methods

A. Subjects

Twelve male endurance-trained runners (mean ± SD: age: 21 ± 1 years; stature: 170.8 ± 3.5 cm; body mass, 58.5

F. Ibrahim et al. (eds.), *3rd International Conference on Movement, Health and Exercise*,
IFMBE Proceedings 58,
DOI: 10.1007/978-981-10-3737-5_25

± 3.5 kg; VO_2max: 58.3 ± 3.3 $mL.kg^{-1}.min^{-1}$) volunteered to participate in this study. All the participants were trained at least 5 days a week for endurance running and they were acclimatized to at least an hour long training and/or competition. All participants were briefed on the measurement procedures. However, the participants were not fully informed about the true purpose of the study. Participants were told that 2 types of solutions were being tested. Early interviews with the participants indicated that none of them practiced mouth rinsing when exercising. Each participant read and signed the consent to participate in this study, and all experimental procedures concerning human subjects were approved by the Universiti Sains Malaysia Ethics Committee.

B. *Preliminary testing*

All participants underwent a maximal graded exercise test on a motorised treadmill (HP Cosmos, Nussdorf, Germany). The graded exercise test consisted four sub-maximal running speeds of 7, 9, 11 and 13 km/hr with 4 min duration for each speed interval. Following the completion of sub-maximal running, the participants were required to actively rest for 5 min on the treadmill at a speed of 3.5 km/hr. Then, the workload was set initially at 12 km/hr with 2% increment in treadmill gradient after every 2 min until exhaustion. Respiratory gas exchange was measured throughout the test using a calibrated gas system (TrueOne 2400, Parvomedics, Utah, USA). The maximal tests were conducted in a climatic chamber set to a thermoneutral environment (20 ± 0.5°C ambient temperature and 40 ± 4% relative humidity).

C. *Familiarisation to running trial*

A week from commencing their first experimental trial, participants underwent a time to exhaustion (TTE) familiarisation trial on a motorised treadmill. Each participant sat on a chair for 5 min then immediately performed a TTE exercise at 70% of VO_2max. Time and distance covered were monitored on the treadmill panel. Right before the commencement of the TTE and at every 15 min interval, the participants rinsed for 10 s and expectorated 25 mL of a solution into a container. This solution was later weighed. This system was to ensure that each participant was familiar with the mouth rinse protocol.

D. *Experimental protocol*

Experimental trials were randomised and undertaken in a double-blind, counter balance, crossover manner. The participants were refrained from rigorous exercise, alcohol and caffeine consumption 24 hrs prior to the test. Each participant recorded their diet 48 hrs. before the first trial and replicated it before subsequent trials.

Participants completed three visits to the laboratory. All main exercise trials were carried out on a motorised treadmill (HP Cosmos, Nussdorf, Germany). Every participant observed an overnight fast and performed the exercise trials at a treadmill speed corresponding to 70% VO_2max. Each exercise trial was divided into three phases in order to induce dehydration, rest and carry out the running performance test.

E. *Exercise induces dehydration (EID)*

On their arrival at the laboratory, subjects emptied their bladder and a urine sample was collected for the determination of urine specific gravity, measured using a refractometer (Atago Co. Ltd, PAL-10s, Tokyo). Then the participant's pre-exercise nude body weight was recorded (Mettler Toledo, Columbus, Ohio, USA). The dehydration included 30 min of running at 60% of VO_2max and 30 min passive rest. This protocol was observed in a climatic chamber set at 35°C and with 70% relative humidity. After the passive rest, participants were towel dried and their post-exercise nude body weight was recorded. A post-exercise urine sample was also collected and analysed. The dehydration protocol induced a -2.1% ± 0.1% change in the total body weight of the participants.

Rest. Immediately after the exercise induced dehydration phase, participants were required to rest for 2 hrs in a temperate conditioned room. Then, they consumed a standardised breakfast that was equivalent to 2.5 kg^{-1} of body weight of CHO [4] and consumed only 50 mL of plain

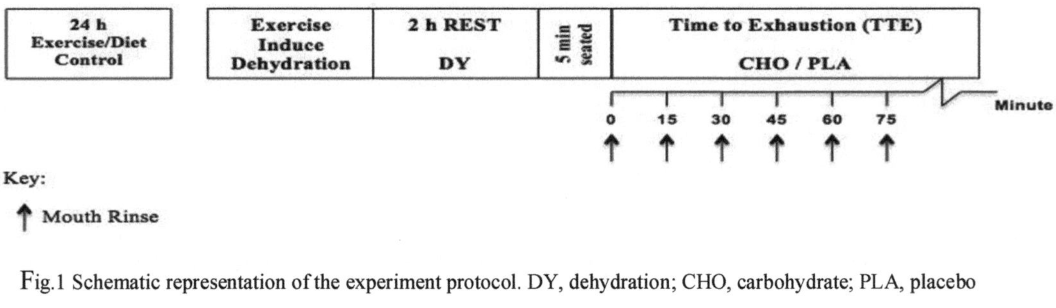

Fig.1 Schematic representation of the experiment protocol. DY, dehydration; CHO, carbohydrate; PLA, placebo

Table 1. Plasma glucose and plasma lactate concentration at pre-exercise and at every 15 min time interval in the TTE duration for carbohydrate mouth rinse (CHO) and placebo mouth rinse (PLA), (mean ± SD). Exh, exhaustion.

	Time to exhaustion (TTE)						
	Pre-exercise	15 min	30 min	45 min	60 min	75 min (n=11)	Exh
Plasma glucose (mmol.L^{-1})							
CHO	5.4 ± 1.0	5.3 ± 0.4	5.2 ± 0.6	5.4 ± 0.6	5.6 ± 0.6	5.7 ± 0.2	6.3 ± 0.3*
PLA	5.1 ± 0.6	4.9 ± 0.5	5.2 ± 0.8	5.3 ± 0.6	5.1 ± 0.3	5.8 ± 0.2	6.1 ± 0.4*
Plasma lactate (mmol.L^{-1})							
CHO	1.5 ± 0.3	2.6 ± 0.3	3.5 ± 0.3*	3.7 ± 0.3*	4.7 ± 0.6*	5.6 ± 0.4*	6.4 ± 0.3*
PLA	1.5 ± 0.3	2.9 ± 0.6	3.7 ± 0.4*	3.6 ± 0.4*	4.7 ± 0.5*	5.7 ± 0.5*	6.5 ± 0.2*

No significant difference between treatments was found. *Significant difference from pre-exercise, $p < 0.05$.

water. Breakfast consisted of a mixture of glycemic index foods, including bread and cereal. Food and water were to be consumed within 30 min of the rest period. No additional water was given to the participants in order to ensure they remained dehydrated before the exercise.

F. Time to exhaustion (TTE)

Participants were instrumented before being ushered into the climatic chamber where they remained seated on a chair for 5 min. Subjects performed the TTE exercise on a treadmill with a constant speed adjusted to 70% of VO_2max at 0% gradient. Participants were asked to cover as much distance as possible without having access to any information about speed and time elapsed.

Oxygen consumption (VO_2) and respiratory exchange ratio (RER) were sampled using calibrated gas system (TrueOne 2400, Parvomedics, Utah, USA) pre-exercise and at every 15 min interval for a 3 min sampling duration. Rectal temperature (T_{re}) was recorded using a disposable thermistor probe (YSI 400 series, Mallinckrodt Medical, Kansas City, Mo., USA) inserted 12 cm beyond the anal sphincter. A data logger (Cole Parmer, Vernon Hills, Ill., USA) was used to record the T_{re}. Skin temperature was measured at four different sites (left shoulder, left chest, right mid-tight and right mid-shin) using iButton temperature sensors (Maxim Integrated Products, Sunnyvale, CA., USA). Mean skin temperature (T_{sk}) was calculated using data from all four sites [16]. T_{re} and T_{sk} were sampled continuously. Participants were asked of their subjective rating of perceived exertion (RPE) on a categorical scale of 6-20 [17] pre-exercise and at every 15 min interval. Heart rate (HR) was continually monitored using an HR monitor (Polar Electro, Kempele, Finland) during the entire exercise period.

At the end of the trial, participants dismounted, removed all clothing and instrumentation; they were towel dried before the post-exercise nude body weight was measured. A post-exercise urine sample was collected for analysis. All trials were conducted in a climatic chamber pre-set at a thermoneutral condition of 20 ± 0.3°C and 40 ± 3% relative humidity. All trials were conducted with a gap of 3 days, in the morning and at the same time of the day. Figure 1 shows a schematic overview of the experimental protocol.

G. Mouth rinse procedure and solutions

At the start and after every 15 min of the TTE, participants were provided with a 25 mL bolus of either a 6% CHO (Sim Company, Penang, Malaysia; 6.0% g/L of glucose) or a placebo (PLA; sucralose, Diabetasol, Jakarta, Indonesia) solution. Both solutions were freshly prepared before each trial and were kept in a chiller at a temperature of 15 ± 0.8°C. Solutions were taste-matched and were the same in colour and appearance. Participants were instructed to rinse the fluids around their mouths for 10 s before expectorating it into a beaker held by an investigator. The beaker was weighed before and after each rinse using an electronic balance (Tanita KD-160, Tokyo, Japan) to determine the volume of rinsed and expectorated solution and to confirm that no fluid had been consumed.

Table 2 Rectal and mean skin temperature at pre-exercise and every 10-min time interval during TTE duration for carbohydrate mouth rinse (CHO) and placebo mouth rinse (PLA), (mean ± SD). Exh, exhaustion.

	Time to exhaustion (TTE)								
	Pre-exercise	10 min	20 min	30 min	40 min	50 min	60 min	70 min	Exh
Rectal temperature (°C)									
CHO	37.7 ± 0.2	37.9 ± 0.2	38.4 ± 0.2^a	38.8 ± 0.2^a	39.0 ± 0.3^a	39.2 ± 0.3^a	39.3 ± 0.3^a	39.4 ± 0.3^a	39.5 ± 0.2^a
PLA	37.4 ± 0.2	37.9 ± 0.2	38.4 ± 0.3^a	38.7 ± 0.3^a	39.0 ± 0.2^a	39.2 ± 0.2^a	39.4 ± 0.2^a	39.5 ± 0.2^a	39.6 ± 0.2^a

Mean skin temperature (°C)									
CHO	29.3 ± 0.8	28.3 ± 0.6	28.1 ± 0.5	28.0 ± 0.4	27.6 ± 0.5[b]	27.3 ± 0.7[b]	26.8 ± 0.8[b]	26.7 ± 0.6[b]	26.6 ± 0.2[b]
PLA	29.3 ± 0.7	28.2 ± 0.7	28.0 ± 0.8	27.7 ± 0.8[b]	27.3 ± 0.8[b]	27.0 ± 0.9[b]	26.8 ± 1.0[b]	26.8 ± 0.5[b]	26.5 ± 0.3[b]

No significant difference between treatments was found.
[a]Significant difference from pre-exercise, $p < 0.05$.
[b]Significant difference from pre-exercise, 10 min and 20 min.

H. Blood collection

Venous blood was sampled via radial vein using a 22G IV catheter insertion (Surflo IV Catheter, Terumo Med Corporation, Eklton, MD., USA) and kept patent by flushing of heparinized saline solution (20 $U.mL^{-1}$) after every sample. A total of 4 mL of whole blood was collected at each sampling point, kept in tubes containing sodium fluoride/ potassium oxalate (BD Vacutainer, Lakes USA), centrifuged at 2500 rpm for 5 min. The plasma was stored at -80°C for later analysis. Plasma lactate was determined using the electro-enzymatic method (YSI 1500 Sport, Yellow Spring, Ohio, USA) while plasma glucose was analysed using UV spectrophotometer (Optima SP-3000 Plus, Tokyo, Japan) using commercially available kits (Randox, Daytona, UK). Blood samples were collected at pre-exercise and after every 15 min during the TTE.

I. Psychological scale

The Feeling Scale (FS) [18] was used to measure the effective dimension of pleasure-displeasure of the participants. The FS consists of a 11-point single item bipolar rating scale that ranges from -5 to +5. Anchors are marked at the "0" point (neutral), ranging from "very good" (+5) to "very bad" (-5) [19]. The participant's perceived active-tion/arousal (energised) level was measured using the perceived activation scale (FAS). FAS is a six-point, single item measure ranging from 1 to 6, with anchors at 1 (low arousal) and 6 (high arousal) [20]. Both FS and FAS scales have been used in previous studies [5, 13, 21] and are easily administered. A 12-point GI comfort scale was used to determine the participant's gastrointestinal comfort with anchors provided at 0 - "neutral", 4 - "uncomfortable", 8 - "very uncomfortable" and 12 - "painful."[5]. The FS, FAS and GI scales were administered during the pre-exercise period and at every 15 min interval during the TTE. The participant's perceived thirst was recorded pre and post exercise using 100 mm visual analogue scale (VAS) where the anchor point was labelled "Not Thirsty" at 0 mm and "Very Thirsty" at 100 mm.

J. Statistical analysis

Statistical analyses were performed using SPSS software package (version 22: SPSS, Chicago IL., USA). All data were reported as mean ± standard deviation (SD). The mean differences in performance (total distance covered and time to exhaustion) were detected using a paired-sample *t*-test. To identify differences in data collected throughout each trial, a repeated-measure factorial ANOVA (trial x time) was employed. Significant main effects for each trial were further analysed using a paired t-test and the Bonferroni adjustment for the number of pair-wise comparisons was used. Significance was accepted at $P < 0.05$.

III. Results

A. Time to exhaustion and total distance covered

The TTE for CHO and PLA trials were 81.2 ± 4.1 min and 76.8 ± 3.9 min, respectively. There was a significant difference in the time performance between CHO and PLA conditions ($F = 5.236$, $p = 0.000$; Figure 2). The runners covered greater distance during CHO (16.0 ± 2.0 km) than during PLA (15.1 ± 2.0 km) rinses. The total distance covered during the CHO condition was 5.6% significantly higher compared to that covered during the PLA condition ($F = 5.119$, $p = 0.001$; Figure 3).

B. Cardio-respiratory response

Figure 4 shows the VO_2 at 15 min time interval during the TTE exercise. There was no significant difference found between trials ($F = 0.527$, $p = 0.666$; Figure 4). Similarly, no significant differences for RER were observed between the CHO and PLA trials ($F = 0.656$, $p = 0.493$; Figure 5). Mean exercise HR increased progressively over time, how ever statistical analyses revealed no significant difference between trials ($F = 0.671$, $p = 0.496$).

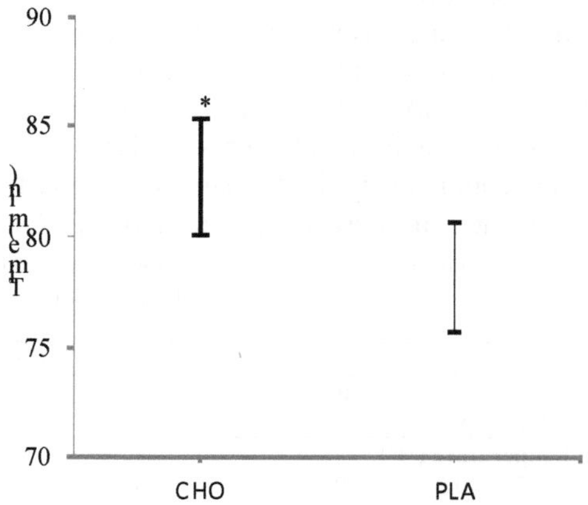

Fig 2 Mean TTE performance in the CHO and PLA trials. * Different from PLA, $p = 0.000$.

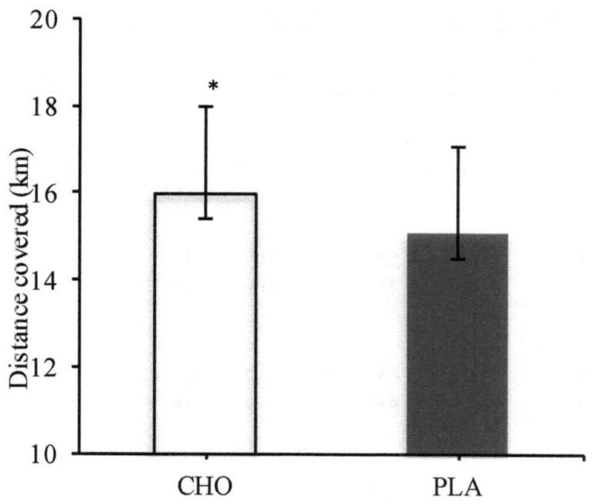

Fig 3 Mean distance covered in the CHO and PLA trials. * Different from PLA, $p = 0.000$.

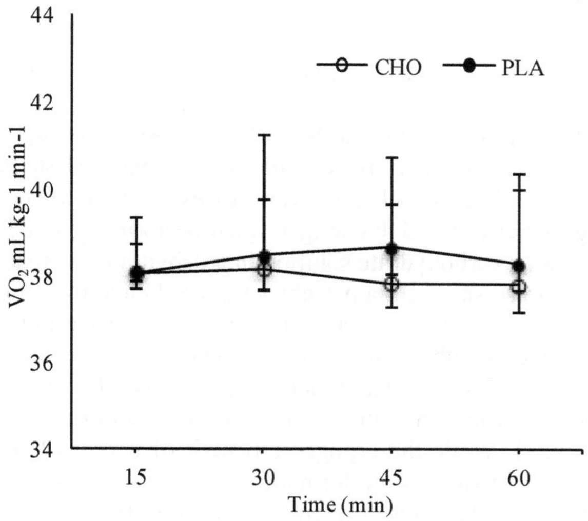

Fig 4 Mean (±SD) of VO_2 at pre-exercise and 15 min time interval of TTE.

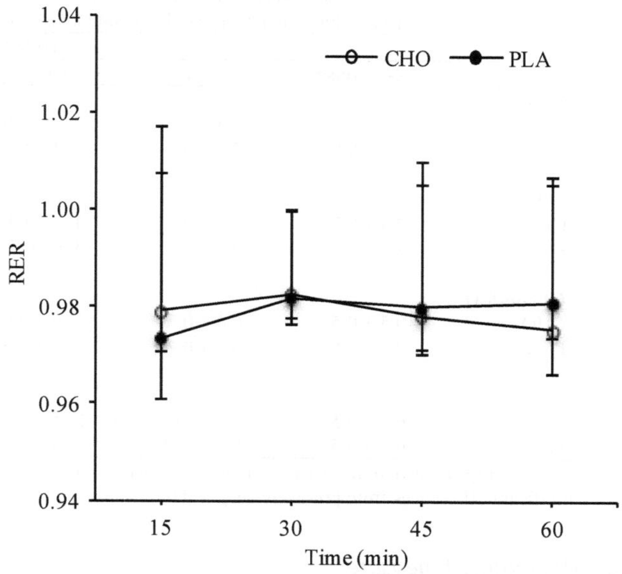

Fig 5 Mean (±SD) of RER at pre-exercise and 15 min time interval of TTE.

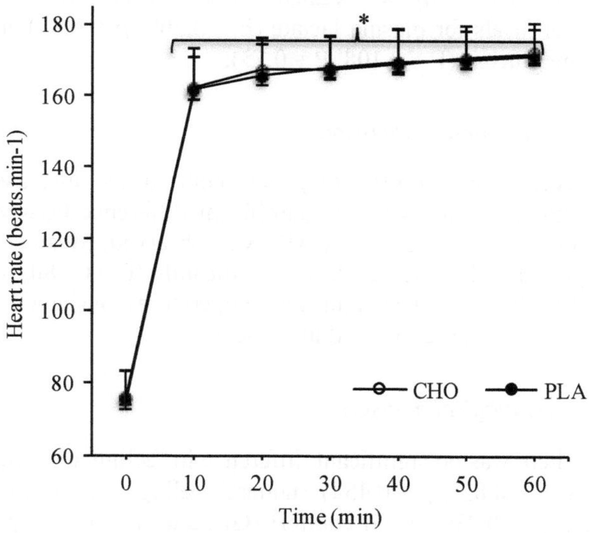

Fig 6 Mean (±SD) of heart rate at pre-exercise and 15 min time interval of TTE. * Significant difference from 0 min in both CHO and PLA trials

Table 3 Mean psychological scores for FAS, FS, GI comfort and RPE at pre-exercise and every 10 min time interval during TTE duration for carbohydrate mouth rinse (CHO) and placebo mouth rinse (PLA), (mean ± SD; $p < 0.05$). Exh, exhaustion.

	Time to exhaustion (TTE)						
	Pre-exercise	15 min	30 min	45 min	60 min	70 min (n=11)	Ext
FAS							
CHO	2.9 ± 0.7	3.3 ± 0.5	3.3 ± 0.6	3.3 ± 0.6	2.9 ± 1.3	2.9 ± 1.3	2.5 ± 0.6
PLA	2.7 ± 0.8	2.8 ± 0.7	3.1 ± 0.5	3.4 ± 0.9	3.0 ± 1.2	3.0 ± 1.2	2.7 ± 0.7
FS							
CHO	2.4 ± 0.7	2.1 ± 0.7	1.3 ± 0.9*	1.5 ± 1.1*	0.8 ± 1.5*	0.9 ± 0.7*	0.9 ± 1.1*
PLA	2.3 ± 0.9	2.3 ± 1.0	1.3 ± 0.5*	1.6 ± 0.7*	1.0 ± 0.7*	0.9 ± 0.9*	0.9 ± 1.2*
GI comfort							
CHO	0.5 ± 0.5	0.6 ± 0.7	1.0 ± 0.9	1.6 ± 0.5*	1.8 ± 0.6*	1.7 ± 0.6*	2.2 ± 1.2*
PLA	0.4 ± 0.5	0.7 ± 0.7	1.2 ± 0.6	1.6 ± 0.7*	1.5 ± 0.5*	1.8 ± 0.4*	2.3 ± 1.1*
RPE							
CHO	6.4 ± 0.5	9.9 ± 1.1*	12.8 ± 1.1*	15.4 ± 1.2*	16.3 ± 0.5*	17.1 ± 0.5*	17.2 ± 0.9*
PLA	6.3 ± 0.5	9.6 ± 1.3*	12.5 ± 1.2*	15.3 ± 1.0*	16.3 ± 0.9*	17.2 ± 0.7*	17.3 ± 0.8*

No significant difference between treatments was found.
*Significant differences from pre-exercise, $p < 0.05$.

C. Haematocrit measure

Plasma lactate and plasma glucose are shown in Table 2. Statistical analysis revealed no significant difference between trials for plasma lactate ($F = 0.309$, $p = 0.87$) and plasma glucose ($F = 0.108$, $P = 0.75$).

D. Thermoregulatory response

Rectal temperature (T_{re}) in both trials increased throughout the trials but no significant difference between treatments ($F = 0.629$, $p = 0.618$) was observed. Mean skin temperature (T_{sk}) did not differ significantly ($F = 0.580$, $p = 0.548$). The mean rectal and skin temperatures, during CHO and PLA trials, are reported in Table 2.

E. Psychological measure

There was no significant difference in feeling activation (FAS: $F = 0.827$, $p = 0.454$), runners' feeling scale (FS: $F = 0.152$, $p = 0.939$) or GI comfort (GI scale: $F = 0.567$, $p = 0.608$). Mean rating perceived exertion (RPE) increased throughout each test. There were no significant differences between trials ($F = 0.17$, $p = 0.921$). However, there was a significant effect of time on both the CHO and PLA trials ($p = 0.000$). Post-exercise thirst perceived ratings (VAS) were significantly higher for both CHO (9.1 ± 1.2) and PLA (8.9 ± 0.9) trials. The mean psychological scores for FAS, FS, GI discomfort and RPE values during TTE are shown in Table 3.

IV. Discussion

This study had examined the effect of mouth rinsing a CHO solution during exercise while in a dehydrated state. The main finding was that the time to exhaustion for the running exercise was 5.4% longer when performing mouth rinsing with a carbohydrate solution in a dehydrated state as compared to rinsing with a placebo solution. In addition, the total running distance was greater when performing carbohydrate mouth rinse. Our findings were consistent with previous studies that reported an improved running performance when performing carbohydrate mouth rinse [3, 5, 6, 22, 23]. While the ergogenic benefit of carbohydrate mouth rinse on running performance remains inconclusive, to our knowledge, the present study was the first to demonstrate the benefit of rinsing a carbohydrate solution on running performance when in a dehydrated state.

The improvement in endurance exercise performance was likely to be linked to the central effect mechanism as previously described by Chambers et al. [22] and Turner et al. [24]. Chambers et al. [22] previously demonstrated that cyclists were able to produce a higher work rate during exercise when rinsing a glucose (sweetener) solution as compared to rinsing a maltodextrin (non-sweetener). Chambers and colleagues [22] further demonstrated that mouth rinsing a glucose solution had activated the supraspinal pathways the brain that is related to motivation and reward during exercise. Another recent study by Turner et al. [24] had also reported activation of brain regions relating to motor control, visual perception and the processing of effective, taste stimuli following carbohydrate mouth rinsing. These findings provided evidence of the ergogenic benefit of mouth rinsing on endurance exercise performance.

In the current study, subjects were 2% dehydrated at pre-exercise and 4.1% - 4.3% at the time to exhaustion. The data from the visual analogue scale for thirst indicated that subjects were extremely thirsty at the end of exercise within both trials. An fMRI study conducted by de Araujo et la., [25] revealed that the level of brain activation is directly associated with the level of thirst among subjects. Brain activation were specifically detected at the anterior insula, medial orbitofrontal cortex and mid-insula (associated with primary taste cortex) which reflected the level of thirst or motivational state of the subjects [25]. Since participants in the present study performed in a dehydrated condition, the presence of CHO solution in the oral cavity and the thirst sensation could have triggered a greater cortical response that may have led to longer running time and distance covered.

To our knowledge, the study by Arnaoutis et al. [12] was the only one that had compared the effect of mouth rinsing to the endurance performance of dehydrated subjects. In this study, the subjects observed an overnight fast (~10 hr) and were dehydrated (2% of body mass) during the exercise sessions. They performed time to exhaustion cycling test at 75% of maximum power output and were given 25 mL of plain water to rinse before and every 5 min of the trial. There were significant exercise improvements in ingestion trial as compared to mouth rinse trial. The use of plain water as a mouth rinse solution may produce a minimal effect that may augment a central effect and possibly enhance performance. In contrast, the finding of the present study revealed that CHO mouth rinsing improved performance of dehydrated subjects as compared to that when rinsing with a PLA, suggesting carbohydrate benefits during exercise.

Although mouth rinsing a CHO solution enhanced endurance performance, there were no differences in psychological response, plasma glucose, plasma lactate, thermoregulatory and cardiovascular response. Rating perceived exertion (RPE) during both trials also seemed to indicate that participants performed with the same maximum effort. We suggested that the cardiovascular strain from the exercise intensity would override any potential ergogenic effects of a carbohydrate mouth rinse. These may the reason that any ergogenic benefit of carbohydrate mouth rinse did not translate into metabolic or psychological changes.

V. Conclusion

In conclusion, mouth rinsing with a CHO solution for 10 s, at every 15 min during exercise in a dehydrated state had improved the running time to exhaustion compared to PLA. We postulated that the improvement in running exercise was likely due to the central mechanism that activates brain areas related to reward and behavioural response. We further believed that the brain activation was greater while rinsing a CHO solution in a dehydrated state. Future research is needed to explore the effectiveness of CHO mouth rinse at a higher level of dehydration (>4%), which is commonly reported among top runners at the end of a marathon.

Acknowledgment

The authors would like to thank the participants for their time and effort. Thank you to all of the people who assisted in recording data during trials.

Conflict of Interest

The authors declare that they have no conflict of interest.

References

1. Carter J M, Jeukendrup A E, Jones D A (2004) The effect of carbohydrate mouth rinse on 1-h cycle time trial performance. Med Sci Sports Exerc 36(12): p. 2107-11
2. Che Muhamed A M, Mohamed N G, Aziz N et al. (2014) Mouth rinsing improves cycling endurance performance during Ramadan fasting in a hot humid environment. Appl Physiol Nutr Metab 39(4): p. 458-64
3. Fares E J, Kayser B (2011) Carbohydrate mouth rinse effects on exercise capacity in pre- and postprandial States. J Nutr Metab 2011: p. 385962
4. Lane S C, Bird S R, Burkde L M et al. (2013) Effect of a carbohydrate mouth rinse on simulated cycling time-trial performance commenced in a fed or fasted state. Appl Physiol Nutr Metab 38(2): p. 134-9
5. Rollo I, Williams C, Nevill M (2011) Influence of ingesting versus mouth rinsing a carbohydrate solution during a 1-h run. Med Sci Sports Exerc 43(3): p. 468-75
6. Fraga C, Velasques B, Koch A J (2015) Carbohydrate mouth rinse enhances time to exhaustion during treadmill exercise. Clin Physiol Funct Imaging
7. Sclafani A (2004) The sixth taste? Appetite, 2004. 43(1): p. 1-3
8. Ataide-Silva T, Ghiarone T, Bertuzzi R et al. (2016) CHO Mouth Rinse Ameliorates Neuromuscular Response with Lower Endogenous CHO Stores. Med Sci Sports Exerc
9. Sinclair J, Bottoms L, Flynn C et al. (2014)The effect of different durations of carbohydrate mouth rinse on cycling performance. Eur J Sport Sci 14(3): p. 259-64
10. Cramer M N, Thompson M W, Periard J D et al. (2015) Thermal and Cardiovascular Strain Mitigate the Potential

Benefit of Carbohydrate Mouth Rinse During Self-Paced Exercise in the Heat. Front Physiol 6: p. 354
11. Watson P, Nichols D, Cordery P (2014) Mouth rinsing with a carbohydrate solution does not influence cycle time trial performance in the heat. Appl Physiol Nutr Metab 39(9): p. 1064-9
12. Arnaoutis G, Kavouras S A, Christaki I et al. (2012) Water ingestion improves performance compared with mouth rinse in dehydrated subjects. Med Sci Sports Exerc 44(1): p. 175-9
13. Rollo I, Williams C (2010) Influence of ingesting a carbohydrate-electrolyte solution before and during a 1-hour run in fed endurance-trained runners. J Sports Sci 28(6): p. 593-601
14. Rollo I, Williams C, Gant N et al. (2008) The influence of carbohydrate mouth rinse on self-selected speeds during a 30-min treadmill run. Int J Sport Nutr Exerc Metab 18(6): p. 585-600
15. Whitham M, McKinney J (2007) Effect of a carbohydrate mouthwash on running time-trial performance. J Sports Sci, 25(12): p. 1385-92
16. Ramanathan N L (1964) A new weighting system for mean surface temperature of the human body. J Appl Physiol, 19: p. 531-3.
17. Borg G A (1982) Psychophysical bases of perceived exertion. Med Sci Sports Exerc 14(5): p. 377-81
18. Hardy C J, Rejeski W J (1989) Not what, but how one feels: The measurement of affect during exercise. J Sport Exerc Psychol 11(3): p. 304-317
19. Ekkekakis P, Backhouse S H, Gary et al. (2008) Walking is popular among adults but is it pleasant? A framework for clarifying the link between walking and affect as illustrated in two studies. Psychology of Sport and Exercise 9(3): p. 246-264.
20. Svebak S, Murgatroyd S (1985) Metamotivational dominance: a multimethod validation of reversal theory constructs. Journal of Personality and Social Psychology 48(1): p. 107
21. Backhouse S H, Biddle S J, William C (2005) Effect of carbohydrate and prolonged exercise on affect and perceived exertion. Med Sci Sports Exerc 37(10): p. 1768-73
22. Chambers E S, Bridge M W, Jones D A (2009) Carbohydrate sensing in the human mouth: effects on exercise performance and brain activity. J Physiol 587(Pt 8): p. 1779-94
23. Rollo I , Williams C (2010) Influence of mouth rinsing a carbohydrate solution on 1-h running performance. Med Sci Sports Exerc 42(4): p. 798-804
24. Turner C E Byblow E, Stiner C M et al. (2014) Carbohydrate in the mouth enhances activation of brain circuitry involved in motor performance and sensory perception. Appetite 80: p. 212-9
25. de Araujo I E, Kringelbach M L, Rolls E T et al. (2003) Human cortical responses to water in the mouth, and the effects of thirst. J Neurophysiol 90(3): p. 1865-76

Author: Ahmad Munir Che Muhamed
Institute: Lifestyle Sains Cluster, Advance Medical and Dental Institute, Universiti Sains Malaysia
City: Kepala Batas, Penang
Country: Malaysia
Email: ahmadmunir@usm.my

The Effects of Releasing Ankle Joint on Pedal Force and Power Production during Electrically Stimulated Cycling in Paraplegic Individuals: A Pilot Study

P.N.F. Hamdan[1], K. Teo[1], N.A. Hamzaid[1], J. Usman[1] and R. Razman[2]

[1] Department of Biomedical Engineering, Faculty of Engineering, University of Malaya, Kuala Lumpur, Malaysia
[2] Sports Centre, University of Malaya, Kuala Lumpur, Malaysia

Abstract— **Previous research has investigated functional electrical stimulation (FES) cycle force and power output (PO) from the perspective of knee and hip joint biomechanics. However, ankle-foot biomechanics and, in particular, the effect of releasing the ankle joint on cycle pedal force and PO during FES cycling in paraplegics has not been widely explored. Therefore, the purpose of this study is to determine whether releasing the ankle joint might influence the peak pedal force and PO during FES cycling in paraplegics. Three complete paraplegics (C7 – T4) participated in this study. All participants performed two sessions of cycling in randomized order. Session 1 and 2 required the participants to cycle in fixed and free-ankle setup, respectively. For each session, the participants performed two sub-sessions of FES cycling. During sub-session 1, the muscles stimulated were upper leg muscles [quadriceps (QUAD) and hamstrings (HAM)]. In sub-session 2, both upper and lower leg muscles [QUAD, HAM, tibialis anterior (TA) and triceps surae (TS)] were stimulated. The normalized peak pedal force and PO of each condition were analyzed. Overall, the normalized peak pedal force and PO during fixed-ankle FES cycling is higher than free-ankle FES cycling. Stimulation of both upper and lower leg muscles during FES cycling provided higher normalized peak pedal force and PO compared to the upper leg muscles stimulated alone. The present pilot study revealed that fixed-ankle FES cycling produced higher normalized peak pedal force and PO than free-ankle FES cycling. Future work involving more paraplegics will be investigated. This finding might serve as a reference for future rehabilitative cycling protocols.**

Keywords— **Functional Electrical Stimulation; Spinal Cord Injury; Cycling; Ankle Movement; Rehabilitation Exercise.**

I. Introduction

Cycling is a popular exercise modality for individuals with spinal cord injury (SCI). The general goal of cycling exercise is to produce the highest possible mechanical power to maximize the merit of health benefits [1]. In SCI populations, such cycling exercise is artificially evoked by functional electrical stimulation (FES), whereby leg muscles are recruited by electrical pulses delivered on the skin surface overlying key muscles [2, 3]. It has been proven to provide benefits including improved muscle strength, endurance, mechanical power output (PO), skin condition, cardiopulmonary fitness, reversal of muscle wasting, blood flow in the legs, reduced incidence of muscle spasms, body composition, bone mass, quality of life, joint health and flexibility, and offsetting some of the secondary complication [2, 4]. However, how the foot is affixed to the pedal has been of interest. A fixed ankle foot orthosis (AFO) or fixed pedal boot is often deployed to affix the foot to the pedal and this has been widely used to also provide shank stability; thus restricting the leg movements in the sagittal plane during cycling. In the standard setup for FES cycling, the ankle joint is immobilised using an orthosis, and stimulation is applied to quadriceps femoris (QUAD), gluteus maximus (GLU), and hamstrings (HAM) using surface electrodes [2, 3].

Researchers have previously sought to elicit maximum PO during FES cycling in order to increase the benefits of cycling during rehabilitation. Berkelmans [5], Sinclair *et al.* [6], Szecsi *et al.* [7], and Duffel *et al.* [8] have reported that the magnitude of mechanical PO produced during FES cycling in individuals with SCI is very low compare to the PO produced during voluntary cycling in able-bodied (AB). The reasons of the low PO magnitude [7] might be due to the inefficiency of artificial muscle activation, the crude control of muscle groups accomplished by stimulation, and muscle atrophy and transformation due to chronic paralysis and disuse. Consequently, several studies have investigated the origins of cycling PO during FES exercise [3, 9].

Ankle positioning during cycling is one of the more important factors for effective pedaling [10, 11], yet this has not received much previous research attention. Theoretically, the PO can be improved by releasing the ankle joint and adding triceps surae (TS) and tibialis anterior (TA) muscles evoked by neurostimulation [4]. Stimulation of the TS and TA has been investigated before in fixed-ankle FES cycling and no remarkable effect on PO was noted, except that it affected only on the cardiovascular and circulatory responses [12]. The stimulation of the TA and TS in a free-ankle setup shows 14% greater PO than the fixed-ankle FES cycling only with the tuning of contact point between the foot and pedal to the relative strength of the ankle plantar flexors [4]. However, Ferrante *et al.* [13] reported that the calf muscle generates limited knee flexion action due to the presence of orthosis that fixed the ankle angle, which may reduce the maximum PO. In another study, Fornusek *et al.* [14] reported that the free-ankle FES

F. Ibrahim et al. (eds.), *3rd International Conference on Movement, Health and Exercise*,
IFMBE Proceedings 58,
DOI: 10.1007/978-981-10-3737-5_26

cycling with the stimulation of the shank muscles (TS and TA) was found safe and increased the ankle excursions that might have improved joint mobility and prevent contractures in persons with paralysis. Taken together, these studies have further shown the importance of investigating maximum PO as a function of ankle movements during FES cycling in paraplegics.

In contrast, a limited number of studies have investigated AFO-constrained ankle movements on the power production during FES cycling in paraplegics. It is an important concern in the rehabilitation systems to elicit maximum pedal force and PO during FES cycling. Therefore, the purpose of this study was to investigate whether a fixed and free-ankle movement might influence cycle peak pedal force and PO during FES cycling in paraplegics. We hypothesize that free-ankle FES cycling might alter the production of peak pedal force and PO, as the biomechanics are affected by the ankle patterns [15].

II. Materials and Methods

A. Participants

Three complete paraplegics (C6 – T4), two males (38.5 ± 14.8 y and 71.0 ± 12.2 kg) and one female (47 y and 82 kg) participated in this study. All participants provided their written informed consent before taking part in the study. The participants had no previous or ongoing record of neuromuscular, musculoskeletal, rheumatological, cardiovascular disorder or orthopaedic lower limb injuries. All the participants were trained with FES cycling for at least 12 weeks. This study was approved by the local Medical Ethics Committee, University of Malaya Medical Centre, University Malaya, Kuala Lumpur, Malaysia (Ref No.: 1003.14(1)).

B. Experimental setup

A FES cycle ergometer (MOTOmed viva2) was utilised in this study. Self-adhesive gel electrodes were placed over the belly of QUAD, HAM, TA, and TS muscle groups. An in-shoe F-scan system (Tekscan Incorporated, Boston, Massachusetts) was placed under the foot of the participants and connected to a "cuff-unit" that linked the foot sensors to a computer via a 10-m cable. Tight socks were applied on the foot to prevent displacement of foot sensor during cycling. For the fixed-ankle FES cycling, the lower legs of each participant were placed on fixed position (FP) AFO that was fixed to the pedal to restrict the ankle joint movement. The seat position from the crank axle was adjusted and recorded for each participant so that the knee extension did not exceed 150-160° at the bottom dead centre (BDC). Motion Capture System (Qualisys) was used to capture the marker placed at the hip, knee, ankle, fifth metatarsophalangeal joints, crank axle and pedal.

C. Data collection protocol

Testing was conducted in two sessions with two sub-sessions for each session. The first session required the participants to perform FES cycling in fixed-ankle setup with FP AFO and the second session required the participants to perform FES cycling with free-ankle setup. Two modes of cycling were performed for each session; passive cycling (without FES induced leg cycling) and FES cycling (FES induced leg cycling). Sub-session 1 consisted of 1 min passive warm up, 2 min FES cycling with QUAD and HAM stimulated, 1 min cool down, and 10 min of resting phase. Sub-session 2 consisted of 1 min passive warm up, 2 min FES cycling with QUAD, HAM, TA and TS stimulated, 1 min cool down, and 10 min resting phase. The order of each session for each participant was randomized. Each session was separated by at least 48 hours. The participants performed cycling at 50 rpm. The stimulation (300 μs pulse width and 30 Hz frequency) was applied by an 8-channels stimulator (RehaStim ScienceMode, HASOMED GmbH, German).

D. Data processing and analysis

The kinetic and kinematic data for each session was recorded in real time at 120Hz by the software [Tekscan Incorporated, Boston, Massachusetts and Motion Capture System (Qualisys)] to store the data into a PC for offline analysis. Only the last 20 s kinetic and kinematic data of each cycling mode for each session was recorded. The peak normalized pedal force and PO during fixed and free-ankle FES cycling were compared and analyzed.

III. Results

A. Pedal force

Fig. 1 showed the normalized peak pedal force during fixed and free-ankle FES cycling. Fixed-ankle FES cycling with both upper and lower leg muscles stimulated showed the highest normalized peak pedal force (87.5 ± 15.1%). Free-ankle FES cycling with upper leg muscles stimulated alone showed the lowest normalized peak pedal force (58.4 ± 11.5%).

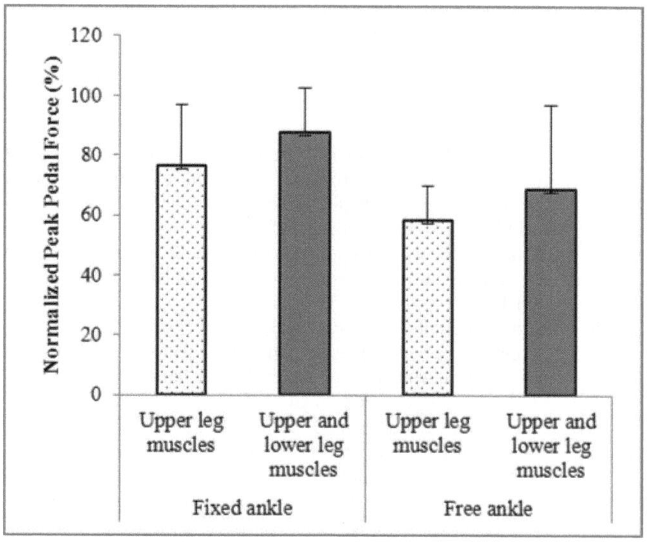

Fig. 1 Normalized peak pedal force (%)

B. Pedal PO

Fig. 2 showed the normalized peak pedal PO during fixed and free-ankle FES cycling. Fixed-ankle FES cycling with both upper and lower leg muscles stimulated showed the highest normalized peak pedal PO (89.9 ± 14.4%). Free-ankle FES cycling with upper leg stimulated alone showed lowest normalized peak pedal PO (54.2 ± 4.3%).

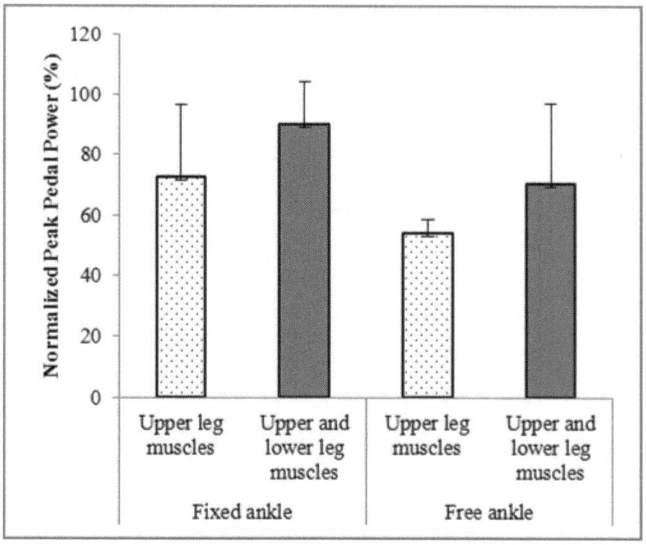

Fig. 2 Normalized peak pedal PO (%)

IV. Discussion

The present study sought to investigate possible differences in normalized peak pedal force and PO generated during fixed and free-ankle FES cycling. To our knowledge, no studies have yet to investigate the effect of releasing ankle joint on the peak pedal force and PO during FES cycling in paraplegics.

A. Pedal force

Pedal force produced in this study was originated from the stimulated muscles during FES cycling. The normalized peak pedal force revealed in the current study was opposed our initial hypothesis. Fixed-ankle FES cycling showed greater normalized peak pedal force compared to the free-ankle FES cycling, either with and without the stimulation of lower leg muscles. This might be due to the similar stimulation angle used for both the fixed and-free ankle FES cycling in this study. Previous research had shown that the crank angles at which the muscle groups are stimulated were altered by the addition of the lower leg muscles stimulation. Therefore, higher normalized peak pedal force could be produced during free-ankle FES cycling if the stimulation angle was adjusted until optimal performance was achieved. It is important to achieve the highest possible pedal force to maximize the merit of health benefits in paraplegics [1].

B. Pedal PO

The present study also reported that the normalized peak pedal PO of fixed-ankle FES cycling with the stimulation of both upper and lower leg muscles was higher than free-ankle FES cycling. The results refuted our initial hypothesis. Ideally, greater PO was generated from the stimulation of both upper and lower leg muscles. In this study, lower leg muscles was stimulated to allow the ankle joint to move in dorsi- and plantarflexion during free-ankle FES cycling. One of the reasons that the current study revealed a lower normalized peak pedal PO during free-ankle FES cycling was due to the power loss at the ankle joint as the participants often experienced muscle spasms during free-ankle FES cycling. Another reason was the drop in cycling cadence observed during free-ankle FES cycling. Changes in cycling cadence could affect the pedal PO production; low cadence produced low pedal PO production [6]. The present study observed that the free-ankle FES cycling produced non-smooth pedaling to the participants. This non-smooth pedaling would affect the cycling cadence which was highly effected the pedal PO production. Therefore, the best stimulation angle was important in order to produce smooth pedaling and thus, producing maximum pedal PO during free-ankle FES cycling.

V. Conclusions

The pedal force and PO found in this pilot study were higher at fixed-ankle FES cycling compared to the free-ankle FES cycling.

Acknowledgment

This research was fully supported by the Ministry of Higher Education, Malaysia and University of Malaya through High Impact Research (HIR), under grant no: UM.C/625/1/HIR/MOHE/ENG/39.

Conflict of Interest

The authors declare that they have no conflict of interest.

References

1. Ragnarsson KT, Pollack S, O'Daniel W et al. (1988) Clinical evaluation of computerized functional electrical stimulation after spinal cord injury: A multicenter pilot study. Arch Phys Med Rehabil 69(9):672-677
2. Bakkum AJT, Groot S De, Woude LHV Van Der et al (2012) The effects of hybrid cycle training in inactive people with long-term spinal cord injury: Design of a multicenter randomized controlled trial. Disabil Rehabil 1-6 DOI 10.3109/09638288.2012.715719
3. Hunt KJ, Fang J, Saengsuwan J et al (2012) On the efficiency of FES cycling: A framework and systematic review. Technol Health Care 20:395-422 DOI 10.3233/THC-2012-0689
4. Van Soest a. J, Gföhler M, Casius LJR (2005) Consequences of ankle joint fixation on FES cycling power output: A simulation study. Med Sci Sports Exerc 37:797-806 DOI 10.1249/01.MSS.0000161802.52243.95
5. Berkelmans R (2008) FES Cycling. J Autom Cont 18(2):73-76 DOI 10.2298/JAC0802073B
6. Szecsi J, Straube A, Fornusek C (2014) A biomechanical cause of low power production during FES cycling of subjects with SCI. J Neuroeng Rehabil 11:123 DOI 10.1186/1743-0003-11-123
7. Duffell LD, Donaldson NDN, Newham DJ (2009) Power output during functional electrically stimulated cycling in trained spinal cord injured people. Neuromodulation 13(1):50-57 DOI 10.1111/j.1525-1403.2009.00245.x
8. Martin JC, Brown NAT (2009) Joint-specific power production and fatigue during maximal cycling. J Biomech 42:474-479 DOI 10.1016/j.jbiomech.2008.11.015
9. Gregor SM, Perell KL, Rushatakankovit S et al (2002) Lower extremity general muscle moment patterns in healthy individuals during recumbent cycling. Clin Biomech 17:123-129 DOI 10.1016/S0268-0033(01)00112-7
10. Pierson-carey CD, Brown DA, Dairaghi CA. (1997) Changes in resultant pedal reaction forces due to ankle immobilization during pedaling. J Appl Biomech 13:334-346
11. Trumbower RD, Faghri PD (2004) Improving pedal power during semireclined leg cycling. IEEE Eng Med Biol Mag 23:62-71 DOI 10.1109/MEMB.2004.1310977
12. Ferrante S, Saunders B, Duffell L, et al (2005) Quantitative evaluation of stimulation patterns for FES cycling, In: 10th Annual Conference of the International FES Society, Montreal, Canada, 2005, pp 2–4
13. Fornusek C, Davis GM, Baek I (2012) Stimulation of shank muscles during functional electrical stimulation cycling increases ankle excursion in individuals with spinal cord injury. Arch Phys Med Rehabil 93:1930-1936 DOI 10.1016/j.apmr.2012.05.012
14. Trumbower RD, Faghri PD (2005) Kinematic analyses of semireclined leg cycling in able-bodied and spinal cord injured individuals. Spinal Cord 43:543-549 DOI 10.1038/sj.sc.3101756
15. Gregor RJ, Broker JP, Ryan MM. (1991) The biomechanics of cycling. Exerc Sport Sci R 19:127-169

Author: Nur Azah Hamzaid
Institute: University of Malaya
Street: Kuala Lumpur
City: Kuala Lumpur
Country: Malaysia
Email: azah.hamzaid@um.edu.my

Effect of High and Low Carbohydrate Meals on Sustained Maximum Voluntary Contraction (MVC) after Prolonged Exercise

T.K. Khong, V.S. Selvanayagam, S.H. Hamzah, P.J. Lim, and A. Yusof

Sports Centre, University of Malaya, Kuala Lumpur, Malaysia

Abstract— **Fatigue is a natural physiological phenomenon where the body's capability to do work is reduced, and in general carbohydrate (CHO) depletion has been attributed to the aetiology of this condition. However, its composition in iso-caloric meals has never been investigated on capacity of sustained force production following a prolonged exercise. Aim: To investigate the effects of high and low CHO meals on sustained maximal voluntary contraction (MVC) capability after a 90-minute run. Methods: Ten (n=10) moderately trained runners (age: 25±3.7 years, and VO2max: 51.42±4.78 ml/min/kg) were prescribed, in a cross-over, randomised, and double blind design, either one of these iso-caloric meals: high CHO meal (1.5 g/kg body weight), or a low CHO meal (0.8 g/kg body weight) prior to the 90-minute run at 65% of VO2max. A 90-second sustained MVC was measured before and after the run (represented at 30, 60, and 90 seconds respectively). Results: MVC dropped significantly (*p*<0.01) after the prolonged running exercise in both groups. The difference in the sustained MVC was significant between the two groups at all time points (*p*<0.01). Conclusion: While both groups completed the 90-minute running task successfully, the high CHO meal allowed a higher sustained force production (MVC) post-exercise, suggesting physiological changes that allows better neuromuscular functions. Theoretically, several factors could be attributed to this phenomenon, such as preservation of fuel source, and/or alteration of brain neurotransmitter concentrations that affect neural drive.**

Keywords— **Muscular Force, Carbohydrate, Sports Nutrition**

I. Introduction

Fatigue would limit sports performance as athletes will not be able to perform at their optimal capability in this condition. This is especially true in prolonged exercise, as this kind of work requires the body to continuously produce muscle force for a long period. Maximum voluntary contraction (MVC) is a common measure in assessing fatigue as shown by the drop in force.

zA widely used acute, pre-competition strategy to avoid fatigue is nutritional aids which are aimed at providing the body with enough energy sources to produce the desired workload. Carbohydrate (CHO) supplementations is vital in preventing hypoglycaemia which generally affects metabolism processes of the muscular system by allowing oxidation of CHO from sources other than muscle glycogen [1]. Blood glucose is also believed to be the main energy source for the central nervous system (CNS), hence a constant supply is required to ensure the CNS functions are not compromised [2]. There is an evidence showing that hypoglycemia during prolonged cycling exercise is associated with the failure to generate the desired force when comparing to a group that is fed CHO drink against placebo [2]. Unfortunately, the study did not compare iso-caloric CHO meals (between high and low CHO), as high fat diet could also influence metabolism during exercise [3]. Controversially, different composition of meals affects substrate utilisation but not prolonged exercise performance [3]. However, little is known on its effect towards sustained force production.

Hence, the aim of this study was to investigate the effect of high and low CHO meals on sustained MVC following a 90-minute running exercise.

II. Methods

A. Participants

Ten (n=10) active male runners volunteered to participate in this study. Subjects were included based on their maximal oxygen consumption level (VO2max), only those who has at least a minimal VO2max of 50 ml/kg/min were selected for this study. Participants' VO2max level were measured on a treadmill using a Cardiopulmonary Exercise Test (CPET) breath-by-breath gas analysis device (COSMED, Rome) before they began participating in familiarisation and intervention tests. Informed consent was obtained prior to beginning of the experimental period.

B. Study Design

Subjects who passed the inclusion test first attended familiarisation session on a week after, where they were familiarised with the flow of the study, including exercise testing procedures and sustained maximal voluntary contraction (MVC) exercise. The subjects also recorded their daily habitual dietary intake and physical activity 3 days prior to the experimental day, and were instructed to eat the same amount of food 3 days prior to the next

F. Ibrahim et al. (eds.), *3rd International Conference on Movement, Health and Exercise*,
IFMBE Proceedings 58,
DOI: 10.1007/978-981-10-3737-5_27

session. Participants underwent two experimental sessions, separated by a week, and after an overnight fast for at least 12 hours. Participants arrived at the lab at the same time on experimental days to avoid the influence of changes in circadian rhythm. Each participant consumed one of two designated iso-caloric meals prescribed in a randomised, double-blind fashion. The meals were: i) high carbohydrate meal (HCHO) consisting of 1.5 g/kg body weight CHO; and ii) low carbohydrate meal (LCHO) consisting of 0.8 g/kg body weight CHO. Protein percentage and calorie were consistent for both meals, and are adjusted to each participants' body weight. Details of test meal is illustrated in Table 1. Throughout this study, subjects were prohibited from consuming alcohol, caffeine or performing any strenuous physical activity prior to exercise testing.

Upon arrival at the lab, participants underwent sustained MVC measurements. They were then provided with the designated meal which they finished consuming within 20 minutes, and additional 1 hour to digest the food. Participants were seated during this one hour with minimal physical activity. They then began to run on a treadmill at a speed equivalent to 65% of their VO2max for 90 minutes. Upon completion of the run, subjects immediately underwent sustained MVC measurements.

C. Designated meals

Table 1. Details of test meal for a 70kg participant.

	HCHO Meal	LCHO Meal
Food	bread, fruit jam, apple, butter, hardboiled egg	bread, butter hardboiled egg,
Macronutrients	618 kcal, 105g CHO, 18g fat, 9g protein	615 kcal, 57g CHO, 39g fat, 9g protein
% of calories from CHO	~68	~37
% of calories from fat	~26	~57
% of calories from protein	~6	~6

D. Sustained MVC measurements

The dominant leg of participant was measured using a dynamometer (CSMi Solutions, USA), where the body was placed in a prone position and the knee joint was placed at an angle of 90o. Maximum isometric plantarflexion was performed for 90 seconds. During the test participants were given visual feedback of elicited force on a monitor, and were verbally encouraged to maintain a maximal effort at all time. Measurements at 30, 60 and 90-sec time points were used for analysis.

E. Statistics

The results were analysed using the Graphpad Prism statistical software. 2-way ANOVA was used to calculate the differences between the time points. A P value of < 0.01 was considered significant. The data is presented in Figure 1.

III. Results

Figure 1 shows the force difference between pre-test and post-test sustained MVC at 3 time points. In the HCHO group, the drop-in force produced was significantly ($P<0.01$) less than LCHO group across all time point.

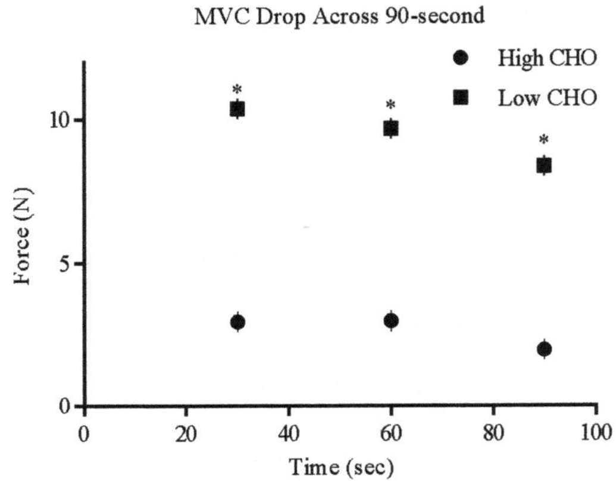

Figure 1: Drop in Sustained MVC after 90-min running between groups. * indicates significant difference ($P<0.01$) between High CHO and Low CHO.

IV. Discussion

Present study is the first to demonstrate that HCHO meal before a 90-minute run is able to preserve post-exercise force production capability better than LCHO meal, despite similar amount of calorie ingested. This indicates that relatively HCHO meal could have induced physiological changes that allows better function to attenuate the loss of muscle force production. Metabolic benefits of CHO to the skeletal muscle are well documented, which include increases in insulin level and subsequently CHO oxidation [4]; as well as enhancing CHO availability and sparing muscle glycogen stores to maintain exercise performance before fatigue kicks in[1]. It should be noted however, a high fat (low CHO) meal have also been documented to improve fat oxidation which could also theoretically preserve muscle glycogen, delaying fatigue [5]. While the effect of CHO and fat compositions on substrate utilisation during prolonged exercise is well documented, they don't seem to be the major limiting factor influencing endurance performance [3]. In relation to force production, however, CHO availability might have been a bigger limiting factor as exogenous CHO could theoretically increase stimulation of glycolytic pathway responsible in force production [6].

Prolonged exercise has also been shown to cause the brain to super-compensate its glycogen availability, indicating the brain's higher demand for CHO[7]. Increased CHO supply to the brain could have preserved force production capability by increasing insulin level, which in turn blunts plasma FFA. High amount of FFA is believed to increase the concentration of free-Tryptophan, a precursor to the serotonin neurotransmitter which causes the sense of fatigue [8].

V. Conclusion

While both groups successfully completed the 90-minute running over two separate occasions, the HCHO meal helped to preserve loss of muscle force production post-exercise, suggesting CHO's beneficial effect on neuromuscular functions. Since changes in muscular force could be attributed to various factors of both central and peripheral sources, future studies combining more parameters including blood and neurophysiological measures with nutritional interventions are vital. This information could provide insights to effects of CHO on fatigue and as strategy to enhance force production.

Conflict of Interest

The authors declare that they have no conflict of interest.

References

1. Coyle EF, Coggan AR, Hemmert MK, Ivy JL (1986) Muscle glycogen utilization during prolonged strenuous exercise when fed carbohydrate. J Appl Physiol 61(1): 165-172.
2. Nybo L (2003) CNS fatigue and prolonged exercise: effect of glucose supplementation. Med Sci Sport Exer 35(4):589-594.
3. Whitley HA, Humphreys SM, Campbell IT et al. (1998) Metabolic and performance responses during endurance exercise after high-fat and high-carbohydrate meals. J of Appl Physiol 85(2): 418-424.
4. Coyle EF, Coggan AR, Hemmert MK et al. (1985) Substrate usage during prolonged exercise following a preexercise meal. J Appl Physiol 59(2): 429-433.
5. Vukovich MD, Costill DL, Hickey MS et al. (1993) Effect of fat emulsion infusion and fat feeding on muscle glycogen utilization during cycle exercise. J Appl Physiol 75(4): 1513-1518.
6. Wax B, Brown SP, Webb HE et al. (2012) Effects of carbohydrate supplementation on force output and time to exhaustion during static leg contractions superimposed with electromyostimulation. J Strength Cond Res 26(6): 1717-1723.
7. Matsui T, Soya H (2013) Brain glycogen decrease and supercompensation with prolonged exhaustive exercise. In Soc Neurosci Public Healt, Springer: pp. 253-264.
8. Meeusen R, Watson P, Hasegawa H et al. (2006) Central fatigue: the serotonin hypothesis and beyond. Sport Med 36:881-909.

Author: Khong Teng Keen
Institute: Sports Centre, University of Malaya
Street: Lembah Pantai
City: 50603 Kuala Lumpur
Country: Malaysia
Email: edwinkhong89@siswa.um.edu.my

Different Instructional Methods Affect the Acquisition and Performance of Sport Stacking

I. N. Mohd Noor and J. P. G. Cheong

Exercise Science Unit, Sports Centre, University of Malaya, Kuala Lumpur, Malaysia

Abstract— **Sport stacking has gained increasing attention from all walks of life, including students, parents, teachers, physical educators, coaches and other practitioners with the emergence of evidence that it is able to facilitate cognitive, physical or perceptual motor skills. However, little is known about how to improve the acquisition and learning of the sport stacking skill. As such, the purpose of this study was to determine the best type of instructional methods (demonstration only and a combination of demonstration with verbal instructions) on the performance of sport stacking in novices. Thirty participants were recruited to participate in this study and were randomly assigned into the demonstration group (DG) or combination of demonstration and verbal instructions group (DVIG). Participants were shown a short video on how to perform sport stacking, whereby DG watched the demonstration with a muted sound, while DVIG watched the video with sound. Following that, participants were given the chance to practice for 15 minutes before completing three trials of the sport stacking cycle whereby total time (seconds) was recorded. The independent sample *t*-test was used to compare the means of DG with DVIG. The findings indicated that there were no differences between the two experimental groups ($p = 0.681$). It appeared that presenting information by way of demonstration or a combination of demonstration and verbal instructions were two possible ways to deliver information concerning how to perform sport stacking for novices. Discussions are presented from the viewpoint of demonstrations providing sufficient and relevant information as well as from the standpoint of verbal instructions being redundant for the learning of sport stacking skills.**

Keywords— **Demonstration, Observational practice, Verbal instructions, Cup stacking, Speed stacking**

I. INTRODUCTION

Role of instructional methods has been repeatedly studied in physical education and sport studies journals, either by use of visual demonstration or verbal instruction. Hodges (2001) stated that both approaches can function as a pre-practice that help to convey a task goal to the learner on how to perform motor skills effectively and may help in selection of strategies that are presumed to be useful to acquisition. Substantial amounts of research have shown that either initial demonstrations or instructions can help in the performance of a novel motor skill and promote learning of a motor task in inexperienced individuals [2, 3].

Nevertheless, the effectiveness of instructional methods usually depends on the type of skills or tasks that are taught and are also related to the existing skills of the learner. Visual demonstration and verbal instructions which supposedly specify the correct or optimal method for executing a skill, might have a minimal learning benefit, when the response of movement is not an existing part of the movement repertoire of learner [4]. Furthermore, tasks become particularly difficult when it involves a skill that is relatively complex and when learners have insufficient intrinsic feedback to enable them to judge how effectively their actions match those of the instructor. In fact, most of the theories regarding instructional methods were only limited to the learning of simple motor skills, rather than complex motor skills [1].

One example of a complex motor skill is a bimanual coordination skill. Bimanual skills involve moving two limbs at the same time, either symmetrically or asymmetrically [5]. There are many sports that involve bimanual coordination skills. Rowing and performing the bench press are examples of symmetrical bimanual skills while serving in volleyball is an example of asymmetrical bimanual task. Another example of an asymmetrical bimanual skill in sports is sport stacking. This sport requires a person to up stack and down stack 12 specially designed cups in a predefined sequence within a fastest time [6]. Sport or cup stacking has become the subject of investigation and is slowly gaining increasing attention in research. So far, six studies have reported the benefits of participation in sport stacking in real world setting, such as improved reading achievement [7, 8], hand-eye coordination [9], limb coordination [1], reaction time [9,11], auditory and visual attention [12], as well as matching in energy expenditure with other physical education activities [13]. Speed Stacks, Inc. [5] had also discovered the efficacy of sport stacking in the acquisition and enhancement of basic motor skills.

However, most previous studies that involve sport stacking has been carried out to measure the effectiveness and influence of sport stacking interventions on cognitive, physical or perceptual motor variables. Attempts to identify motor learning variables that facilitated the improvement of cup stacking performance was ineffective with only one study investigating the effects of practice conditions on cup

F. Ibrahim et al. (eds.), *3rd International Conference on Movement, Health and Exercise*,
IFMBE Proceedings 58,
DOI: 10.1007/978-981-10-3737-5_28

stacking performance. The authors of this study manipulated the amount and distribution of practice and found that undergraduate students achieved faster stacking times when practicing according to a distributed practice schedule [14]. There were no studies that investigated the role of presenting information from a motor learning perspective. As such, the purpose of this study was to identify instructional methods which were most effective and suitable for learning the skills of sport stacking by novices. More specifically, this study aimed to make comparisons between demonstration and a combination of demonstration and verbal instructions on the performance of sport stacking.

II. METHODS

A. Participants

A convenience sample of thirty undergraduate students (n=30) that consisted of ten students each from the first, second and third year of studies, respectively, were recruited to participate in this study (mean age = 22.2 ± 1.6 years). None of the participants had any prior experience in sport stacking. All participants were treated in accordance with the ethical standards of the responsible committee on human experimentation (institutional and national) and with the Helsinki Declaration.

B. Experimental groups

Participants were randomly assigned (using a random number generator) into the Demonstration group (DG) or the Demonstration and Verbal Instruction group (DVIG). A short video on how to play sport stacking in a correct and easy way was shown to both experimental groups during the learning phase of the study. The only thing that differentiated these two groups was the sound of the video. DG learned sport stacking by only observing the movement demonstrated by the person in the video while the voice of the demonstrator was muted. Meanwhile, DVIG learned sport stacking by observing the movement demonstration, together with verbal instructions by the person in the video.

C. Instrumentation and Measurements

A seven and a half minutes video, which comprised of sport stacking lessons obtained from the Speed Stack website, was used. A set of cups (Speed Stacks, Inc.) that consisted of 12 cups made specifically for sport stacking was used during the practice and testing phases of this study. A stop watch was used during the testing phase to record the total time (in seconds) required to complete the stacking task and the time performance to complete a cycle stack.

D. Procedures

Participants were required to attend one session that took approximately 40 minutes. This session included the learning, practice and testing phases. At the beginning of the session, a participant information sheet was distributed and an informed consent was obtained. Subsequently, participants were briefed about the procedures of the study before being randomly assigned into one of the two experimental groups (DG or DVIG).

As all participants had no prior experience in sport stacking, it was not possible to conduct a pre-test because the participants did not have any clue how to perform the cup stack. As such, each group immediately underwent 15 minutes of the learning phase of sport stacking whereby the video containing the demonstration, with or without verbal instructions, were shown to the respective groups. The video consisted of information on how to up stack and down stack the cups, including the 3-6-3 stack, 6-6 stack and the 1-10-1 stack. After the learning phase, the participants were given the opportunity to practice sport stacking for 15 minutes. Feedback on whether the cup stacking sequences were correct or incorrect was given to all participants in this practice phase. Finally, the participants were tested on a complete cycle of sport stacking. They were given three attempts to perform sport stacking and the fastest time among the three trials was taken.

E. Statistical Analyses

All statistical analyses were performed using the Statistical Package for Social Sciences (SPSS for Windows, version 22.0; SPSS Inc., Chicago, IL). Measures of central tendency and spread of the data are presented as means and standard deviations. The independent sample *t*-test was used for mean comparisons to identify the magnitude of the differences in time performance of sport stacking between DG and DVIG. The significance level was set at $p \leq 0.05$.

III. RESULTS

The means and standard deviations of time performance to complete a cycle stack for both DG and DVIG groups, are presented in Table 1. A graphical representation of the means and standard deviations are shown in Figure 1. There was a slight difference in mean value between DG and DVIG, where the mean difference was -0.73 seconds.

From the independent samples t-test, there appeared to be no significant differences on time performance of sport stacking between DG and DVIG ($t = -0.415$, $p = 0.681$).

Table 1 Means and standard deviations of time performance of sport stacking between two experimental groups.

Groups	Means	Std. Deviation
Demonstration	30.40	5.356
Demonstration and Verbal Instructions	31.13	4.257

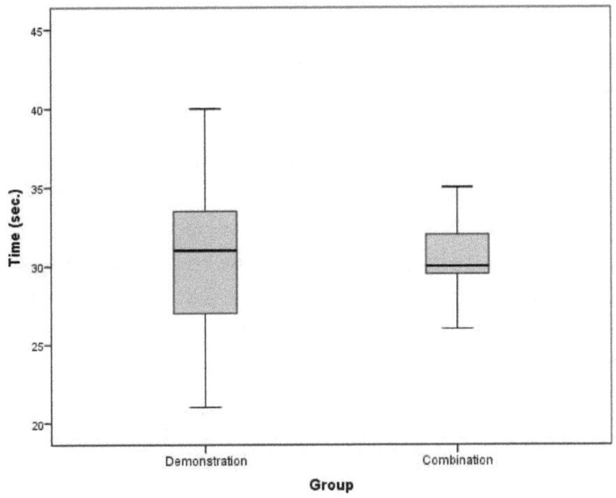

Figure 1 Boxplot showing the time scored to complete a cycle stack by the Demonstration and Demonstration with Verbal Instructions groups, respectively.

IV. Discussion

The purpose of this study was to identify instructional methods which are most effective for acquisition of the skill of sport stacking. The two instructional methods studied were learning by way of demonstration and learning by way of a combination of demonstration and verbal instructions.

The results indicated that there were no significant differences detected between the two groups. It appeared that participants from DG and DVIG performed relatively similarly on time performance regardless of the type of instructional methods used. Previous investigations into the effectiveness of learning by demonstration were equivocal, with some studies finding that demonstration is better than other types of instructions on some occasions, while on other occasions, it was not found to be better [5]. Taking into account investigations specifically into complex skills, one recent study comparing verbal instructions and demonstration found that providing verbal instructions did not facilitate learning of figure skating skills [15]. The findings of this study consisting of the complex skill of sport stacking echoed the results of the study of the complex skills of figure skating.

From this study, it appeared that visual demonstration as an instructional method in sport stacking provided relevant and sufficient information for executing the sport stacking skills without the help of verbal instruction. It has been said that `when the information to be conveyed through demonstration is redundant, there is, by definition, no information transmitted to the learner' [16]. In this case, it appeared that information conveyed through the demonstration was clear and useful and managed to be transferred to the learner. In fact, it is noteworthy that although significant differences were not detected between DG and DVIG, there was a small difference in the mean value between these two experimental groups, whereby the DG obtained a slightly faster mean time to complete a full cycle stack compared to DVIG.

Interpretation of the results could also come from the viewpoint of verbal instructions being redundant to the learning of sport stacking skills, hence the DVIG not benefitting from the verbal instructions. Firstly, it is possible that the information for the DVIG group received too much information due to the combination of both instructional methods. Processing the given information was necessary from a cognitive as well as motor perspective, and could have affected the participants at the time of learning which led to a slightly inferior time performance of sport stacking. One way around this is to perhaps present the verbal instructions implicitly as implicit motor learners have been found to not need to progress through the initial cognitive stage of learning [17].

Secondly, a study by Hinds and colleagues cautioned the use of verbal instructions given by experts because the information provided was viewed as too conceptual and organised [18]. Perhaps this was the case in this study as the video viewed by the participants was of an expert performer who has many years of experience and had won numerous sport stacking competitions thus far. Therefore, it has been suggested that when giving verbal instructions, the information should consist of more concrete and specific pieces of information, with minimal concepts and interrelationships among them [5].

In conclusion, presenting information by way of demonstration or a combination of demonstration and verbal instructions appeared to have the same effect on sport stacking performance of novices. For future directions, it is recommended that investigations on the use of demonstration and verbal instructions featuring an individually based movement template or less skilled performers are carried out. Less skilled models have been shown to be just as good as 'expert models' and only demonstrate the 'correct' performance, where the learners must find out what does and does not work by themselves.

Acknowledgment

The authors would like to thank the lecturers of the VHEE 3170 Research Project in Exercise Science class for their advice and guidance throughout this research project.

Conflict of Interest

The authors declare that they have no conflict of interest.

References

1. Hodges, NJ (2001) The role of instructions and demonstrations in learning a coordination skill. Doctoral dissertation, University of British Columbia, Vancouver.
2. Carroll, WR, & Bandura, A (1982). The role of visual monitoring in observational learning of action patterns: Making the unobservable observable. J Motor Behav 14(2):153-167.
3. Swinnen, SP (1996) Information feedback for motor skill learning: A review. Adv Motor Learn Control 37-66.
4. Hodges, NJ, & Franks, IM (2002). Modelling coaching practice: the role of instruction and demonstration. J Sports Sci 20(10): 793-811.
5. Magill, R, & Anderson, D (2016) Motor learning and control: Concepts and applications (11th Ed.). McGraw-Hill Education, Singapore.
6. Speed Stacks, Inc. (2001) Cup stacking with Speed Stacks: See it, believe it, teach it! Speed Stacks, Colorado.
7. Davis, S (2003) Cup Stacking and its Effect on Reading Scores in Sixth Grade Elementary School Students. Doctoral dissertation, The University of North Carolina at Pembroke, North Carolina.
8. Uhrich, TA, & Swalm, RL (2007) A pilot study of a possible effect from a motor task on reading performance. Perceptual Motor Skills, 104(3):1035-104.
9. Udermann, BE, Murray, SR, Mayer, JM, & Sagendorf, K (2004) Influence of cup stacking on hand-eye coordination and reaction time of second-grade students. Perceptual Motor Skills, 98(2):409-414.
10. Rhea, CK (2004) Changes in upper limb coordination and kinematics following a five week instructional unit in cup stacking. Master's thesis, Barry University, Florida.
11. Li, Y, Coleman, D, Ransdell, M, Coleman, L, & Irwin, C (2011) Sport stacking activities in school children's motor skill development. Perceptual Motor Skills, 113(2): 431-438.
12. Mortimer, J, Krysztofiak, J, Custard, S, & McKune, AJ (2011) Sport stacking in auditory and visual attention of Grade 3 learners. Perceptual Motor Skills, 113: 98-112.
13. Murray, S, Udermann, BE, Reineke, DM, & Battista, RA (2009) Energy expenditure in sport stacking. The Phys Educ, 66:180-185.
14. Gibbons, E, Hendrick, JL, & Bauer, J (2007) Distribution of practice on cup stacking performance. AAHPERD National Conference, Maryland, USA.
15. Haguenauer, M, Fargier, P, Legreneur, P, Dufour, AB, Cogerino, G, Begon, M, & Monteil, KM (2005) Short-term effects of using verbal instructions and demonstration at the beginning of learning a complex skill in figure skating. Perceptual Motor Skills, 100(1): 179-191.
16. Newell, KM (1981) Skill learning. Human Skills, 35(3): 203-225.
17. Masters, RS, & Poolton, JM (2012).Advances in implicit motor learning. Skill acquisition in sport: Research, theory and practice 59-75.
18. Hinds, PJ, Patterson, M, & Pfeffer, J (2001) Bothered by abstraction: the effect of expertise on knowledge transfer and subsequent novice performance. J Appl Psych, 86(6): 1232-1243.

Author: Jadeera Phaik Geok Cheong, PhD
Institute: Exericse Science Unit, Sports Centre, University of Malaya
Street: Lembah Pantai
City: Kuala Lumpur
Country: Malaysia
Email: jadeera@um.edu.my

Monitoring the Walking Pattern of Lower Limb Prosthetic Users Using Mobile Accelerometer Apps

M. S. Mohktar[1,2] and N.R. Hasan Basri[1]

[1] Department of Biomedical Engineering, Faculty of Engineering, University of Malaya, Malaysia
[2] Centre for Innovation in Medical Engineering, University of Malaya, Malaysia

Abstract— **Physical activities are any movement that required muscles to work and consequently use more energy as compared to resting. Physical activities are the best parameter in determining a person health. This is especially important for the health of lower limb prosthetic users who are in progress of restoring their walking and standing ability by wearing the prosthetic leg. The objective of this study is to develop an accelerometer apps that can identify the walking pattern of lower limb prosthetic users. The Smartphone accelerometer application function is to detect and to collect acceleration data while user's performing physical activities such as walking on flat surface. As a results an accelerometer application on android smartphone has been developed. The magnitude of acceleration data from the accelerometer are used to identify three participants; normal, transtibial and transfemoral prosthetic users walking patterns.**

Keywords— **Accelerometer apps, Prosthetic user**

I. Introduction

Physical activities are any movement that use our body muscle to work and use more energy as compared to resting. Example of physical activities are walking, standing, and any movements that use muscle to work. Physical activities have different level that determine whether ones have a vigorous or moderate lifestyle and are the best parameter for determining health. A lifestyle with adequate or active physical activities translates a healthy lifestyle [1].

Amputation is a process of removing limb or body part as a treatment or preventative method to avoid more severe disease to occur. Amputee is a person who had lost one or more of his or her limb and has undergone amputation surgery. There is also congenital case of amputated limb which occur since born. There are several levels of amputation which cover upper limb and lower limb. Upper extremities amputations have several part which are below elbow and above elbow amputation. For lower extremities, amputations are divided into below knee amputation and above knee amputation. There are specific names of the amputation according to the location of the bone or joint that is amputated [2]. Subjects with amputation show disorders in blood circulation and metabolism, reduced physical functioning capacity, and reduced tolerance capability [3].

In medical field, acceleration have become an important parameter to assess patient health condition through their activities. The acceleration measured can determine patient physical health condition to diagnose or improve the rehabilitation treatment of the patient. Accelerometer is one of the best tools to detect the physical activity level of human form sedentary to vigorous level of activities. Accurate and precise data of the accelerometer is also the reasons it is the best tools to measure the acceleration of physical activities. Accelerometer is a device that measure acceleration of an object per unit time [4]. It measures the rate of change of velocity of an object when it is moving.

The objective of this study is to develop a smartphone accelerometer apps that can measure the walking acceleration of lower limb prosthetic users. The collected acceleration data are used to visualized the pattern. Hence, the walking pattern of the prosthetic users can be monitored.

II. Methodology

A. Sensor

The sensor used in this study was an own developed accelerometer application for android smartphones. The accelerometer sensor detects the translational acceleration; the X, Y and Z axis acceleration data. The axes define the movement; Z-axis detect forward movement, Y-axis detect upward and downward movement and X-axis detect horizontal movement [5].

The apps also calculate the total acceleration data which is the combination of the translational acceleration with gravity acceleration. Eq. 1 shows the total acceleration [5]:

$$\text{Sum of Vector Magnitude (a)} = \sqrt{X^2 + Y^2 + Z^2} \qquad (1)$$

Eq 1. Sum of Vector Magnitude (Total Acceleration)

The data from the accelerometer then were stored in the secure digital (SD) card and later transferred to laptop for analysis.

F. Ibrahim et al. (eds.), *3rd International Conference on Movement, Health and Exercise*,
IFMBE Proceedings 58,
DOI: 10.1007/978-981-10-3737-5_29

For data analysis, the total acceleration (a) is subtracted with the gravity acceleration (g) to get the output acceleration data (Oa) on the limb. Eq. 2 was used to calculate the ouput acceleration [6]:

$$\text{Output acceleration } (O_a) = a - g \quad (2)$$

Eq. 2. Output acceleration acted on limb (Oa)

The sensor accelerometer application was developed using Java Development Kit and Android Development Tools (ADT).

B. Testing

The apps was tested on lower limb prosthetic users and normal subject. The apps was installed in an android smartphone for testing. The acceleration data were stored in the smartphone's SD card. An experiment was conducted on subjects and sample data were collected from the accelerometer apps. Then, the collected data were visualized to show the pattern of acceleration of the participants.

C. Participant

Male participants were recruited from Prosthetic and Orthotic laboratory patients who came to be study subjects. Ethics approval was obtained from the Department of Biomedical Engineering and subjects were asked to sign consent forms.

The participants involve in the experiment were transtibial and transfemoral prosthetic users and a normal person. The age of the participants is all above 18 years old and the participant with lower limb prosthesis must be able to walk with their prosthesis. The muscle grades of the recruited participants were between 3 to 4. Table 1 shows the height and the weight of the participants.

Table 8 Participants Height, Weight and Body Mass Index (BMI)

Participants	Height (m)	Weight (kg)	BMI (kg/m^2)
Transtibial	1.63	120	45
Transfemoral	1.65	79	29
Normal	1.52	58	25

D. Study Protocol

The participants were required to wear pants. First, the accelerometer apps was switched ON, then the smartphone was placed vertically in the left pocket (as all the prosthetic users recruited were amputated at left leg). Participants were asked to walk for about 2 minutes on a flat surface. The accelerometer apps was switched OFF and the recorded data were observed. The participants were required to repeat the procedure for three times and the recorded data were averaged.

III. Results and Discussion

The design of the accelerometer apps is shown in fig. 1. It displays the X, Y, Z acceleration data and the total acceleration value. The data were recorded in SD card of the smartphone.

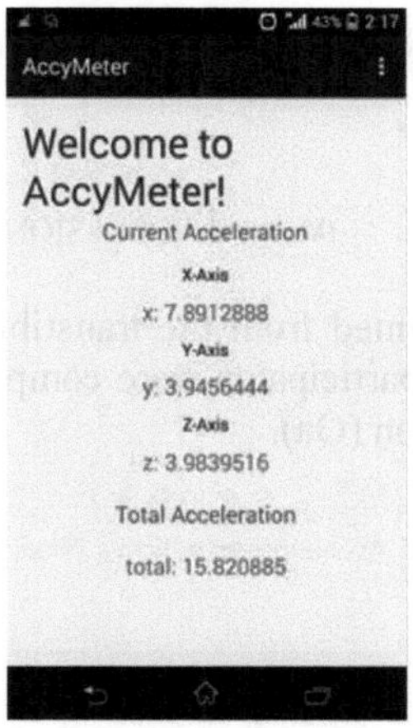

Fig.1 Screen Display of AccyMeter App

Fig. 2 shows the raw acceleration data of the X, Y and Z axis during walking obtained from one of the participant.

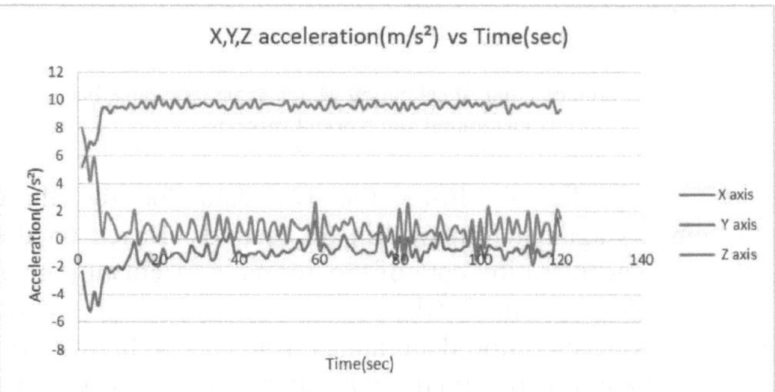

Fig. 2 Raw acceleration data of Transtibial participant obtained from accelerometer apps

The raw data; X, Y, and Z acceleration were used to calculate the total acceleration (sum of vector magnitude) by using Eq. 1., then the output accelerations were calculated using Eq. 2. The output acceleration that represents the real acceleration acted on the limb of the prosthesis were plotted against time. Fig. 3 shows the graph of the transfemoral participant.

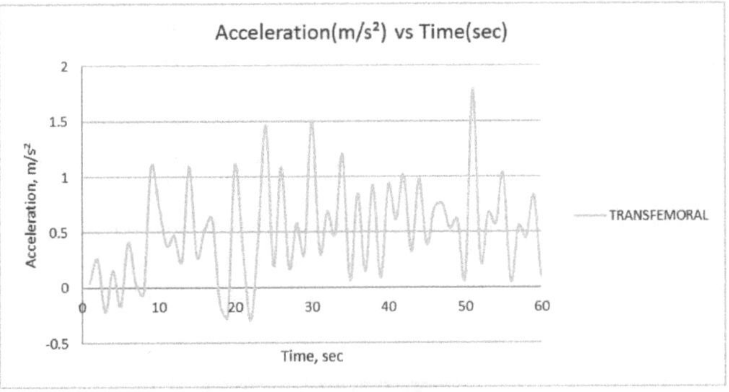

Fig. 3 Output acceleration graph against time for the Transfemoral participant.

IV. DISCUSSIONS

The data obtained from the transtibial, the transfemoral and the normal participants were compared in terms of the output acceleration (Oa).

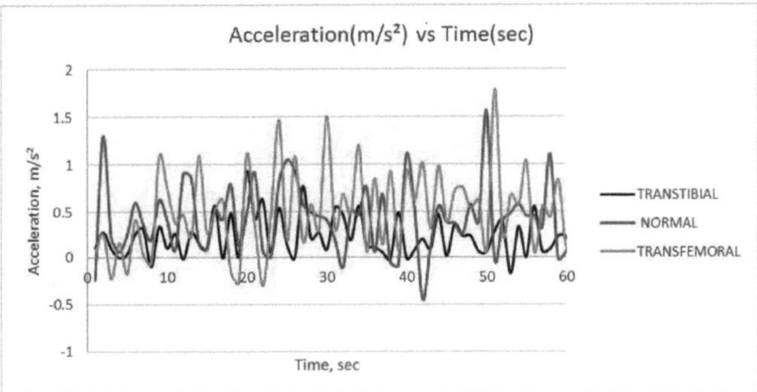

Fig. 4 Graph of Total Acceleration versus Time for Transtibial, Transfemoral and Normal subjects.

Graph in fig. 4 shows the *Oa* data for the three participants against time within 60 seconds. Based on the graph, the acceleration data of the normal participant ranges from -0.48 to 1.6m/s^2 whereas the transtibial prosthetic participant acceleration data ranges from -0.2 to 0.9ms^2 and the transfemoral prosthetic participant acceleration data ranges from -0.38 to 1.8ms^2. The transfemoral acceleration data were nearly similar to the normal participant's data. The negative values of the acceleration for all three participants were due to decrease in speed during walking.

The normal participants' acceleration data in this study were slightly higher compared to the normal acceleration presented in Oberg et. al, as they recorded 1.2 to 1.4 m/s^2 for male and 1.1 to 1.3 m/s^2 for female (Öberg, Karsznia, & Öberg, 1993). This is probably because of the ethnicity differences in the relationship of the gait speed between the study participant and the participants recruited by Oberg et. al [7, 8].

The transtibial participant's walking pattern showed that the participant walk slower as compared to the other participants walking speed and in lower range of acceleration. This might be due to the weight of the transtibial participant that was greater than the other two participants. Based on observation, the transtibial participant performed more stance phase and less swing phase. He had to decrease his step length or cadence to stabilize his walking. This finding is in agreement with Del Porto et. al, stating that there is a strong link between weight and balance impairments [9].

The acceleration data showed that the transfemoral participant had a good control of knee and ankle joint while walking using the prosthesis. His walking pattern was similar to the normal participant walking pattern. However, the range of acceleration were slightly higher. Moreover, based on observation of the transfemoral participant's walking pattern, the participant step further with his prosthetic leg compared to his intact leg. He reduces the swing phase of his intact leg and performed earlier stance phase during walking. He also tends to lean towards his prosthetic leg during stance phase for stability. His gait pattern was similar to many transfemoral prosthetic users which usually had uneven step length, lateral bending toward prosthetic limb and increase in walking speed [10].

Most of the gait analysis study of prosthetic users were conducted in traditional and expensive gait laboratories [11]. This study provided the proof of concept of the capacity of an android smartphone apps to monitor the walking patterns of the prosthetic users. Thus, it could expand the study of the prosthetics user's gait pattern during them performing daily life routine, which is an important especially for rehabilitation and health monitoring purposes.

Moreover, with its affordable price, smartphone is a widely used electronic device [12]. Thus, the healthcare service provider can monitor the gait of the prosthetic users almost immediately after the patients being released from the hospital. Likewise, the monitoring could also be applied to prosthetic users who lives in the rural area.

V. CONCLUSION AND FUTURE WORK

As a conclusion, an android smartphone accelerometer app was successfully developed in this study. The apps is capable in measuring the acceleration data on lower limb prosthesis participants and on a normal participant. The walking patterns among the participants were displayed, observed and compared.

Furthermore, the developed smartphone apps may help in the post treatment assessment among the prosthetic users. The apps could be used for data collection and processing thru the daily lives of the prosthetic users.

Conflict of Interest

The authors declare no conflict of interest.

Acknowledgment

This research is supported by the MOSTI Sciencefund Project (SF-017-2013).

References

1. F. J. Penedo and J. R. Dahn, "Exercise and well-being: a review of mental and physical health benefits associated with physical activity," Current opinion in psychiatry, vol. 18, pp. 189-193, 2005.
2. R. Hoile, "Amputation: Surgical Practice and Patient Management," bmj, vol. 312, p. 984, 1996-04-13 07:00:00 1996.
3. S. F. Kurdibaylo, "Cardiorespiratory status and movement capabilities in adults with limb amputation," J Rehabil Res Dev, vol. 31, pp. 222-35, Aug 1994.
4. J. M. Broderick, J. Ryan, D. M. O'Donnell, and J. Hussey, "A guide to assessing physical activity using accelerometry in cancer patients," Support Care Cancer, vol. 22, pp. 1121-30, Apr 2014.
5. J. R. Kwapisz, G. M. Weiss, and S. A. Moore, "Activity recognition using cell phone accelerometers," SIGKDD Explor. Newsl., vol. 12, pp. 74-82, 2011.
6. R. Takeda, S. Tadano, A. Natorigawa, M. Todoh, and S. Yoshinari, "Gait posture estimation using wearable acceleration and gyro sensors," J Biomech, vol. 42, pp. 2486-94, Nov 13 2009.
7. B. Xu, D. K. Houston, S. S. Gropper, and C. A. Zizza, "Race/Ethnicity differences in the relationship between obesity and gait speed among older Americans," J Nutr Elder, vol. 28, pp. 372-85, Oct 2009.
8. T. Oberg, A. Karsznia, and K. Oberg, "Basic gait parameters: reference data for normal subjects, 10-79 years of age," J Rehabil Res Dev, vol. 30, pp. 210-23, 1993.
9. H. Del Porto, C. Pechak, D. Smith, and R. Reed-Jones, "Biomechanical effects of obesity on balance," International Journal of Exercise Science, vol. 5, p. 1, 2012.
10. H. B. Skinner and D. J. Effeney, "Gait analysis in amputees," Am J Phys Med, vol. 64, pp. 82-9, Apr 1985.
11. A. Muro-de-la-Herran, B. Garcia-Zapirain, and A. Mendez-Zorrilla, "Gait analysis methods: an overview of wearable and non-wearable systems, highlighting clinical applications," Sensors, vol. 14, pp. 3362-3394, 2014.
12. R. LeMoyne, T. Mastroianni, and K. Montoya, "Implementation of a smartphone for evaluating gait characteristics of a trans-tibial prosthesis," Conf Proc IEEE Eng Med Biol Soc, vol. 2014, pp. 3674-7, 2014.

Author: Mas Sahidayana Mohktar
Institute: Centre for Innovation in Medical Engineering, University of Malaya
Street: Lembah Pantai
City: 50603, Kuala Lumpur
Country: Malaysia
Email: mas_dayana@um.edu.my

Diet and Bone Status in Eumenorrheic Female Athletes

N.S. Ahmad[1], M. S. Hamid[2], J.P.G. Cheong[1] and S.H. Hamzah[1]

[1] Sports Centre, University of Malaya, 50603 Kuala Lumpur, Malaysia
[2] Sport Medicine Unit, Faculty of Medicine, University of Malaya, 50603 Kuala Lumpur, Malaysia

Abstract— **Various risk factors for low bone mass density (BMD) have been reported, such as diet restriction, unbalance body composition and menstrual irregularities. This is apparent in female athletes who have a tendency to seek for a low body weight by adopting chronic energy deficits (low calories intake or malnutrition) practices. The aim of this study was to determine the relationship between energy intake, nutrient consumption amount, energy availability, body weight with BMD status. Eighty-five moderately trained female athletes aged 18–29 years took part in this study. Body fat percentage and BMD were measured using the Tanita Weighing Scale and 200-CM Furuno Ultrasound bone densitometer, respectively. Data on physical activity, menstrual cycle status, and eating attitude were obtained using questionnaires. Dietary variables were assessed from a prospective combination of 24-hour diet recall and 3-day food diary of macronutrients and micronutrients. Results showed that more than half (53%) of the female athletes have low BMD, with z-scores ranging between -2.00 and -0.03 (below the normal range). The mean (SD) of energy intake of female athletes was 1291 (304) kcal/day which is below than recommended nutrient intake (RNI). Similarly minerals (Calcium and Phosphorus) intake did not achieve the RNI. Low energy intake was significantly (p≤0.05) correlated with low eating attitude score (EAT-26). Pearson Correlation also showed that low BMD was significantly associated with low body weight. The common factors of low BMD among eumenorrheic athletes are an insufficient intake of energy intake and bone building nutrients (Calcium, Vitamins D, Vitamin C and Zinc). Higher BMD could be achieved through actively encouraging high-risk group of athletes to focus on good dieting practice.**

Keywords— **Energy intake, Nutrient, Exercise, Physical activity**

I. Introduction

Adequate nutrition is important not only for maintaining good health and nutritional status of women but also for strengthening their bone health. Mostly, nutrient intake of female athletes is below energy requirement [1]. Energy intake less than 2000 kcal/day cannot support the nutritional demands and boost extra energy for the intense exercise of female athletes [2]. Previous studies have reported that low energy intake of1939 kcal/day was discovered among female college athletes [1]. Similarly, elite cross-country skiers consumed 1988 kcal/day [3] and elite skaters with 1491 kcal/day [4]. These findings are made worse when the daily total energy expenditure is reported to be significantly higher than energy intake [1].

Low energy intake combined with high energy expenditure lead to negative energy balance or known as low energy availability. Low energy availability may be the result of inadequate intakes of macronutrients, such as carbohydrate, essential amino acids, and fatty acids, or mineral and vitamins, such as calcium and vitamin C, which are essential for bone health [2].

Márquez & Molinero [5] found that if adolescents or adult women do not get enough nutrition and become underweight (low BMI) and had a low body fat percentage, then will disturb menstrual cycle and bone density. Imbalance nutritional status, menstrual dysfunction, and low BMD are known as 'the female athlete triad' (FAT) [6]. Therefore, priority should be placed on the adequate nutritional intake of athletes because unbalance diet is the key contributing factor for menstrual irregularities and low BMD. The present study was conducted to determine the relationship between energy intake, energy availability, nutrients consumption amount and body weight on BMD status.

II. Methods

A. Participants

Healthy female athletes aged between 18 to 29 years old who trained or exercised three times per week for the last five years were invited to participate in this study. Participants were explained about the objectives of the study and the procedures involved. All participants were required to sign an informed consent form prior to participation.

The sample size was calculated using G-Power version 3.1.9 and based on a previous study conducted by Ismail et al. (1997). The power of the study was set at 80%, with a 95% confident interval and effect size was 0.99. The calculated sample size was 85 participants with 20% drop out rates considered. A total of 85 female athletes were recruited from all sports. This study was approved by the Human Ethical Committee, University of Malaya (UM.TNC2/RC/H&C/UMREC – 43).

F. Ibrahim et al. (eds.), *3rd International Conference on Movement, Health and Exercise*,
IFMBE Proceedings 58,
DOI: 10.1007/978-981-10-3737-5_30

B. Measurement

Body composition and anthropometry assessment: Bio-impedance analysis (BIA) (Tanita, Japan) was used to estimate body composition, and in particular body weight, body mass index and percentages of body fat. Meanwhile, height was measured by using a portable stadiometer (SECA, Germany).

Bone Densitometer: Bone density was measured using the Furuno CM 200 Ultrasound Bone Densitometer. This equipment is very convenient for osteoporosis screening and easy to operate and only takes 3-10 seconds for the measurement.

Estimated Energy Expenditure: Basal Metabolic Rate was measured by using COSMED Quark C-PET. The test was performed with a flow dilution canopy hood for about 20 minutes. The BMR value was multiplied by PAL of either by 1.4 (sedentary/mild) or 1.7 (moderate) or 2.0 (strenuous) activities based on Physical Activity Record.

BMR x PAL= Estimated Energy Expenditure [7]

C. Dietary Intake measures

24-hour food recall: For 24-hour food recall questionnaire, participants were required to recall and record the food and beverages that were consumed within the last 24hours. A food recall kit (bowl size- L, M, and S, plate, teaspoon, tablespoon, and scoop) was provided to help the participants to determine the amounts of foods consumed.

3-day food record: Participants were also required to complete a 3-day food diary (two weekdays and one weekend) at home. A 3-day food record is designed to get an accurate description of a typical daily diet. Participants were required to record all types of food and beverage and the amounts as accurate as possible. Dietary intake of energy and macronutrients of 24-hour food recall and 3-day food dairy were calculated using Nutritionist Pro software.

BMD results	No of participants	Z- score (SD)
z-score > 0	40	1.06 (0.99)
z-score < 0	45	-0.59 (0.06)

D. Questionnaires

Eating Attitudes Test-26 (EAT-26): The EAT-26 was used to assess "eating disorder risk". Screening for eating disorders is based on the assumption that early identification can lead to earlier treatment, thereby reducing serious physical and psychological complications or even death. Individuals who scored 20 or more on the test have a high risk of eating disorder.

Menstrual History Questionnaire: Menarche age, cycle frequency, cycle flow were determined from six questions in this questionnaire. Menarche age is years since menarche, Cycle flow is the average length of menstrual flow and Cycle frequency/year is average menstrual cycles experienced each year.

E. Statistical analyses

Data were analyzed using Statistical Package for the Social Sciences (SPSS) version 22.0. Data were described descriptively and a normality test was performed using the Kolmogorov- Smirnoff test. Continuous data such as training activity, energy intake, eating attitude, menstrual status, and bone status were presented in mean and standard deviation. The Pearson correlation was used to assess the association between bone measurement with energy intake, nutrient consumption amount, and body weight.

III. Results

A. Participants characteristic, bone and menstrual status

A total of 85 participants completed the study. Table 1 summarizes the participant characteristics. Participants exercise in average 11 month per year, physical activity level (PAL) of 1.81 and trained 4 day per week.

Table 9 Physical characteristic of the participants (n=85)

Variables	Mean (SD)
Age (year)	21 (3)
Body weight (kg)	56 (9)
Height (cm)	159 (0.06)
Body mass index	22 (3)
Body Fat (%)	15 (6)
Years training (month/year)	11 (4)
Frequency of training / week	4 (2)
Physical activity level	1.81 (0.2)

SD=standard deviation

Table 2 Bone status data of the participants (n=85)

BMD results	No of participants	Z- score (SD)
z-score > 0	40	1.06 (0.99)
z-score < 0	45	-0.59 (0.06)

SD=standard deviation

Table 2 showed that 53% of participants had z-score less than zero which indicated low bone density. Another 47% showed normal bone density. The menstrual history questionnaire classified all participants as eumenorrhoeic. No significant different between participants' age of menarche (12.3 ± 1.2 years old), cycle frequency (12 ± 1.0 times/year) and cycle flow (6.6 ± 1.7 days/ month).

B. *Dietary intake and energy expenditure*

Dietary energy intake and macronutrients are presented in Table 3. The energy intake of the female athletes was 1291 (304) kcal/day which was lower than RNI requirement. Carbohydrate and protein intake found to be lower as compared to the RNI. Meanwhile fat was slightly higher than RNI value. As for selected mineral intake, calcium and phosphorus were observed to be lower than RNI. Participants also showed high energy expenditure (1807 (311) kcal /day) compared to their energy intake which can produced low energy availability (29 (4) kcal/day/kg FFM)

Table 3 Energy intake, Macronutrients, and Micronutrients intake compared to Recommended Nutrient Intake (RNI) (n=85)

Nutrients	Amount	RNI
Energy Intake (kcal/day)	1291 (304)	2000
Macronutrients		
Carbohydrate (g/day)	167 (43)	209
Protein (g/day)	50 (15)	55
Fat (g/day)	55 (17)	50
Micronutrients		
Calcium (mg/day)	416 (164)	800
Phosphorus (mg/day)	106 (43)	550

SD = standard deviation

C. *Eating Attitude Test-26*

The present results also showed the mean (SD) for Eating Attitude test (EAT-26) among female athletes was 21 (3) score representing a high risk of eating disorder. There was significant correlation of high EAT-26 score with low energy intake. The higher score showed that majority of the participants were fear of gaining weight, not satisfy of current body image and conscious of food selection.

D. *Relationship of body weight and dietary intake with BMD*

The association of body weight and BMD were showed in Fig 1. Low body weight was significantly correlated with low BMD (r=0.3) which revealed that low body weight increases the chances of low BMD. However, there was no significant correlation between low energy intake with low BMD (r=0.147). Nevertheless, it was found that low energy intake significantly ($p < 0.05$) correlated low body fat percentage (r= 0.9) and low body fat percentage also significantly associated low body weight (r=0.7).

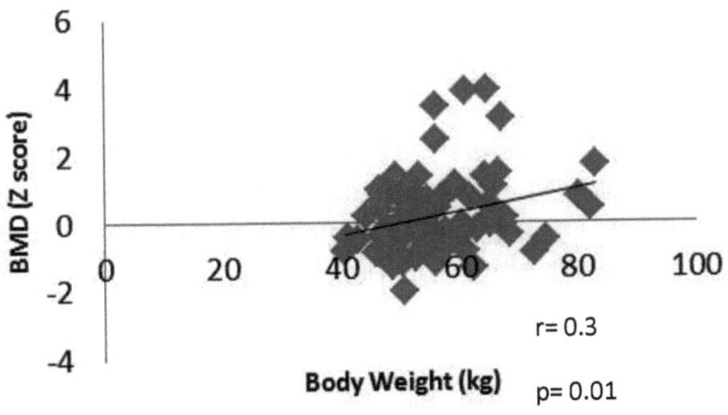

Fig. 1 The relation between Body Weight (kg) and bone mineral density (BMD) at the calcaneus bone (n=85

IV. DISCUSSION

Female athletes in this study undergo intense exercise training session which is resulting in high energy expenditure coupled with low energy intake. This condition will cause negative energy balance or low energy availability to occur. High pressure to obtain optimal training bodyweight and long-term of low energy availability can cause loss of BMD [8].

Low energy intake can be associated with low intake for some macronutrients (carbohydrate and protein) and minerals (Calcium and Phosphorus). In Malaysia poor energy intake, food selection and dieting habit are the main factors of insufficient protein and carbohydrate intake [9]. The present study also found that the mean carbohydrate and protein intake showed a lower value than RNI which contribute to low energy intake body fat and body weight. Meanwhile, micronutrients intake of calcium and phosphorus which are important for bone development are also below the RNI, especially for the active individual. In fact, a lower incidence of low bone density was found among women taking calcium and phosphorus supplementation regularly as opposed to those who did not, 18.7% vs. 29.3% (p=0.036) [10]. Several studies revealed that improving energy balance leads to improvement on overall nutritional status thus returning the athlete to normal bone health especially in female athletes [11, 12].

Consequently from poor eating attitude, 85% of participants had low body weight due to a low fat percentage. It can be speculated that low fat percentage. Dieting behavior and lack of knowledge on energy needs often leads to energy deficit and increased the risk of bone mass loss. The results are consistent with earlier literature who found a strong relationship of eating disorder, body composition, and diet behavior [15]. Thus, the long-term effect of eating disorder might have a negative impact on body health. Overall, the present study can be concluded that low energy availability, inadequate body fatness, body weight and high exercise stress negatively affect the bone status in female athletes.

V. CONCLUSIONS

Several factors were found to contribute to low BMD status. It is recommended to a nutritionist, coach, and sports scientist to monitor and provide a guideline for proper calories intake suit with exercise demand to prevent low energy availability issues. As a conclusion, a sufficient amount of energy, macronutrient and bone-building intake can contribute to optimal body weight, body fat and bone mass.

ACKNOWLEDGMENT

We would like to express our sincerest gratitude to the participants for their commitment and support.

CONFLICT OF INTEREST

The authors declare that they have no conflict of interest.

REFERENCES

1. Shriver, L.H., N.M. Betts, and G. Wollenberg, *Dietary intakes and eating habits of college athletes: are female college athletes following the current sports nutrition standards?* Journal of American College Health, 2013. **61**(1): p. 10-16.
2. McLEAN, J.A., S.I. Barr, and J.C. Prior, *Dietary restraint, exercise, and bone density in young women: are they related?* Medicine and science in sports and exercise, 2001. **33**(8): p. 1292-1296.
3. Papadopoulou, S.K., et al., *Body composition and dietary intake of elite cross-country skiers members of the greek national team.* Asian journal of sports medicine, 2012. **3**(4): p. 257.
4. Dwyer, J., et al., *Eating attitudes and food intakes of elite adolescent female figure skaters: a cross sectional study.* Journal of the International Society of Sports Nutrition, 2012. **9**(1): p. 53.
5. Márquez, S. and O. Molinero, *Energy availability, menstrual dysfunction and bone health in sports; an overview of the female athlete triad.* Nutr Hosp, 2013. **28**(4): p. 1010-1017.
6. Joy, E., et al., *2014 Female athlete triad coalition consensus statement on treatment and return to play of the female athlete triad.* Current sports medicine reports, 2014. **13**(4): p. 219-232.
7. Ismail, M., W. Wan-Nudri, and H. Zawiah, *Energy expenditure studies to predict requirements of selected national athletes.* Mal J Nutr, 1997. **3**: p. 71-81.
8. Nichols, D.L., C.F. Sanborn, and E.V. Essery, *Bone density and young athletic women.* Sports Medicine, 2007. **37**(11): p. 1001-1014.
9. Manore, M.M., *Dietary recommendations and athletic menstrual dysfunction.* Sports medicine, 2002. **32**(14): p. 887-901.
10. Lim, P., et al., *Bone health in urban midlife Malaysian women: risk factors and prevention.* Osteoporosis international, 2005. **16**(12): p. 2069-2079.
11. Loucks, A.B., B. Kiens, and H.H. Wright, *Energy availability in athletes.* Journal of sports sciences, 2011. **29**(sup1): p. S7-S15.
12. Doyle-Lucas, A.F., J.D. Akers, and B.M. Davy, *Energetic Efficiency, Menstrual Irregularity, and Bone Mineral Density in Elite Professional Female Ballet Dancers.* Journal of Dance Medicine & Science, 2010. **14**(4): p. 146-154.

Author: Sareena Hanim Hamzah
Institute: Sport Centre, University of Malaya
Street: Jalan University
City: Kuala Lumpur
Country: Malaysia
Email: sareena@um.edu.my

Does Imagery Facilitate a Reduction in Movement Variability in a Targeting Task?

N.A. Mohammed Suberi[1], R. Razman[1] and N. Callow[2]

[1] Exercise Science Unit, University of Malaya, Kuala Lumpur, Malaysia
[2] School of Sport, Health, and Exercise Sciences, University of Wales, Bangor, United Kingdom

Abstract— **It is widely accepted that imagery is an effective strategy for improving skilled performance, motivation and rehabilitation of sports injuries. Previous empirical studies showed that the decrease in variability contributed to a consistency of movement which led to performance improvement. However, there are no studies examining the role of visual imagery in movement variability during target aiming task. Thus, the objective of this study was to discover the possible moderating role of imagery on movement variability at the starting point of a target aiming task. It was hypothesized that visual imagery perspectives would induce a significantly more consistent movement pattern than a control group. Thirty-six participants that passed the Vividness of Movement Imagery Questionnaire-2 for imagery ability screening were randomly assigned to three groups; External Visual Imagery (EVI), Internal Visual Imagery (IVI) and a control group. In a seated position, with their elbow bent at 90 degrees, participants performed a no-vision pullback aiming task on a sliding track. The movement was tracked by an 8 camera 3D motion capture system. Both imagery groups received the relevant imagery script between blocks from the second block onwards, while the control group answered a set of math questions during the break. A two-way Mixed ANOVA was conducted to examine the main effect and interaction for the dependent variable (p<.05). Movement variability was not significantly different between external visual imagery, internal visual imagery, and the control group. However, there was a visible reduction in variability. The results of this study suggest that imagery does not give a direct effect on the consistency of the movement patterns in a repeated task. Instead, the increase in consistency of movement possibly stems from the learning effect of continuous practice.**

Keywords— **Visual imagery perspective, Internal visual imagery, External visual imagery, Movement variability, Consistency**

I. Introduction

There are several previous studies done to examine the importance of movement variability in sports performance and injuries. For example, shoulder alignment towards the stance are needed for reduced detrimental variability in golf [1], consistency in elbow and wrist joints were important in release boundaries of basketball shooting [2], and, the reduction of variability in a distal joint movement is greater of skilled players on standing throw with run-up during the acceleration phase in team handball [3]. In motor control, there are different connotations use to enlighten the meaning of movement variability. The conventional approach assumes the variability as noise in the nervous and classified it as dysfunctional [4]. On the other hand, according to the dynamical systems theory, variability is thought to have a useful role to human movement [5]. Variability shows multiple patterns of movement, providing flexibility and adaptive strategies for each changing condition. Indeed, the lack of variability of movement has been shown to contribute to the abnormal sensory cortex and thus interfere with motor function [6]. In other words, variability serves up huge amounts of information to the sensory system, which in turn can prevent injury in movement. In addition, with sophisticated measurement, the significant potential of variability as a signal for flexibility and adaptability in biological human are now apparent [7]

In general, movement variability encompasses the variation that exists in motor performance resulting from multiple repetitive movements of a particular task. However, extreme variability in movement pattern is often assumed as noise and is detrimental to performance. Therefore, this noise should be reduced to optimize performing a movement. Various methods have been explored to increase the consistency of the movement, including psychological approach [8].

Research has shown the role of imagery practice in sports performance. It is used in many fields including cognitive psychology [9], neuropsychology, neurophysiology and neuropathic technique for rehabilitation [10]. Also, it has been reported to increase motor learning and motor control [11]. Imagery helps an athlete for a wide variety of cognitive and motivational purposes that can occur at a general or specific level [12]. It has been shown to be worthwhile when used for learning and performance for strategies of play [13], goal achievement [14], mental control, sports confidence [15] and a decline in arousal and stress [16].

In many brain imaging studies, researchers have shown the same brain areas are activated during unconscious planning and execution of movement whenever the person represent an action through imagery [11]. Therefore, when the sensory input interacts with motor imagery, prospective sensory feedback is created [17]. For example, Naito and colleagues [18] proposed that tendon vibration (proprioceptive input) of wrist extensor was raised by the

F. Ibrahim et al. (eds.), *3rd International Conference on Movement, Health and Exercise*,
IFMBE Proceedings 58,
DOI: 10.1007/978-981-10-3737-5_31

motor imagery of wrist flexion. This shows the shared processes between imagery and execution, primed the action execution.

Previous studies show that visual imagery perspective can help in improving performance. Visual imagery has been divided into two perspectives; internal visual imagery and external visual imagery. Internal visual imagery involves the participant imagining the scene as looking through their eyes and mentally rehearsing the spatial and temporal conditions. Conversely, external visual imagery involves participants imagining from a third perspective and allows them to "see" the movement from the outside.

It would be of interest to explore the possible different effect of both types of visual imagery on movement variability, especially in achieving a targeted task. Therefore, the aim of this novel study was to discover the possible moderating role of visual imagery perspective on the consistency of movement patterns at the start of a target aiming task.

II. METHODS

A. Participants

Forty-one collegiate level students volunteered for the study. Participants' imagery ability was measured using the Vividness of Movement Imagery Questionnaire-2 [19]. Based on previous studies, the participant who scored below 36 on each subscale of the VMIQ-2 were found to be able to produce images at least moderately clear and vivid [20]. Thus, only 36 participants who met the visual imagery criteria from the screening were accepted into the study. Participants were randomly assigned to three groups: External Visual Imagery, Internal Visual Imagery and a Math's-control group (M age = 20.75 ±1.8 years, Male N=18, Female N=18). All participants were required to fill a written informed consent before participation. Ethical approval was granted by the ethics board of the University of Malaya (UM.TNC 2/RC/H&E/UMREC-62) on research involving human subjects. *Procedure*

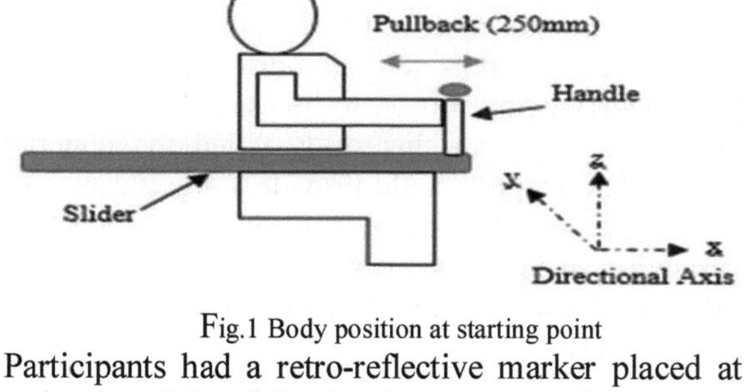

Fig.1 Body position at starting point

Participants had a retro-reflective marker placed at the lateral epicondyle of the dominant arm, and the movement of elbow position in X, Y and Z direction in space was tracked by an 8 camera Qualisys 3D motion capture system. In a seated position, with their elbow bent at 90 degrees, participants performed a no-vision pullback aiming task on a sliding track. There were 7 blocks of 3 trials. The target distance was set at 250mm. Standard deviation (SD) of discreet elbow position at the start point of the task was used as the score for the variability in each block. Lower values represent lower movement variability for that block and vice versa.

Before the first trial of the six blocks following the pre-test block, the IVI and EVI groups were given an IVI and EVI script to read through at his or her own pace and image the task. Before the second and third trials in each block, they were asked to image the task from their respective perspective without reading the script in 30-second. The control group has answered a set of math questions to take their minds off the task during the break.

B. Statistical analyses

Data was analyzed using Statistical Package for the Social Sciences (SPSS) version 22.0. Data is presented descriptively, and a normality test was performed using the Kolmogorov-Smirnoff test. A two-way mixed ANOVA (3 group x 7 blocks) was conducted on the movement variability scores.

III. RESULT

A total of 36 participants completed the study. It was found that movement variability was not significantly different between external visual imagery, internal visual imagery, and control group. The main analysis on elbow position in the X and Y and Z direction at start point revealed that no significant interaction in X direction; ($F_{10,165}$ = 1.045, p = .409), in Y direction; ($F_{9,155}$ = .789, p = .632), and in Z direction; ($F_{10,163}$ = 1.189, p = .302). No significant main effect for block in X; ($F_{5,165}$ = 1.103, p = .362), in Y; ($F_{5,155}$ = 1.842, p = .112), and in Z; ($F_{5,163}$ = 2.23, p = .054) and no significant main effect for group in X; ($F_{2,33}$ = .181, p = .835), in Y; ($F_{2,33}$ = .175, p = .840), and in Z; ($F_{2,33}$ = 1.924, p = .162). Although there was no significant difference across block for elbow position in the X, Y and Z, interestingly, the result did show there was a visible decline in movement variability with the progress from block 1 to block 7 (Fig. 2). Mean and SD of group for X, Y and Z direction for blocks are presented in Table 1, Table 2 and Table 3.

Table 1 Mean SD score for X direction in blocks

Group	Block 1	2	3	4	5	6	7
EVI	2.17 ±1.26	3.46 ±2.10	2.50 ±0.51	2.39 ±1.28	2.30 ±0.75	1.81 ±0.64	2.44 ±1.88
IVI	2.75 ±1.93	2.74 ±1.61	2.78 ±2.16	2.61 ±1.24	2.65 ±1.46	2.34 ±1.38	2.25 ±1.09
Math control	2.85 ±1.89	2.37 ±1.64	2.97 ±1.61	2.85 ±1.58	2.23 ±1.34	2.97 ±1.24	1.70 ±0.45

Table 2 Mean SD score for Y direction in blocks

Group	Block						
	1	2	3	4	5	6	7
EVI	4.10 ±1.78	2.68 ±0.39	2.52 ±0.79	4.07 ±3.01	3.25 ±2.79	2.97 ±1.61	3.69 ±2.90
IVI	3.83 ±2.52	3.54 ±2.22	2.83 ±1.58	3.05 ±1.79	3.37 ±2.66	3.20 ±2.01	3.44 ±1.18
Math control	4.65 ±2.56	3.72 ±2.18	3.74 ±1.74	2.79 ±1.69	3.57 ±2.76	2.39 ±1.41	4.02 ±2.54

Table 3 Mean SD score for Z direction in blocks

Group	Block						
	1	2	3	4	5	6	7
EVI	5.50 ±2.84	3.95 ±1.49	4.50 ±2.64	4.47 ±2.38	3.51 ±2.21	4.11 ±1.91	4.24 ±2.17
IVI	4.39 ±1.47	4.63 ±2.16	4.85 ±2.56	4.40 ±2.02	2.76 ±1.44	3.39 ±2.16	3.11 ±1.26
Math control	4.33 ±1.71	3.61 ±3.36	2.83 ±1.89	2.93 ±1.13	3.90 ±2.84	3.06 ±1.43	3.07 ±1.68

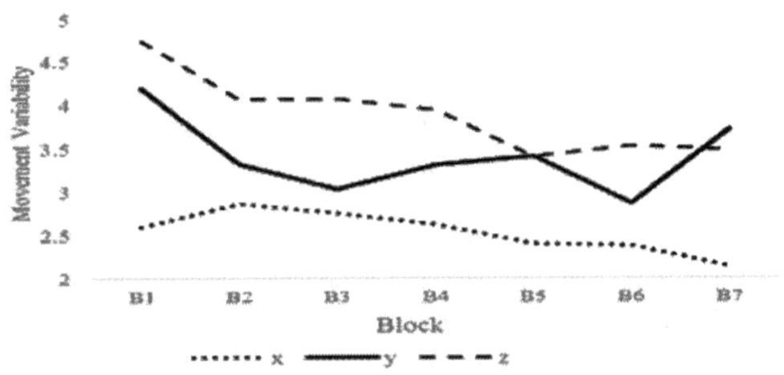

Fig. 2 Movement variability as a role of block

IV. Discussion

This study aimed to examine the role of imagery intervention on the consistency of movement within a repeated aiming task. The primary analysis shows there were no significant interaction and main effect in group and block over all trials. Therefore, the result failed to clarify the evidence for supporting the hypothesis that internal visual imagery or external visual imagery may promote an increase in consistency of movement patterns.

Despite the fact there were no significant differences in the consistencies of movement between the groups, there was a clear trend of declining discreet variability values with progress blocks of trials (Fig. 2). A visible decline was seen for elbow position in the X, Y, and Z direction from block 1 to block 7. Interestingly, this does not only apply to the treatment group but was also obvious for the control group. Thus, it suggests the decline in variability in the block of trials was not affected by the imagery intervention but is instead driven by another factor (s).

One possible explanation is that the decline in variability across all groups as the blocks progressed may be associated with the learning effect of practice from repeated trials. The repetition task leads to improved memory and strengthens the motor skills execution of the real world target. Thus, the decline in the variability of the movement is probably because the participants had a learning effect from the repeated trials - which in turn lead to a more consistent movement pattern at the start of the task. This possible explanation is in line with Domkin et al. [21], who reported that further decrease in variability would be noticeable following a prolonged bout of training.

The results are in line with a previous finding on an arm pointing movement [22], which states the decline in the variability of movement of the arm is affected by the constant practice.

V. Conclusion

Variability in human movement is necessary for efficient function of the motor system. On the other hand, too much variability is described as noise and disturbs motor funtion. Despite imagery being proven to be useful for a variety of purposes, the results of this study failed to support the hypothesis that imagery could provide a significant impact on the consistency of movement. Conversely, a decline in the variability of the movement found here is more likely related to the learning effect of practice within the repeated target aiming task.

Acknowledgment

This research was supported by the University of Malaya Postgraduate Research Grant (PPP) -Research PG092-2014B.

Conflict of Interest

The authors declare that they have no conflict of interest.

References

1. Langdown BL, Bridge MW, Li FX (2013) Address position variability in golfers of differing skill level. Int J Golf Sci 2(1):1–9.
2. Button C, MacLeod M, Sanders R, Coleman S (2003) Examining movement variability in the basketball free-throw action at different skill levels. Res Q Exercise Sport, 74(3):257–269.
3. Wagner H, Pfusterschmied J, Klous M, von Duvillard SP, Müller E (2012) Movement variability and skill level of various throwing techniques. Hum Mov Sci 31(1):78–90 DOI 10.1016/j.humov.2011.05.005
4. Newel KM, Corcos DM (1993) Issues in variability and motor control. In Newel KM & Corcos DM (Eds.), Variability and motor control (pp. 1–12) Champaign, IL: Human Kinetics.
5. Bartlett R, Wheat J, Robins M (2007) Is movement variability important for sports biomechanists? Sports Biomech 6:224–243 DOI 10.1080/14763140701322994
6. Harbourne RT, Stergiou N (2009) Movement variability and the use of nonlinear tools: principles to guide physical therapist practice. Phys Ther DOI 10.2522/ptj.20080130
7. Müller H, Sternad D (2009) Motor learning: changes in the structure of variability in a redundant task. Adv Exp Med Bio. 629:439–456 DOI 10.1007/978-0-387-77064-2_23
8. Stergiou N, Decker LM (2011) Human movement variability, nonlinear dynamics, and pathology: Is there a connection? Hum Mov Sc. 30(5):869–888 DOI 10.1016/j.humov.2011.06.002
9. Frank C, Land WM, Popp C, Schack T (2014) Mental representation and mental practice: Experimental investigation on the functional links between motor memory and motor imagery. PLoS ONE 9(4) DOI 10.1371/journal.pone.0095175
10. Rebecca H, Evans L, Callow N (2008) Imagery use during rehabilitation from injury: A case study of an elite athlete. Sport Psychol 22:405–422
11. Munzert J, Lorey B, Zentgraf K (2009) Cognitive motor processes: The role of motor imagery in the study of motor representations. Brain Res Rev 60(2):306–326 DOI 10.1016/j.brainresrev.2008.12.024
12. Nordin S, Cumming J (2008) Types and functions of athlete's imagery: Testing prediction from the applied model of imagery use by examing effectiveness. Int J Sport Exerc Psychol 6:189–206.
13. MacIntyre T, Moran A (1996) Imagery use among canoeists: a worldwide survey of novice, intermediate, and elite slalomists. J Appl Sport Psychol 8: S132.
14. Callow N, Hardy L (2001) Types of Imagery Associated with Sport Confidence in Netball Players of Varying Skill Levels. J. Appl Sport Psychol 13(1):1–17 DOI 10.1080/104132001753155921
15. Callow N, Hardy L, Hall C (2001) The effects of a motivational general-mastery imagery intervention on the sports confidence of high-level badminton players. Res Q Exerc Sport 72(4):389–400.
16. Mellalieu SD, Hanton S, Thomas O (2009) The effects of a motivational general-arousal imagery intervention upon pre-performance symptoms in male rugby union players. Psychol Sport Exerc 10(1):175–185 DOI 10.1016/j.psychsport.2008.07.003
17. Annet J (1996) On knowing how to do things: a theory of motor imagery. Cogn. Brain Res 3:65–69.
18. Naito E, Kochiyama T, Kitada R, Nakamura S, Matsumura M, Yonekura YSN (2002) Internally simulated movement sensation during motor imagery activate cortical motor areas and the cerebellum. J Neurosci 22(May 1):3683–3691
19. Robert R, Callow N, Hardy L, Markland D, Bringer J (2008) Movement imagery ability: development and assessment of a revised version of the vividness of movement imagery questionnaire. J Sport Exerc Psychol 30(2):200–221.
20. Roberts R, Callow N, Hardy L, Woodman T, Thomas L (2010) Interactive effects of different visual imagery perspectives and narcissism on motor performance. J Sport Exerc Psychol 32(4):499–517.
21. Domkin D, Laczko J, Jaric S, Johansso H, Latash ML (2002) Structure of joint variability in bimanual pointing tasks. Exp Brain Res 143(1):11–23 DOI 10.1007/s00221-001-0944-1
22. Darling WG, Cole KJ, Abbs JH (1988) Kinematic variability of grasp movements as a function of practice and movement speed. Exp Brain Res 73:225–235.

Author: Nor Azlina Mohammed Suberi
Institute: Sports Centre, University of Malaya
Street: Lembah Pantai
City: 50603 Kuala Lumpur
Country: Malaysia
Email: zlinas4352@gmail.com

Reliability of a Pendulum Apparatus Designed to Test the Taekwondo Electronic Body Protector

R.W.L. Chong and R. Razman

Exercise Science Unit, Sports Centre, University of Malaya, Kuala Lumpur, Malaysia

Abstract— **The sport of Taekwondo has advocated the use of the Protector and Scoring System (PSS) in all of its major competitions to help reduce scoring errors from human judges. The main purpose of this study is to determine the reliability of a custom built mechanical pendulum apparatus designed to test the effectiveness of the PSS. To determine the apparatus reliability, an accelerometer was placed at the end of the pendulum to record the peak acceleration of the pendulum in the Y axis. Two sets of data, consisting of 50 trials each were collected on separate days, using two different testers but using the same exact hardware settings. It was found that there was no significant difference between the two sets of data ($p > 0.05$). The standard error of the apparatus was similar for both sets at 0.002ms^{-2}. The Bland-Altman plot showed evenly distributed points in the scatterplot graph which indicates that the two sets of data are in agreement together with the linear regression analysis. The data from this study suggest that the custom built apparatus is a reliable instrument. The peak acceleration of the pendulum for each trial was highly consistent. Thus, the apparatus should be able to serve its purpose of measuring the reliability and accuracy of the Taekwondo electronic body protector scoring system.**

Keyword— **Reliability, Pendulum, PSS, Taekwondo**

I. Introduction

Taekwondo is now an Olympic sport since the Sydney 2000 Olympic Games. The sport had evolved from being a subjective sport into an objective one that utilized cutting edge technology on the body pad – known as the Protector and Scoring System (PSS). All points are given to exponents based on sensors reading whereby it is able to detect impact force. By using the PSS points can only be given to exponents when the sensors detect a minimal impact threshold, which are set differently for each weight division. The World Taekwondo Federation (WTF) had introduced the Daedo TK-Strike PSS into the Olympic Games for the first time during the London 2012 Olympic Games and it will again be used as an official scoring system again in this upcoming Rio 2106 Olympic Games.

There are not many studies that had looked at the reliability and validity of the PSS. To date, there seem to be only two studies that had been conducted before this, both of which had found that the PSS was unreliable. Both also used different test methods. In theory, PSS utilises a pre-set minimal impact threshold to record a scoring point, but it was found that the electronic body protector routinely scored points even with low impact [1]. The PSS was tested using an elastic spring steel rod in a mannequin foot which acts as a pre-loaded catapult, mimicking a taekwondo kick. Although the spring steel method was similar to a kick to a certain extent, the spring steel used can gradually degrade in strength and flexibility due to the possibility of fatigue loading and spring relaxation [2]. This might mostly likely have reduced the impact consistency.

In the only other study of the PSS, it was found that the PSS showed poor accuracy, reliability and linearity which is essential for a Taekwondo scoring tool [3]. The study was done using a drop test method with three different drop heights. The main drawback of that study was that the weights that were dropped on the PSS remained on it after each and every drop. This procedure is dissimilar to the momentary contact that occurs in an actual Taekwondo kick. Kicks are retracted immediately after each kick, meaning that the impulse time of the study was not applicable in the real world.

The sole purpose of the PSS is to eliminate biasness and encourage transparency of scoring during competition in conjunction with the Olympic spirit. Since this system is now an official scoring system of the Olympic Games, the validity and reliability of the PSS is extremely important as to provide fair play to every exponent. However, although the PSS was used since the London 2012 Olympic Games, there are surprisingly minimal evidence to support its validity and reliability. In order to test the PSS, one must first create a dependable test apparatus. The test rig must be first tested for its reliability to provide reliable data collection for testing the PSS in the future. The main purpose of this study is to find out the reliability of a customized gravity driven mechanical pendulum specifically created to test the PSS.

II. Research Design

The mechanical pendulum is built accordingly as per Figure 1 to try to mimic a kick hitting on the PSS, in the most consistent manner possible. It is made up of metal plates and tubular mild steel sections. Two sealed ball bearings were used to provide low friction rotational movement for the pendulum arm. While, the base of the

F. Ibrahim et al. (eds.), *3rd International Conference on Movement, Health and Exercise*,
IFMBE Proceedings 58,
DOI: 10.1007/978-981-10-3737-5_32

apparatus was securely bolted on to a concrete floor. The frame consists of four poles with the height of 1.4 meter each, and slanted at a 15 degrees angle. The base is 0.63 meter wide and 0.41 meter long and it was bolted onto the concrete floor of the lab. The length of the pendulum was 0.94 meter. A metal plate was welded in between the front and back frame and has an adjustable clamp to hold the PSS in place. The pendulum was held in place using an electromagnetic lock that is located on an extended tubular holding arm. The pendulum was released with a flick of a switch. The mechanical rig's pendulum swings at the same distance and velocity which produces the same amount of kinetic energy for a specific weight. The pendulum was purely accelerated by gravity without having other external forces acting on it, wind resistance was negligible. More importantly, because of the 15 degree slant, the pendulum will not 'stick' to the back plate but instead swing back after the impact.

To test on the apparatus's reliability, a calibrated Crossbow Accelerometer GP series (Crossbow, USA) with a sensitivity of 0.08 Volt was placed on top of the pendulum to record the peak acceleration of the pendulum in the Y Axis (Figure 1). For this study, the pendulum was released freely, without additional weights and without placing a PSS – meaning that the pendulum hits the back plate directly. Two sets of data, consisting of 50 trials each were collected on separate days, using two different testers but using the same exact hardware settings. Data was recorded using Powerlab 4/25(ADinstrument, Australia) and Powerlab Chart software. The released height of the pendulum was checked after every two trials to ensure that it was exactly the same height.

Data was analysed using SPSS software Version 23. Paired t-test, linear regression analysis followed by Bland-Altman test were used to analyse the differences and agreement between the two data sets.

III. Results

A total of 100 trials were recorded. Each set of data had 50 trials and Table 1 shows the mean peak acceleration of the pendulum, standard deviation, standard error of the apparatus, mean differences between trials of set one and two; and significance value of mean differences between the sets.

From Table 1 the standard error of the apparatus was similar for both sets at 0.002ms^{-2}. It was found that there is no significant difference between the two sets ($p > 0.05$). Points of differences between trials against mean of Sets were plotted using Bland-Altman method as in Figure 2.

Table 1: Mean ± SD peak acceleration and Standard Error.

	Set 1 (S1)	Set 2 (S2)	(S1-S2)
Mean peak acceleration of pendulum (ms^{-2})	2.17 ± 0.014	2.18 ± 0.013	-0.0038
Standard Error (ms^{-2})	0.002	0.002	
P value			0.205

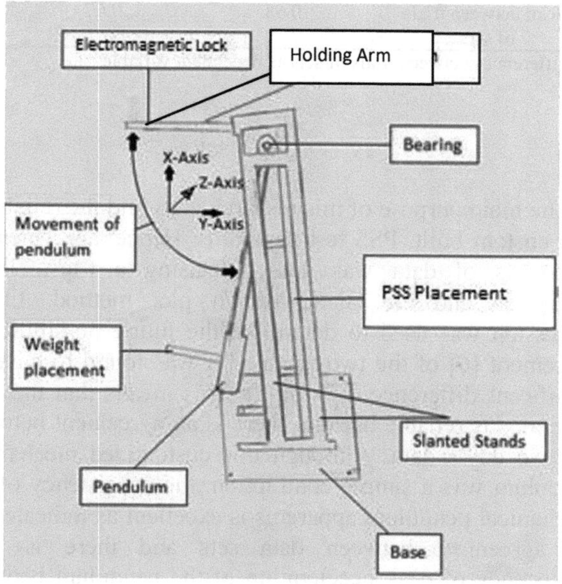

Fig. 1: Customized Pendulum Apparatus

The Bland Altman plot showed evenly distributed points in the scatterplot graph. Bland-Altman statistical analysis was chosen as the statistical analysis method in this study because it is able to find out the agreement between two different measurements [4]. According to the Bland Altman plot, there is an agreement and no proportional biasness

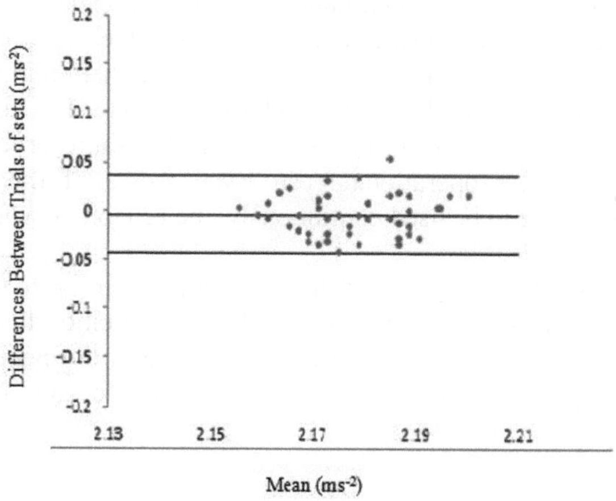

Figure 2 Dietary intake and energy expenditure

between the two data sets because of the evenly distributed points in the scatterplot graph [5].

Linear regression was analysed on both sets and found that there are no significant differences between the two sets ($p > 0.05$) as in Table 2.

Table 2: Coefficient of linear regression

Predictor variable	β Estimate[a]	*P* value
Mean between trials of sets	0.68	0.67

*Differences between trials of sets as dependent variable

IV. Discussion

The main purpose of this research is to find the reliability of a custom built, PSS test apparatus. Hence, agreement of both sets of data was analysed using a t-test, linear regression and the Bland-Altman plot method. Linear regression was used to determine the future possibility of agreement [6] of the two sets and it was found to have no significant difference ($p > 0.05$). This means that the PSS apparatus is reliable because there is an agreement between the two day's data. Although this customized mechanical pendulum was a simple contraption, the consistency of the mechanical pendulum apparatus is excellent as indicated by the agreement between data sets and there is high consistency of peak acceleration of the pendulum between each individual trial.

The mechanical pendulum is a reliable test apparatus possibly because it is purely gravity driven; and gravity is constant. Under the constant gravitational force, acceleration of the pendulum will also remain consistent for any number of given repetitions. By using this apparatus, one does not need to worry about the degradation of strength of the pendulum because it does not need external forces to move the pendulum such as for example when using the steel spring apparatus [1]. Steel spring will degrade in strength and flexibility after a certain number of repetitions [2] and it will need to be replaced. This will cause more time in terms of resetting the same procedure and a higher operational cost. Even worse, if the same spring were to be used continuously, the reliability would be questionable.

Although the concept of this apparatus is similar to one of the previous study which is the drop test method that also uses gravitational force - the impact force of the pendulum is only momentary due the slant angled design of the apparatus as opposed to a ball being stuck on the PSS [3]. This method is much more similar to an execution of a kick whereby kicks are retracted immediately right after contact.

The mechanical pendulum is much more convenient to be used as a test apparatus compared to the other two method mentioned [1] [3]. Extra weights can be added to the pendulum during a test to produce higher kinetic energy during impact. Besides that, because it has adjustable clamps, it can also be used to test all sections of the body protector. At the same time, with a solid metal surface, it will also reduce the absorption of kinetic energy during impact as well as increase the durability of the apparatus.

V. Conclusion

The current study has shown that the customized pendulum test apparatus is a reliable measurement tool. Thus, it should be able to serve its purpose of measuring the reliability and accuracy of the Taekwondo electronic body protector.

REFERENCES

1. Ramazanoglu, N (2013) Transmission of impact through the electronic body protector in taekwondo. International Journal of Applied Science and Technology 3(2): 1-7.
2. Salvange, P. S (2012) Design of helical coil compression spring: A review. International Journal of Engineering Research and Applications 2(6): 513-522.
3. Tasika, N (2013) Reliability & linearity of an electronic body protector employed in taekwondo games: A preliminary study. Journal of Human Sport and Exercise 8(3): 622-632 Doi:10.4100/jhse.2013.8.Proc3.08.
4. Zaki, R., Bulgiba, A., Ismail, R., et al (2012) Statistical methods used to test for agreement of medical instruments measuring continuous variables in method comparison studies: A Systematic Review. Plos One 7(5) Doi:10.1371/journal.pone.0037908.
5. Giavarina, D (2015) Understanding Bland Altman analysis. Biochemia Medica Biochem Med 25(2): 141-151. Doi:10.11613/bm.2015.015.
6. Zou, K. H., Tuncali, K., Silverman, S. G (2003) Correlation and simple linear regression 1. Radiology 227(3): 617-628.

Author: Ryan Chong Wy Lunn
Institute: University of Malaya
Street: Lembah Pantai
City: Kuala Lumpur
Country: Malaysia
Email: cp86ae@yahoo.com

Comparison of Vertical Jump Height Using the Force Platform and the Vertec

P.Y. Cheah, J.P.G. Cheong, R. Razman and N.E. Zainal Abidin

Exercise Science Unit, Sports Centre, University of Malaya, Kuala Lumpur, Malaysia

Abstract— **Vertical jump height is often being measured in many ways to evaluate the ability of an individual to jump, and to provide an estimation of lower limb muscle power. The various jump tests and apparatus each have their advantages and disadvantages. The main purpose of this study was to compare and explore the relationship between the two more popular test apparatus - the force platform and the Vertec. Sixty-nine university students, 32 females and 37 males between the ages of 20 to 23 (M age 22.01±0.66 years) were recruited to participate in this study. They were required to perform countermovement jumps on the force platform and using the Vertec. Both jump tests were run on the same day, with a 10-minute interval between tests. Results showed that the jump height values obtained from the force platform were significantly lower compared to jumping with the Vertec. The mean jump height for the force platform was 0.36m ± 0.094 and for the Vertec was 0.55m ±0.120 respectively. Although the mean jump height differs significantly, both the force platform and Vertec displayed a significant correlation (r =0.826, *p*<0.01). It was concluded that both devices are suitable to be used to measure jump height because those who obtained a high jump height at the Vertec also obtained a high value of jump height on the force platform. An important difference between both devices was the presence of a target of achievement for the Vertec.**

Keywords— **Jump height, Vertec, Force Platform, Countermovement Jump**

I. Introduction

Generally, every individual has the ability to jump as it is one of the fundamental motor skills that is acquired at a very young age. However, there are many factors that influence the capability of an individual to jump and attain the maximum jump height. The vertical jump test is frequently used to assess athletic skill by comparing the jump performance of starters vs. nonstarters [1] and to measure lower body power [2]. It also appeared to be a valuable measurement in identifying lower-limb functionality for the non-athletic population [3, 4]. While there are many types of devices that can be used to measure vertical jump height, two types of devices, the force platform and the Vertec, was investigated in this study.

A force platform is a rectangular metal plate with a piezoelectric or strain gauge transducer attached at each corner to give an electric output that is proportional to the force on the plate. It is a device that measures the force applied by an individual onto the force plate. There are few methods that can be used to measure the jump height using the force platform, including the flight time method, impulse-momentum method and the work-energy method [5]. Linthorne [5] found that both flight time and impulse momentum methods obtained via the force platform are able to provide a good estimation of jump height.

Alternately, the Vertec consists of several plastic swivel vanes arranged in half inch (0.0127m) increments which are attached to a telescopic metal pole that can be adjusted to a participants' standing reach. The test requires a participant to use their hand to displace the vanes with an overhead swinging motion at the peak of their vertical jump. The highest displaced horizontal swivel vane determines the maximum jump height. Vertical jump height is calculated by taking the difference between the fully extended standing reach measurement and the highest displaced horizontal swivel vane [6]. The Vertec is relatively inexpensive, require minimal training for operation, do not require extensive data analysis, and provide immediate results. [3]

In previous studies, jump height has been measured using Just Jump, contact mat, and Myotest accelerometers [2, 3, 7, 8, 9, 10, 11]. One study by Erich and colleagues [11] found 27% higher jump height when assessed by the Vertec, compared to the force platform. Another study conducted by Isaacs [7] compared the Vertec to the Just Jump system (VJM) in children aged 7 to 11 years old and found a significant difference ($r = 0.83$, $p < 0.001$) between vertical jump height scores measured with the VJM and the Vertec Jump height in the VJM is determined by measuring flight time [3], and the flight time is subsequently determined by microswitches located within the mat, which are sensitive to the lift-off of the feet from the mat and to the landing of the feet back on to the mat [6]. Participants jumping on the mat focuses purely on the jump and do not have the knowledge of results about the height of jump. Meanwhile, jumping using the Vertec would involve the coordination of jumping and swiping the vanes simultaneously as participants have a target to aim and reach. However, previous research has yet to investigate the relationship between the Vertec and the force platform on jump performance. Thus, the purpose of this study is to compare jump height and determine the relationship between the force platform and the Vertec.

F. Ibrahim et al. (eds.), *3rd International Conference on Movement, Health and Exercise*,
IFMBE Proceedings 58,
DOI: 10.1007/978-981-10-3737-5_33

II. Methods

A. Participants

Sixty nine university students, 32 females and 37 males between the ages of 20 to 23 (M age 22.01±0.66 years), from the University of Malaya were recruited to participate in this study. All the participants were healthy and had no injuries during the data collection period. They were briefed on the procedures of the study and required to sign an informed consent form. The study was approved by the University of Malaya Research Ethics Committee (UM.TNC2/RC/H&E/UMREC – 116) and procedures of this study were carried out in accordance with the guidelines regarding the use of human participants.

B. Instruments

The force platform was a rectangle metal piece (0.5m x 0.5m). It was set to 1200Hz for the capture rate. Gen 5 AMTI-NetForce (Watertown, MA, USA) was used in this study. According to Linthorne [5], jump height can be calculated using the flight time method where;

$$v_{to} = \frac{g t_{flight}}{2}$$

$$y_{flight} = \frac{v_{to}^2}{2g}$$

Where V_{to} is the vertical take-off velocity, t flight is duration of participant in air and y_{flight} is the jump height of the participant.

The Vertec (Sports Imports, Columbus, OH, USA) consists of 48 plastic vanes with three colours (white, blue and red). Every red vane indicates 6 inch, blue indicates 1 inch and white indicates 0.5 inch. The difference between standing height and the highest displaced vane measurement is the value of the jump height.

C. Procedures

Participants were required to perform countermovement jumps on two jumping test apparatus, which were carried out in a laboratory on the same day. The rest period between tests was 10 minutes. Height and weight of the participants were recorded. They were briefed on the procedures and the protocols of the tests. Prior to testing, participants warmed up by cycling on an ergometer at an average of 60 RPM for five minutes. Participants then proceeded to either jumping on the force platform followed by jumping with the Vertec, or vice-versa, the order by which was determined randomly. Three trials were given for each jumping device and the highest score for each was used for analysis. Only countermovement jumps were recorded, and trials with the wrong technique were repeated. A countermovement jump, starts from a standing position, followed by a squat with arms swinging backwards and then taking off the ground with the arms swinging up forward and upwards (see Figure 1 for an illustration).

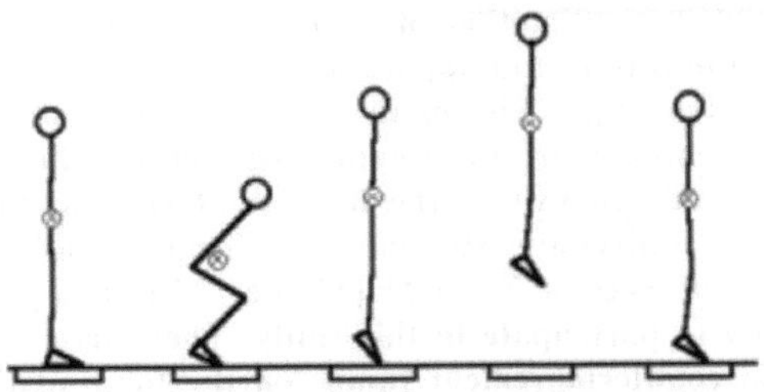

Figure 1. Countermovement Jump
(Adapted from Linthorne [5])

For the force platform jump test, participants stepped on the force platform when instructed and executed a countermovement jump. If they landed outside the platform on descend, the trial was repeated. Rest interval between trials was one minute.

For the Vertec jump test, the plastic vanes were adjusted according to the participants' maximum standing reach by extending their arm straight over the head (standing height). The dominant side was closest to the vanes. Participants then performed a maximal countermovement jump and displaced the vanes with their dominant hand. The highest displaced vane was recorded.

D. Data Analyses

To compare the vertical jump height between jumping on the force platform and jumping with the Vertec, the highest score for each of the jumps on each jumping device was recorded. Data was analyzed using IBM Statistical Package for the Social Science (SPSS) software. Data distribution was assessed using Kolmogorov Smimov's test for normality. Paired sample t-test was used to determine if there were significant differences between the two types of jump test apparatus with p level set at .01. Relationship between the two tests were analyzed using the Pearson product moment correlation coefficient. Means and standard deviations of jump height (m) were obtained using descriptive statistics.

III. Results

The mean jump height on the force platform was 0.36_ 0.094m while for the Vertec, it was 0.55±0.120m. There were significant differences between jump height for the force platform and Vertec, t (68) = -23.414, $p<0.01$, whereby participants jumping on the force platform showed a significantly lower value compared to those jumping with the Vertec. In terms of relationship there was a significant positive correlation (r =0.82, $p<0.01$). By convention, r = .70 to .90 is considered as a high positive correlation [12]. Figure 2 represented the relationship between the force platform and Vertec for jump height performance.

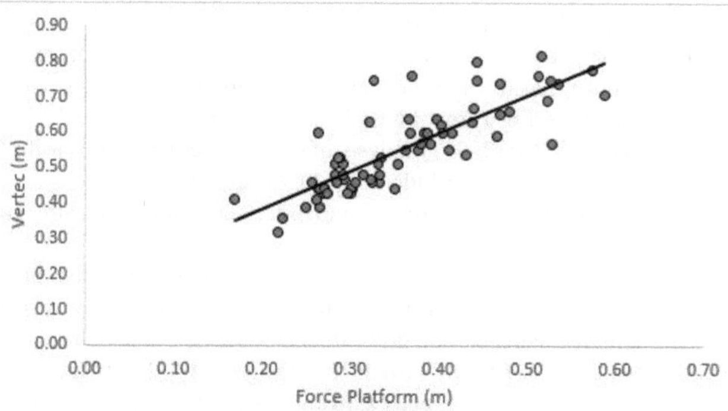

Figure 2: Scatter plot of jump height between force platform and Vertec

IV. Discussion

The purpose of this present study was to compare the difference in jump height between the force platform and the Vertec, as well as to look at the relationship between the two devices.

From the results obtained, participants' jump height was varied between the Vertec and the force platform, with mean scores for the Vertec significantly higher than the force platform. A previous study by Magnúsdóttir and colleagues [9] comparing three devices to measure jump height showed similar results whereby jump height results from the contact mat and accelerometer results that relied on flight time calculation showed systematically lower values compared to the Vertec. Another study had also found 27% higher jump height values when assessed by the Vertec, compared to the force platform [11].

Higher values from the Vertec could probably be due to extraneous factors such as target of achievement, motivation, knowledge of results, technique of jumping, and coordination. Jump height is increased when performers are given an external focus of instructions, relative to an internal focus or no focus instructions [8]. As seen from the results, an external focus which is the Vertec vanes seem to influence and drive the participants to jump higher and break their previous record. Participants are constantly motivated to reach for a higher vane as the knowledge of results exist. On the other hand, participants jumping on the force platform are unable to obtain the knowledge of results or any feedback of their performance.

Different motor coordination needs of the two types of jumps could also play a role. On the force platform, participants will jump with both arms swinging simultaneously throughout the swing as they are only focused on the jump. There is no need to reach for a target. Some participants tend to tuck their legs in during the jump and some would have their legs fully extended. Motor coordination patterns using the Vertec is different whereby participants need to coordinate the timing of the hand to swat the vanes of the Vertec while jumping. In order to reach higher, participants would put more effort to explode off the ground and at the same time reach as high as possible to displace a higher vane on the Vertec.

Position and movement of the limbs could well affect the center of gravity position and indirectly affect the jump height. Raising both arms or both arms and both legs, moves the center of gravity closer to the head thus closer to the hands. Lifting of arms causes another body part to move downwards in order for the center of gravity to continue moving along the parabolic path [13]. When jumping on the force platform, participants may raise their arms and legs, therefore causing the head and the trunk to go lower in order to compensate for the jump movement. This causes the jump height to be lower compared to jumping with only one hand raised (as in the Vertec), or jumping without lifting of the legs. Figure 3 illustrates the height of jump when different limbs are raised during the jumping phase.

Figure 3: Three different vertical jump techniques result in different jump heights but with the same height of the center of gravity. (Adapted from McGinnis [13])

Despite having distinct differences in the jump height between the force platform and Vertec, both these devices showed a strong positive correlation. When the participant was able to obtain a very high jump height on the force platform, the value obtained using the Vertec appeared to be high as well. Conversely, when jump performance was low on the force platform, the scores obtained from the Vertec were also low. This outcome displays a consistency in the evaluation between two different types of test. Looking through the past studies, Leard and colleagues [10] had also found similar results to this present study, whereby the correlation between Just Jump (contact mat), Vertec (jump and reach) and the criterion reference (3-camera motion analysis system) was statistically significant, although the mean scores among the three devices measuring vertical jump height were significantly different.

These results of this study provide evidence that although both tests are distinctly different from one another because jumping on the force platform does not require the participant to reach a target and involve a different pattern of coordination, the jump heights between the two are closely related. Future research should possibly compare the differences between jumping on the force platform without a target and jumping on the force platform with a target that is specific to a sport skill.

V. Conclusion

The present study showed that the jump height on the force platform and Vertec were significantly different yet were strongly correlated. Previous studies with the same concept gave similar results as what was found. It appeared that either one of these tests are suitable and valid to be used as a jumping assessment for athletes as well as sedentary people. Depending on the availability of the test apparatus, both the Vertec and the force platform would provide researchers with valid results.

Acknowledgment

The authors would like to acknowledge funding received from University of Malaya's grant number BK041-2015 "Contextual Interference Effects on the Acquisition and Retention of Power on the Lower Limbs". The authors also thank all the participants who volunteered and spent their time participating in this study.

Conflict of Interest

The authors declare that they have no conflict of interest.

References

1. Black, W, Roundy, E (1994) Comparison of size, strength, speed, and power in NCAA division 1-A football players. J Strength Cond Res 8: 80–85.
2. Harman, EA, Rosenstein, MT, Frykman, PN, Rosenstein, RM, Kraemer, WJ (1991) Estima-tion of human power output from vertical jump. J Appl Sport Sci Res 5: 116–120.
3. Nuzzo, JL, Anning, JH, Scharfenberg, JM (2011) The reliability of three devices used for measuring vertical jump height. J Strength Cond Res 25(9): 2580–2590.
4. Pijnappels, M, van de Burg, JCE, Reeves, ND, van Dieen, JH (2008) Identification of elderly fallers by muscle strength measures. Eur J Appl Physiol 102: 585–592.
5. Linthorne NP (2001) Analysis of standing vertical jumps using a force platform. Am J Phys 69(11): 1198-1203.
6. Klavora, P (2000) Vertical-jump test: A critical review. Strength Cond J 22(5): 70–74.
7. Isaacs, LD (1998) Comparison of the Vertec and Just Jump systems for measuring height of vertical jump by young children. Percept Mot Skills 86: 659–663.
8. Wulf G, Dufek JS, Lozano L, Pettigrew C (2010) Increased jump height and reduced EMG activity with an external focus. Hum Mov Sci 29: 440–448.
9. Magnusdottir A, Porgilsson B, Karlsson B (2014) Comparing three devices for jump height measurement in a heterogeneous group of sub-jects. J Strength Cond Res 28(10): 2837-2844.
10. Leard JS, Cirillo MA, Katsnelson E, Kimiatek DA, Miller TW, Trebincevic K, & Garbalosa JC (2007) Validity of two alternative systems for measuring vertical jump height. J Strength Cond Res 21(4):1296-9.
11. Erich JP, Tyler VZ, Brad W, William PE (2010) Comparison of jump height values derived from a force platform and Vertec. Proc. 28th Conference of the International Society of Biomechanics in Sports, Marquette, Michigan, USA, 2010, pp 572-575.
12. Mukaka, MM (2012) A guide to appropriate use of correlation coefficient in medical research. Malawi Med J 24(3): 69-71.
13. McGinnis, PM (2013) Biomechanics of sport and exercise. Champaign, IL: Human Kinetics.

Author Pei Yi Cheah,
Institute: Sports Centre, University of Malaya
Street: Lembah Pantai
City: 50603 Kuala Lumpur
Country: Malaysia
Email: peiyicheah@gmail.com

Fabric-Based Sensor for Applications in Biomechanical Pressure Measurement

N. Mohd Alias[1], M.Y. Ahmad[1,2] and N.A. Hamzaid[1]

[1] Department of Biomedical Engineering, Faculty of Engineering, University of Malaya, Malaysia
[2] Center for Innovation in Medical Engineering (CIME), Department of Biomedical Engineering, Faculty of Engineering, University of Malaya, Malaysia

Abstract— **Wearable pressure sensor is an important element in human gait and biomechanical investigations. Many studies have been carried out to ensure its reliability and comfort to the user even when the sensor has to be operated under extreme conditions such as uneven surface with high curvature area, high humidity environment, and wide range of dynamic pressure. However, the existing pressure sensors that could endure such extreme conditions usually involve complex fabrication processes and often incur high cost of materials. Therefore, in this paper we present a novel pressure sensor made of low cost fabric materials with simple fabrication process. The sensor is constructed from a conductive fabric using piezoresistive technique. A unit area of our pressure sensor is made of a thin layer of piezoresistive material and it is placed in between two layers of conductive fabrics. To vary sensing sensitivity and to cater wide range of pressure, our proposed sensor uses special mesh layers which are placed in between the piezoresistive layer and the conductive layers. This way, wide range of dynamic pressure can be captured. To test the proposed sensor, five units of pressure sensors have been designed with variety of mesh layering techniques. These sensors are tested using 0- 3 kg load and the data are analyzed graphically. Our experimental results on the resistive versus pressure response show a similar trend as compared with the existing popular pressure sensor such as flexiforce in term of force versus resistance curve. The use of mesh layer allows measurement of higher magnitude of pressure. Since the performance of our sensor can be tailored to exhibits the performance of the exiting commercial sensors therefore the proposed technique can be an alternative affordable solution to that of the existing expensive sensors.**

Keywords— **Pressure sensor, Wearable sensor, Biomechanics, Piezoresistive.**

I. INTRODUCTION

For prosthetic users, especially running athletes, measurement of surface pressure is highly important. A layer of pressure sensor can be useful in measuring physical irritations that may appear on the surface of human skin and underlying soft tissues [1]. The measurement are useful for adaptive technology to control the effect of high pressure, shear stress, abrasive relative motions and other physical irritations on the skin machine interface [2]. A Part from it, pressure sensor is also useful in biomechanics and gait investigations. For better performance, a highly flexible and comfortable pressure sensor is demanded in biomechanics investigation that can operate within complex interface due to geometry of human body structure [3]. For this reasons, many studies have been carried out to ensure the pressure sensor is reliable and comfort to the user even when the sensor has to be operated under extreme conditions such as under uneven surface with high curvature areas, high humidity environments, and wide range of dynamic pressure [4]. However, the existing pressure sensors that could endure such extreme conditions are usually involving complex fabrication process and often incur high cost of materials [5].

In recent years, pressure sensing units can now be implemented onto fabric surfaces. Simple fabrication process, low cost and reliable performance are among desirable features anticipate for wearable pressure sensor in biomechanics and gait investigation. Different types of material can be utilized along with conductive fabrics that eventually performs as pressure sensor. Such technology can be used to measure pressure on human body surface [6]. Fabric based pressure sensor technology showed an encouraging results in providing measurements for gait performance [7]. However, the existing fabric pressure sensor is too sensitive and limited range of pressure measurement normally in the range of 0-1N/m^2. Therefore, the objective of this paper is to investigate a method to improve the pressure measurement range using off the shelf material and fabrics.

II. SENSOR DESIGN

A. Group A sensor

The construction of fabric-based pressure sensor is divided into two groups: Group A and group B. Group A sensor consists of two outer layers (top and bottom) made of a common fabric from neoprene with the dimensions of 25 x 25 x 0.01 mm. Neoprene is a non-conductive fabric consisting of synthetic rubber material designed with high flexibility and can maintain its properties even in the presents of high humidity. These two layers function to protect the conductive materials located underneath of each of the outer layers. These conductive materials act as electrodes that transport current into and out of the sensor. The conductive materials are stretchable type of conductive fabric made up from 78% Nylon, 22% elastomer and 99% pure silver. Finally a layer of velostat is placed in between

F. Ibrahim et al. (eds.), 3rd International Conference on Movement , *Health and Exercise*,
IFMBE Proceedings 58,
DOI: 10.1007/978-981-10-3737-5_34

the two electrodes thus forming piezo resistive effects. Velostat is carbon thin approximately 4mm infused plastic (polyolefin) that has high resistivity in the range of 500 Ω per cm. The resistance of the velostat changes when pressure is applied. Three types of group A pressure sensors are fabricated by modifying the number of velostat layers. Type A1 has single layer of velostat, type A2 has double layers and type A3 has three layers.

B. Group B sensor

The materials, arrangement and shape of the internal layers of sensor Group B exactly similar as in sensor Group A. The only different is that in sensor group B an additional mesh layer of fabric is added between the velostat and conductive fabric as illustrated in figure 1. Different types of materials, opening area and texture of mesh layers are employed to investigate the property of the sensor. We find out that, different types of mesh layer varies the sensitivity of the sensor and more importantly it changes the range of pressure measurement. Figure 1 illustrates the layer arrangement of group B sensor. In this study, we fabricated two types of group B sensors namely B1 and B2. The only difference between these two sensors is on the type of mesh fabric used. B1 uses square shaped mesh fabric made of polyester with opening unit square of (0.36 mm x 0.40mm), thickness of 0.10mm. B2 uses diamond shaped mesh fabric made of nylon and it has opening unit square of (2.55 mm x 5.68mm), thickness of 0.54 mm.

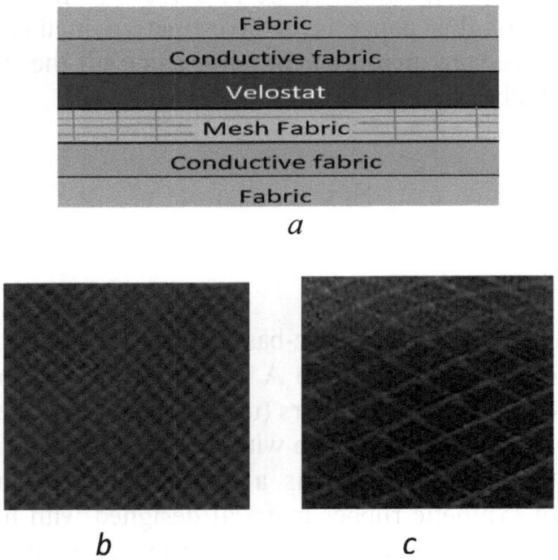

Figure 1 –a: Described the layers of group B sensor, b: polyester mesh fabric that are used for Sensor type B1, c: nylon mesh fabric that are used in sensor type B2

C. Sensor Design

All the constructed sensors are designed with "Mickey Mouse" shape (figure 2), with pin terminals located on its top. At this stage, the design is aimed to sense equally distributed pressure. The sensing area of the sensor is 12.6 cm^2 following the overlapping area between the conductive fabric and velostat. A bigger area of velostat as compared to the conductive fabric is used to avoid short circuit.

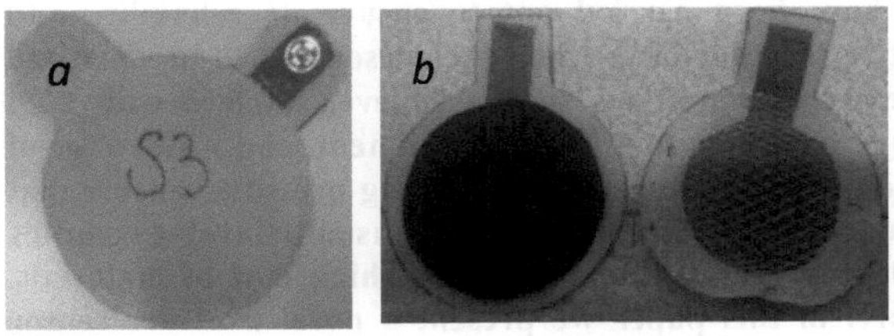

Figure 2 – a. "Mickey Mouse" shape fabric-based pressure sensor with sensing area of 12.6 cm2. b. Internal design of the sensor.

III. Experimental setup

A bench testing are performed by using power generator to supply voltage into the sensor. The input and the output voltages are recorded using an oscilloscope. A circular plastic probe with an area 7.07 cm^2 height 1cm was used to apply a uniform force perpendicular to the surface of the sensor. Different level of applied forces is achieved using a standard set of circular metal blocks having their specific weight ranging from 0 to 30N. With these metal blocks, the applied force can be varied between 0-30N. To ensure accuracy a calibrated electronic weighting platform is employed to verify the actual force. Pressure acting on the sensor is obtained by the well-known equation (Pressure = Force applied/Contact area). The experiment was repeated 5 times for every five units of sensor samples. The experiment setup is illustrate in figure 3.

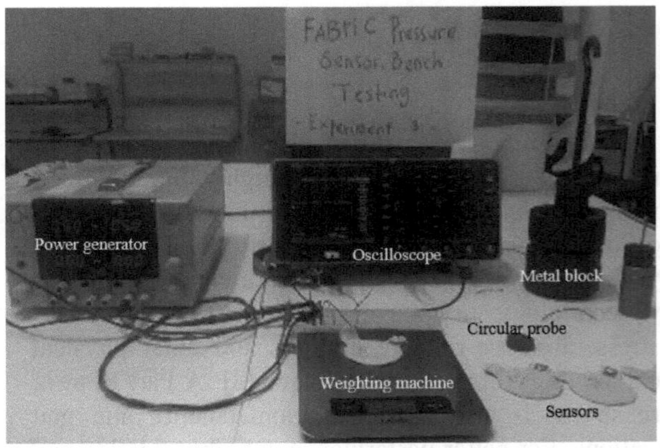

Figure 3 Experimental setup for sensor bench testing

The wiring connection of the experimental setup can be simplified as in figure 4. A series resistor (R) with the value of 10kΩ is connected in series with the fabric sensor to form a simple voltage divider. The input voltage (V_1) is set to 5.2V while. The variations of output voltage (V_3) are recorded from oscilloscope. The changes of sensing resistor that appears across the sensor terminals can be calculated as:

$$R_s = \frac{V_1 - V_3}{V_3}(R)$$

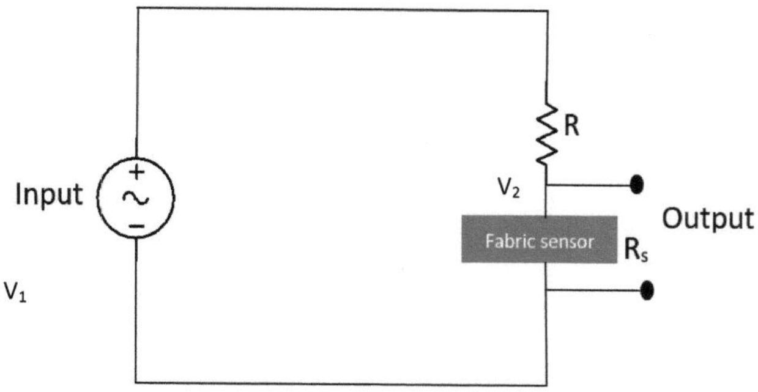

Figure 4 The Simplified test circuit Fabric-based pressure sensor

IV. Result and Discussion

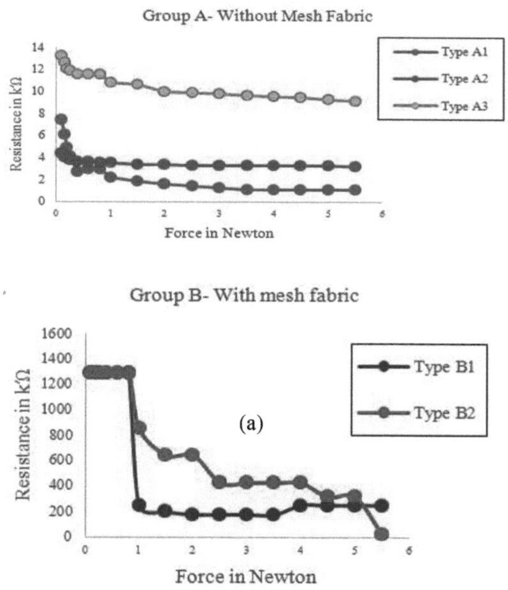

(b)

Figure 5 Results of fabric-pressure Based on characterization of Group A- without Mesh Fabric while Group B- with Mesh fabric.

Results in Figure 5 indicate that all the fabricated sensors have the ability to detect the changes of pressure. The use of additional layers of velostat in sensors A2 and A3 increases the overall sensor resistivity however their sensitivity remains unchanged. The response curve for group A is suitable for measurement in the range of 0 to 1N.

As for the group B sensors, the use of mesh fabric modifies the resistivity and also the sensitivity. It only becomes responsive when the force exceeds 1N. In group B, B2 is the most suitable sensor that can perform over a high range of force starting from 1N and above up to 5N.

Based from the above results, the technique of using mesh layer can be an effective approach to increase the range of sensing measurement which is desirable for biomechanics investigations. Our preliminary findings have revealed that it is possible to custom made simple, low cost, soft and flexible pressure sensors using the existing fabric-based materials.

V. Conclusions

Different sensors have been designed and categorized based on the force sensitive resistor principle. The effect of piezoresistive layer(s) and additional mesh fabric layer were investigated. It was found that different number of layers of piezoresistive material affects the initial resistivity of the sensor. The use of mesh fabric help to modify the sensitivity of the sensor as well as increase the range of measurement. The sensor have high potential to be used in pressure ulcer investigations in prosthetic socket and have many others application especially in sport biomechanics and orthopedic devices

Acknowledgment

The authors gratefully acknowledge the funding from Postgraduate Research Grant (PPP) PG167-2015B and University of Malaya Research Grant (UMRG Programme) - AET (Innovative Technology (ITRC)) RP009D-13AET from University of Malaya.

Conflict of Interest

The authors have no conflict of interest in the development of fabric-based pressure sensor.

References

1. S. Patel, H. Park, P. Bonato, L. Chan, and M. Rodgers, “A review of wearable sensors and systems with application in rehabilitation,”

pp. 1–17, 2012.
2. D. Tsiokos, G. T. Kanellos, G. Papaioannou, and S. Pissadakis, "Fiber Optic Based Pressure Sensing Surface for Skin Health Management in Prosthetic and Rehabilitation Interventions," *Biomed. Eng. - Tech. Appl. Med.*, pp. 245–268, 2012.
3. M. Lowe, A. King, E. Lovett, and T. Papakostas, "Flexible tactile sensor technology: bringing haptics to life," *Sens. Rev.*, vol. 24, no. 1, pp. 33–36, 2004.
4. E. a Al-Fakih, N. A. A. Osman, A. Eshraghi, and F. R. M. Adikan, "The capability of fiber Bragg grating sensors to measure amputees' trans-tibial stump/socket interface pressures.," *Sensors (Basel).*, vol. 13, no. 8, pp. 10348–57, Jan. 2013.
5. C. Oquigley, M. Sabourin, S. Coyle, J. Connolly, J. Condall, K. Curran, B. Corcoran, and D. Diamond, "Characteristics of a piezo-resistive fabric stretch sensor glove for home-monitoring of rheumatoid arthritis," *Proc. - 11th Int. Conf. Wearable Implant. Body Sens. Networks Work. BSN Work. 2014*, pp. 23–26, 2014.
6. L. G. Harris, "Design and Fabrication of a Piezoresistive Tactile Sensor for Ergonomic Analyses," 2014.
7. O. Tirosh, R. Begg, E. Passmore, and N. Knopp-Steinberg, "Wearable textile sensor sock for gait analysis," *Proc. Int. Conf. Sens. Technol. ICST*, pp. 618–622, 2013.

Author: Mohd Yazed Ahmad
Institute: University of Malaya
Street: Department of Biomedical, Faculty of Engineering
City: 50603 Kuala Lumpur
Country: Malaysia
Email: myaz@um.edu.my

A Pilot Study on Physical Performance Measures: What is Better for Knee Osteoarthritis Patients, Orthosis or Gait Modifications?

S.J. Khan[1], S.S. Khan[1], J. Usman[1,2], A.H. Mokhtar[3] and N.A. Abu Osman[1,2]

[1] Department of Biomedical Engineering, University of Malaya, Kuala Lumpur, Malaysia
[2] Centre of Applied Biomechanics, University of Malaya, Malaysia
[3] Sports Medicine Unit, Faculty of Medicine, University of Malaya, Kuala Lumpur, Malaysia

Abstract— **Background: Knee osteoarthritis (kOA), the most commonly occurring type of osteoarthritis in the world and one of the key contributors to global burden of disease significantly affects the physical functionality of the patient. The two conservative techniques of kOA include orthoses (knee braces and wedged insoles) and gait modification techniques (toe-in and toe-out gait). Previous studies assessing these two types of techniques do not present a comparison between their effects on physical functionality of the patient. Methods: Five OARSI (Osteoarthritis Research Society International) recommended performance-based tests to measure physical function of kOA patients: (1) 30s chair stand test (2) 40m fast-paced walk test (3) stair climb test (4) timed up and go (TUG) test (5) 6min walk test were applied on a total of 5 kOA patients randomly (Age: 59.2 ± 11.09 years, BMI: 25.94 ± 3.77 kg/m2, Kellgren-Lawrence Grades 1,2 and 4). The test conditions included normal gait, toe-out gait, toe-in gait, use of laterally wedged insoles and use of knee brace. The study was a single visit study assessing immediate effects, however, patients were given enough time and training to get used to the interventions. Data from the tests were analyzed using Repeated-Measures ANOVA. Results: No significant differences were observed among the five test conditions for 30sec chair stand test, 40m fast-paced walk test, TUG test and 6min walk test ($p > 0.05$). Pairwise comparison showed that toe-out differed significantly from toe-in ($p = 0.042$), insoles ($p = 0.037$), and brace ($p = 0.011$) while toe-in differed significantly from toe-out ($p = 0.042$), insoles ($p = 0.035$) and brace ($p = 0.012$). Conclusions: In general, orthoses and gait modifications seemed to have similar physical performance measures for kOA patients. The results however, indicated that gait modifications are affecting stair ascent of the kOA patients differently as compared to orthoses. Larger sample size may reveal a more accurate influence of the test conditions used in this study.**

Keywords— **Physical performance measures, Toe-out gait, Toe-in gait, Knee brace, Lateral wedged insoles.**

I. Introduction

Joint degeneration can be understood more comprehensively by taking into account the mechanical factors involved. Joint loadings, angles and moments all need to be in delicate balance in order to maintain a healthy knee and can hence prevent gradual degradation [1]. One of the major consequences of knee joint degradation is knee osteoarthritis.

Knee osteoarthritis (kOA) is the most commonly occurring type of osteoarthritis in the world largely because of its wearisome mobility and load-bearing characteristics during gait [2, 3].

It has been ranked one of the largest contributors to global disability, with the highest prevalence in Asia Pacific high income region [4]. There has been an alarming increase of 64.8% in Years of Life Lived with Disability (YLDs) due to kOA from 1990 to 2010 [5]. Due to this high prevalence and YLDs, there is a growing research interest in its prevention, therapy and causative factors.

The management of kOA goes through two streams: first is to slow down its progression through different types of interventions and second, which is considered as the gold standard is the replacement of the diseased knee joint. The design of any intervention aimed at slowing down the progression of kOA includes decreasing knee joint load, decreasing pain and increasing physical performance of the patient. Gait retraining methods that include toe-out gait and toe-in gait have been tested for load and pain reduction but not for physical performance [6-8].

Literature reports a shift in preference towards performance-based tests over self-reported physical function tests [9-12]. This is so because of the biasness of perception present in the scores in self-reported function tests. The physical performance tests, on the other hand, report the actual abilities of a person, rather than their perception of their abilities. Currently, there exist no standardization for performance-based tests, therefore we have included the OARSI (Osteoarthritis Research Society International) recommended performance-based tests to measure physical function of kOA patients [13]. Three of these five tests (30-second chair test, 40m fast-paced walk test and stair climb test) are the three core tests, while the other two (timed up and go test and 6-minute walk test) are additional tests.

The objectives of the study were (1) evaluating the immediate effects of toe-in gait and toe-out gait on physical performance measures in kOA patients (2) comparing these effects with those of orthoses (knee brace and lateral wedged insoles).

F. Ibrahim et al. (eds.), *3rd International Conference on Movement, Health and Exercise*,
IFMBE Proceedings 58,
DOI: 10.1007/978-981-10-3737-5_35

II. Materials and Methods

A. Sample Information

Five knee osteoarthritis patients (2 males, 3 females) were recruited for this pilot study through University of Malaya Medical Centre and community advertisement. The mean and standard deviation (SD) values (Mean ± SD) of the sample taken were: Age: 59.2 ± 11.09 years, BMI: 25.94 ± 3.77 kg/m^2

B. Inclusion Criteria

The inclusion criteria for the subjects included medial knee osteoarthritis patients (Kellgren-Lawrence grades 1-4), aged 40 to 70 years, having a BMI of less than 30 kg/ m^2 (non-obese [14]) and having no other physical disability. The participants were excluded on the basis of any neurological or musculoskeletal disorder or inability to adopt toe-in and toe-out gait pattern.

C. Ethical Approval

Approval was obtained from University of Malaya Medical Centre (UMMC). All participants provided written informed consent for the study.

Interventions

D. Orthoses

The orthoses used in this study were knee brace (KB) and lateral wedged insole (WI). The knee brace (Donjoy OA Adjuster™ 3, USA) worked on 4 points of leverage system. The lateral wedged insole (Salfordinsole™, UK) was full-length with 5◦ inclination. Both of these orthoses are designed for medial compartment knee osteoarthritis.

Gait Modifications

The study deals with the effects of two gait modification techniques that deal with foot progression angle. The participants walked with straight foot, self-selected toe-out angle (TO: making V shape with their feet) and toe-in angle (TI: making A shape with their feet). The participants were given practice sessions for these gait modifications with toe-out angle of 30 ° and toe-in angle of -10°. During experiment the participants were allowed to modify/ self-select their toe angles per their comfort level.

E. Data Collection

The participants were provided with shoes (Supercloud, Adidas, UK) in order to avoid influence of different shoes. They were briefed about interventions and study protocol and were given adequate practice time.

We followed OARSI (Osteoarthritis Research Society International) recommended performance-based tests to assess the physical function in people with KOA, described briefly through Table 1 and Figure 1. These tests were performed in a randomized order for five test conditions: natural walk (N) and each of the four interventions (TO, TI, KB, WI).

F. Statistics

Shapiro-Wilk test was applied to the data to assess normality. Differences among base-line and intervention values were evaluated using repeated measures ANOVA ($p < 0.05$) with pairwise comparison (LSD method) using IBM SPSS version 20 (SPSS Inc., USA).

III. Results

Figure 2 represents mean values of outcome variables for all test conditions. Repeated measures ANOVA showed insignificant differences for 4 out of 5 performance based tests.

Significant differences were found only for stair climb test [$F(4, 16) = 5.532$, $p = 0.020$]. Pairwise comparison showed that TO differed significantly from TI ($p = 0.042$), WI ($p = 0.037$), and KB ($p = 0.011$) while TI differed significantly from TO ($p = 0.042$), WI ($p = 0.035$) and KB ($p = 0.012$).

IV. Discussion

Knee joint replacement is considered to be the gold standard in the treatment of kOA. The main rationale behind opting for this treatment technique by patients is to increase the physical performance of the patients [15]. This need of improving physical performance should be considered while devising other treatment and prevention protocols for kOA also. This led us to evaluate gait modification techniques in terms of physical performance of kOA patients and comparing them with those of other conservative treatment techniques (orthoses).

A qualitative observation made from the results is that the orthoses have an edge in improving physical capabilities of the kOA patients during 30-second chair stand test, TUG test, 40m fast-paced walk test and stair climb test. The 6-minute walk test showed the knee brace to be decreasing the physical performance the most. Since this test involves longer period of walking, the patients reported difficulty and discomfort in carrying the brace reporting it to be "heavy" and "bulky" as compared to other interventions. But since these differences

Table 11: Brief description of the five standard performance-based tests used in the study.

Performance Test	Protocol	Equipment Used	Outcome Variable
30-second Chair Stand Test	The participant is asked to perform a sit-to-stand activity for a period of 30sec without any support	Chair without arms with a seat height of 44cm	Number of chair stands
Timed Up and Go (TUG) test	The participants are asked to stand up from a sitting position, walk 3 meters at normal walking speed, turn around, walk the same path to the chair and sit down on it.	Chair with arm rest. Seat height of the chair was 44cm and arm-rest height was 20cm from seat.	Time in seconds
6-minute walk test	Participant is asked to walk on a defined path in 6 minutes.	Flat walking area with length = 10m, width = 3m and arc-length =1m. Total perimeter = 30m	Distance in meter
Stair Climb Test	Participant is asked to ascend and descend a staircase, without using handrails.	Staircase with 10 steps. Each step with height = 18cm and width =35cm	Time in seconds
40m (4 x 10m) Fast-paced walk Test	Participants are asked to walk 4 times on a 10m walkway as fast as they can.	Walkway with a length of 10m	Walking speed in meter/ second

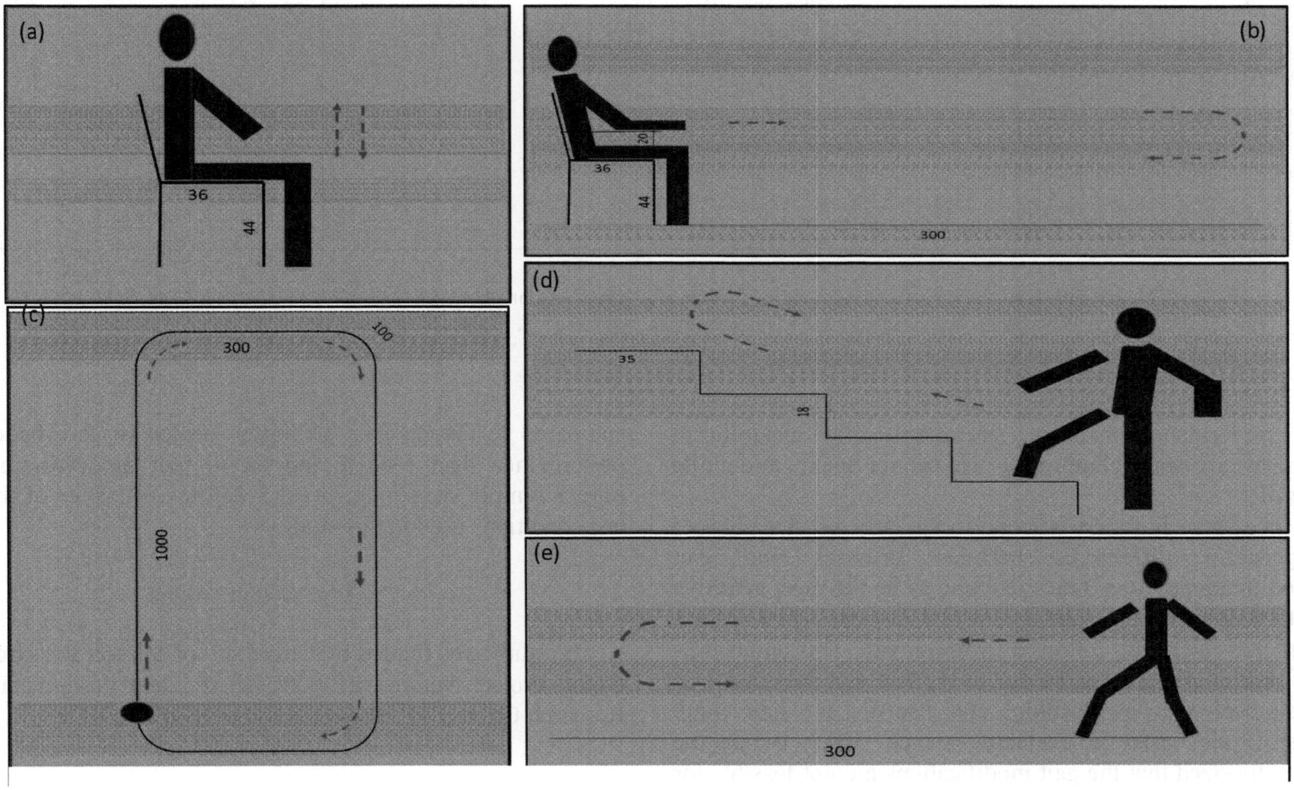

Figure 1 Schematics of performance-based tests (a) 30s chair stand test (b) time up and go (TUG) test (c) 6min walk test (d) stair climb test (e) 40m fast-paced walk test. Grey arrows show direction of movement or progression. All dimensions are given in centimetres.

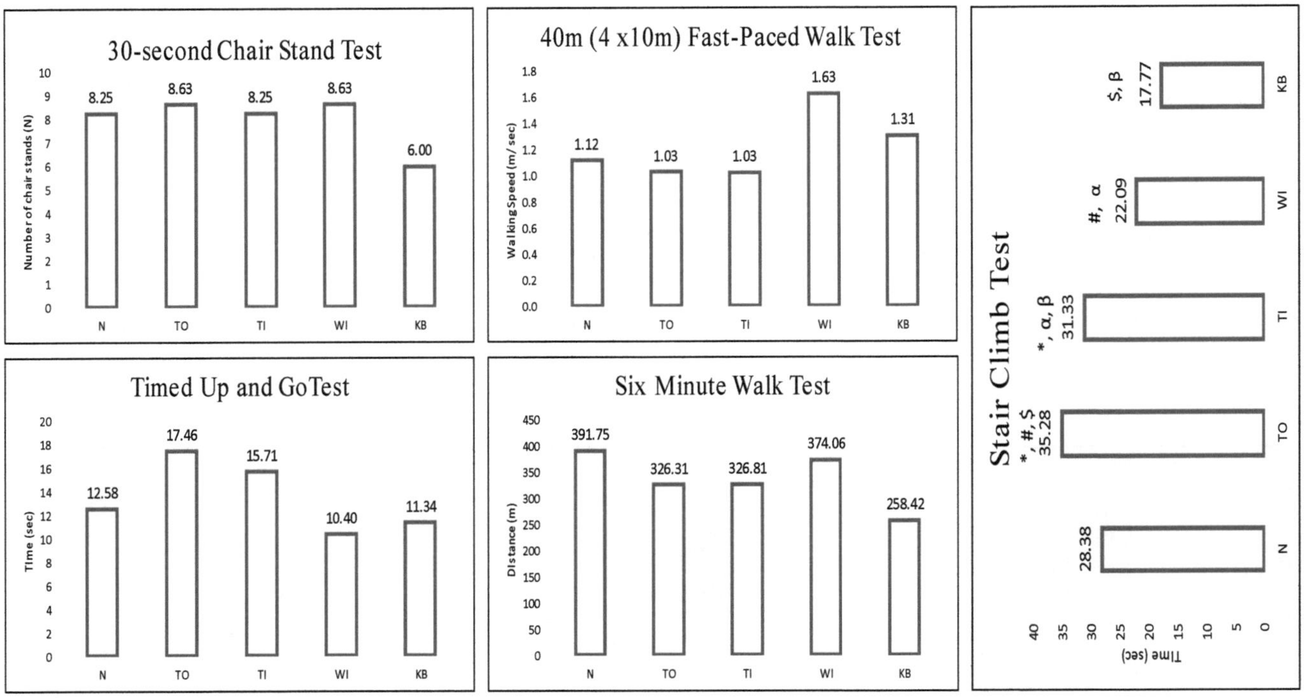

Figure 2: Mean values representing outcome variables of performance-based tests for all test conditions. Where N is natural gait, TO is toe-out gait, TI is toe-in gait, WI is lateral wedged insoles and KB is knee brace. Symbols (*, #, $, α and β) represent significant differences between test conditions.2

in performance measures are statistically insignificant except for stair climb test, we cannot make a definite verdict.

The stair climb test is found to the only test that showed significant differences between orthoses and gait modifications. Knee brace is found to be the most effective intervention during stair ascent and descent while toe-out gait was the worst, followed by toe-in gait. This uniqueness of stair climb test may be due to the fact that the other four tests deal with level-walking and also that the stair climb test is also used to evaluate balance [13]. It may be hypothesized that the gait modifications are not feasible for the activities that require higher postural stability and lower risk of fall. Since kOA patients already have compromised balance, this observation can be important in recommending gait modification techniques during stair ascent and descent [16].

V. Conclusions

Gait modification techniques did not seem to have any considerable effect on physical performance measures as compared to natural gait. Orthoses seemed to give better performance than toe-out and toe-in gait modifications. Larger sample sizes may reveal a definite influence of the test conditions used in this study

Acknowledgment

The study was funded by University of Malaya Research Grant, Project Number RP-020C-13AET and Postgraduate Research Grant (PPP) Project Number PG072-2015B.

Conflict of Interest

The authors declare having no conflict of interest.

References

1. Moyer, R.F., et al., Osteoarthritis Year in Review 2014: mechanics - basic and clinical studies in osteoarthritis. Osteoarthritis and Cartilage, 2014. 22(12): p. 1989-2002.
2. Badley, E.M., Arthritis in Canada: what do we know and what should we know? . J Rheumatol Suppl 2005: p. 32-41.
3. Lawrence, R.C., D.T. Felson, and C.G. Helmick, Estimates of the prevalence of arthritis and other rheumatic conditions in the United States. Arthritis Rheum, 2008. 58: p. 26-35.
4. Cross, M., et al., The global burden of hip and knee osteoarthritis: estimates from the global burden of disease 2010 study. Annals of the rheumatic diseases, 2014: p. annrheumdis-2013-204763.
5. Vos, T., et al., Years lived with disability (YLDs) for 1160 sequelae of 289 diseases and injuries 1990–2010: a systematic analysis for the Global Burden of Disease Study 2010. The Lancet, 2012. 380(9859): p. 2163-2196.
6. Hunt, M.A. and J. Takacs, Effects of a 10-week toe-out gait modification intervention in people with medial knee osteoarthritis: a pilot, feasibility study. Osteoarthritis and Cartilage, 2014. 22(7): p. 904-911.
7. Simic, M., et al., Altering foot progression angle in people with medial knee osteoarthritis: the effects of varying toe-in and toe-out angles are mediated by pain and malalignment. Osteoarthritis and Cartilage, 2013. 21(9): p. 1272-1280.
8. van den Noort, J., et al., The effectiveness of voluntary modifications of gait pattern to reduce the knee adduction moment. Hum Mov Sci, 2013. 32(3): p. 412-24.
9. Parent, E. and H. Moffet, Comparative responsiveness of locomotor tests and questionnaires used to follow early recovery after total knee arthroplasty. Archives of physical medicine and rehabilitation, 2002. 83(1): p. 70-80.
10. Stratford, P.W. and D.M. Kennedy, Does parallel item content on WOMAC's pain and function subscales limit its ability to detect change in functional status? BMC Musculoskeletal disorders, 2004. 5(1): p. 1.
11. Stratford, P.W. and D.M. Kennedy, Performance measures were necessary to obtain a complete picture of osteoarthritic patients. Journal of clinical epidemiology, 2006. 59(2): p. 160-167.
12. Stratford, P.W., D.M. Kennedy, and D.L. Riddle, New study design evaluated the validity of measures to assess change after hip or knee arthroplasty. Journal of clinical epidemiology, 2009. 62(3): p. 347-352.
13. Dobson, F., et al., OARSI recommended performance-based tests to assess physical function in people diagnosed with hip or knee osteoarthritis. Osteoarthritis and Cartilage, 2013. 21(8): p. 1042-1052.
14. Organization, W.H., Obesity: preventing and managing the global epidemic. 2000: World Health Organization.
15. Kennedy, D.M., et al., Assessing stability and change of four performance measures: a longitudinal study evaluating outcome following total hip and knee arthroplasty. BMC Musculoskelet Disord, 2005. 6: p. 3.
16. Khalaj, N., et al., Balance and risk of fall in individuals with bilateral mild and moderate knee osteoarthritis. PloS one, 2014. 9(3): p. e92270.

Author: Juliana Usman
Institute: University of Malaya
Street: Jalan University
City: Kuala Lumpur
Country: Malaysia
Email: juliana_78@um.edu.my

Kinematics Analysis of a 3DoF Lower Limb Exoskeleton for Gait Rehabilitation: A Preliminary Investigation

M. A. Zakaria, A. P.P. Abdul Majeed, I. M. Khairuddin, and Z. Taha

Faculty of Manufacturing Engineering, Universiti Malaysia Pahang

Abstract— **Robotics have been engaged to address the shortcomings of conventional rehabilitation therapy as well as the ever increasing demand for rehabilitation services. This paper presents the kinematics of a 3DoF lower limb exoskeleton restricted to the sagittal plane. The Denavit-Hartenberg representation, as well as the geometrical solution approach, are employed to obtain the forward and inverse kinematics of the exoskeleton, respectively. A simulation study is performed to validate the proposed model.**

Keywords— **Exoskeleton kinematics, Exoskeleton forward kinematics, Exoskeleton inverse kinematics, Exoskeleton D-H representation, Lower limb kinematics.**

I. Introduction

The World Health Statistics 2014 reported that about 8 percent of Malaysia's population is well over 60 years old [1]. The Malaysian Ministry of Healths 2011 report recorded that almost 11 percent and 7.2 percent of children between the age from one month to 18 years old were identified with physical and cerebral palsy disabilities, respectively [2], [3]. The report also suggests that there is an average increase of 300 percent of stroke patients annually. More often than not, the aforementioned statistics are affected by gait disorders [4]. Gait is essentially the capability of a person to maintain balance and assume the upright position apart from ones ability to initiate and sustain rhythmic stepping [5]. Gait abnormalities may originate from cerebellar disease, neuromuscular disease, cardiac disease, cognitive impairment, stroke, brain or spinal injury or even other general circumstances that may bring about this condition [6], [7].

The demand for rehabilitation services is on the rise globally primarily due to the growing number of ageing society as well as the aforementioned contributing factors. Studies have evidently shown that through continuous and repetitive locomotion activity, patients mobility may be improved [8]–[10]. The assistance of at least two physical therapists is required in conventional rehabilitation therapy to facilitate this form of activity [9]. Nonetheless, this category of therapy is deemed cost demanding and laborious to the therapist. This scenario has led the research community as a whole to address the shortcomings of conventional rehabilitation therapy as well as the increasing demand for gait rehabilitation through the engagement of robotics.

In this paper, the kinematics of the lower limb exoskeleton is presented. The development of the mathematical modelling is discussed in this article as well as the simulation results of the modelling. The kinematics model deals with the analytical study of the lower limb motion with respect to its fixed reference frame (e.g., the body) without taking into consideration the forces or torques that generate the motion. In particular, the lower limb kinematics developed in this paper relates the relation between the joint variables with the position and orientation of the lower limb exoskeleton.

II. Mathematical modelling of the robot

In this study, the Denavit-Hartenberg (D-H) algorithm is used to construct the exoskeleton axis and its relevant parameters. Although the D-H method provides a long and complex mathematical solution, nonetheless, it is a systematic method to develop sophisticated robotic modelling. It is worth mentioning that there are other methods available to develop such model, however, it is beyond the scope of this paper. The readers are encouraged to refer to [11] for further detailed elaboration of the other techniques.

The method involved are as follows:

1. The lower limb exoskeleton is shown in Figure 1. Fromthe figure, the D-H parameter may be assigned to the degree of freedom of interest. The resultant D-H parameter is then depicted in the figure.
2. Then, the D-H parameters are tabulated as in Table 1 to find its corresponding values.
3. Obtain the inverse kinematic solution using the inverse transform technique and geometrical solution.

Table 1 D-H Lower Limb Robotic Leg Parameters

Link, i	Joint angle (degree), θ_i	Link offset (meter), d_i	Link length (meter), a_i	Twist angle (degree), α_i
1	$\theta 1$	0	1.5	0

F. Ibrahim et al. (eds.), *3rd International Conference on Movement, Health and Exercise*, IFMBE Proceedings 58,
DOI: 10.1007/978-981-10-3737-5_36

2	$\theta 2$	0	0.5	0
3	0	0.2	0	
4	$\theta 4$	0.2	0	0

A. Transformation Matrix

The robot transformation matrix, A can be obtained by substituting the D-H parameters into the Equation 1.

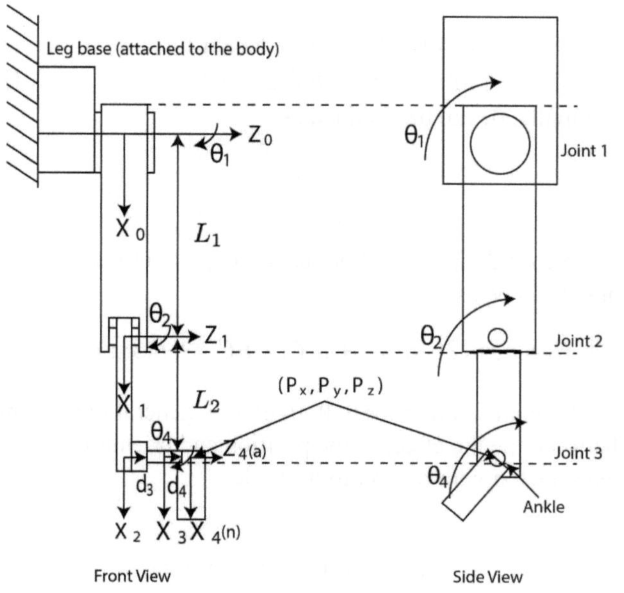

Fig. 1. Robotic Leg with D-H Representation

$$^{i-1}A_i = \begin{bmatrix} \cos\theta_i & -\cos\alpha_i \sin\theta_i & \sin\alpha_i \sin\theta_i & a_i \cos\theta_i \\ \sin\theta_i & \cos\alpha_i \cos\theta_i & -\sin\alpha_i \cos\theta_i & a_i \sin\theta_i \\ 0 & \sin\alpha_i & \cos\alpha_i & d_i \\ 0 & 0 & 0 & 1 \end{bmatrix} \quad (1)$$

and the total transformation matrix with respect to the robotic base can be defined as

$$^0T_4 = {}^0A_1\,{}^1A_2\,{}^2A_3\,{}^3A_4$$

where,

Therefore, the total transformation matrix between the base and lower limb ankle may be expressed as

$$^0T_4 = \begin{bmatrix} C_{1,2,4} & -S_{1,2,4} & 0 & L_2C_{1,2} + L_1C_1 \\ S_{1,2,4} & C_{1,2,4} & 0 & L_2S_{1,2} + L_1S_1 \\ 0 & 0 & 1 & -d_3 - d_4 \\ 0 & 0 & 0 & 1 \end{bmatrix} \quad (3)$$

Where,

$C_{i,j,k} = \cos(\theta_i + \theta_j + \theta_k)$ and $S_{i,j,k} = \sin(\theta_i + \theta_j + \theta_k)$

B. Inverse Kinematic

The inverse kinematics is essential for control system development. It allows the user to determine the required joint angles command in order to reach specific position or configuration. The inverse kinematics utilises the ankle required position Px,Py,Pz as inputs to determine the joint angles, θi.

To calculate the desired angle of the exoskeleton, the general orientation, and position of the represented end effector can be defined as

$$^0T_4 = \begin{bmatrix} n_x & s_x & a_x & p_x \\ n_y & s_y & a_y & p_y \\ n_z & s_z & a_z & p_z \\ 0 & 0 & 0 & 1 \end{bmatrix} \quad (4)$$

Where the vector n, s , a are the normal vector, sliding vector and approach vector that represents the robotic end effector (ankle) orientation. Meanwhile, p represents the robot's position with respect to its base.

By using trigonometric relationship and equating Equation 3 and Equation 4, angle θ_2 may be computed as:

$$P_x^2 + P_y^2 = (L_2C_{1,2} + L_1C_1)^2 + (L_2S_{1,2} + L_1S_1)^2 \quad (5)$$

$$= L_1^2 + L_2^2 + 2L_1L_2C_{12}C_2 + 2L_1L_2S_{12}S_2 \quad (6)$$

$$= L_1^2 + L_2^2 + 2L_1L_2C_2 \quad (7)$$

$$C_2 = \frac{P_x^2 + P_y^2 - L_1^2 - L_2^2}{2L_1L_2}$$

$$\theta_2 = \cos^{-1}\left(\frac{P_x^2 + P_y^2 - L_1^2 - L_2^2}{2L_1L_2}\right)$$

(2) rearrange the equation above yields,

(8)

(9).

To solve for θ_1, we can use trigonometric identities to expand the equation:

$$L_2C_{1,2} + L_1C_1 = P_x \quad (10)$$

$$L_2S_{1,2} + L_1S_1 = P_y \quad (11)$$

$$P_x = L_1C_1 + L_2C_2 - L_2S_1S_2 \quad (12)$$

$$= (L_1 + L_2C_2)C_1 - L_2S_1S_2 \quad (13)$$

$$P_y = L_1S_1 + L_1S_1C_2 - L_2C_1S_2 \quad (14)$$

$$= (L_1 + L_2C_2)S_1 - L_2C_1S_2 \quad (15)$$

C_1 can be obtained as:

$$C_1 = \frac{P_x + L_2S_1S_2}{L_1 + L_2C_2} \quad (16)$$

Subtitute C_1 into the P_y in Equation 15 and rearrange the equation gives:

$$S_1 = \frac{-L_2S_2P_x + P_y(L_1 + L_2C_2)}{(L_2S_2)^2 + (L_1 + L_2C_2)^2} \quad (17)$$

S_1 can be substituted into Px equation which gives:

$$C_1 = \frac{P_x(L_1 + L_2C_2) + P_yL_2S_2}{(L_2C_2)^2 + (L_1 + L_2C_2)^2} \quad (18)$$

To consider all the exoskeletons rotation quadrant, the atan2 function should be considered. Finally, the θ1 equation can obtained as such as

$$\theta_1 = atan2(-L_2S_2P_x + P_y(L_1 + L_2C_2), P_x(L_1 + L_2C_2) + P_yL_2S_2) \quad (19)$$

To solve for θ_2,

$$tan_{1,2,4} = \frac{S_{1,2,4}}{C_{1,2,4}} = \frac{n_y}{n_x} \quad (20)$$

From this equation, θ_2 can also be found as:

$$\theta_2 = atan2(n_y, n_x) - \theta_1 - \theta_4 \quad (21)$$

Besides, the θ4 can be calculated using the inverse transform technique. The total homogenous transformation matrix can be rewritten back as:

$$^2A_3^{-1}\ {}^1A_2^{-1}\ {}^0A_1^{-1}\ {}^0T_4 = {}^3A_4 \quad (22)$$

Equating the resultant of Equation 22 with $^3A_4$,

$$^3A_4 = \begin{bmatrix} \cos\theta_4 & -\sin\theta_4 & 0 & 0 \\ \sin\theta_4 & \cos\theta_4 & 0 & 0 \\ 0 & 0 & 1 & -d_4 \\ 0 & 0 & 0 & 1 \end{bmatrix} \quad (23)$$

Then yields,

$$\cos\theta_4 = n_xC_{1,2} + n_yS_{1,2} \quad (24)$$

$$\sin\theta_4 = -n_xS_{1,2} + n_yC_{1,2} \quad (25)$$

Finally, θ_4 can be obtained as:

$$\theta_4 = \tan^{-1}\frac{-n_xS_{1,2} + n_yC_{1,2}}{n_xC_{1,2} + n_yS_{1,2}} \quad (26)$$

To consider all the robot quadrant operations, the operator atan2 can be used

$$\theta_4 = atan2(-n_xS_{1,2} + n_yC_{1,2}, n_xC_{1,2} + n_yS_{1,2}) \quad (27)$$

The final solution for $\theta_1,\theta_2,\theta_4$ gives a general relationship between the the effect of the position and orientation of the exoskeleton with respect to its angle.

III. Results validation

To verify the developed kinematics model, MATLAB is used to simulate the mathematical modelling. The verification is made by giving the input signal to the kinematics model and analysing the output computed from the mathematical model. Three cases are presented to simulate the working principles of exoskeletons kinematics.

A. Forward Kinematic

Case 1:

Simulation of the robot with angle θ1 moving, whilst θ2 and θ4 are constant to simulate exoskeletons "kick" movement. Figure 2 and Figure 3 depicts the input given and the obtained result, respectively. From the result, we can expect the sinusoidal output of the robots position as the joint angle θ2 is set to constant. The Z axis position remains constant as the exoskeletons movement is restricted to the sagittal plane. The maximum magnitude for Px is 1.5 meter, which is the total length of the link L1 and L2 (1.0 meter and 0.5 meter). From Figure 3, the start position for the robotic leg is at rest position (see Figure 1, Px = 1.5 meter, Py = 0 meter and Pz = 0.4 meter) and the values changes gradually due to the motion input from θ1.

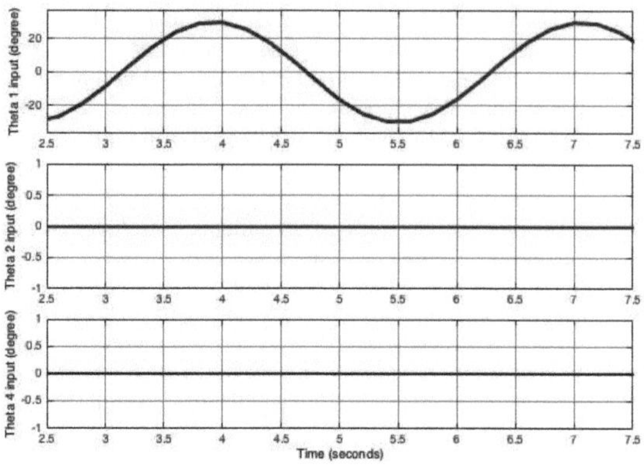

Fig. 2. Case 1: The robot input angle (degree)

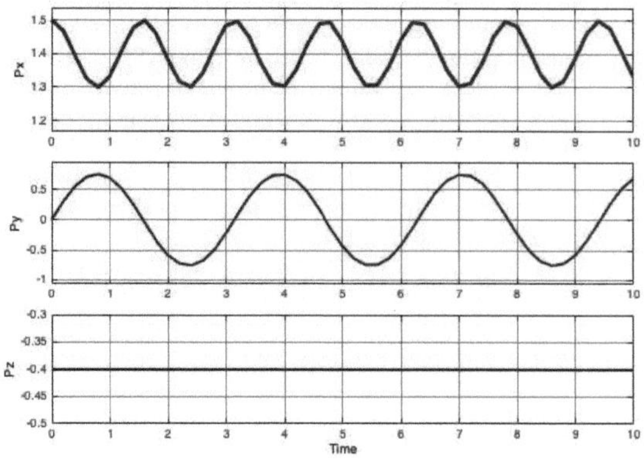

Fig. 3. Case 1: The robot output position (meter)

Case 2:

Moving both θ1 and θ2 angles to simulate leg movement. In order to simulate the robotic simultaneous leg movement, θ2 should be less than 0 (to mimic the human walking angle). Therefore, a saturation function is added to the θ2 sinusoidal input to limit its value. The case 2 input signal and output positional response is shown in Figure 4 and 5, respectively. It is apparent that when θ1 and θ2 are varied, the position output of Px and Py are coupled between the two input angles which in turn generates a non-uniform motion. The Pz axis remain constant since there is no movement along the Z-axis.

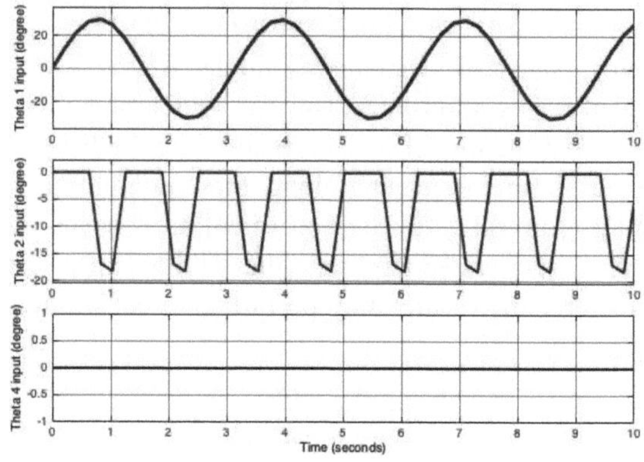

Fig. 4. Case 2: The robot input angle (degree)

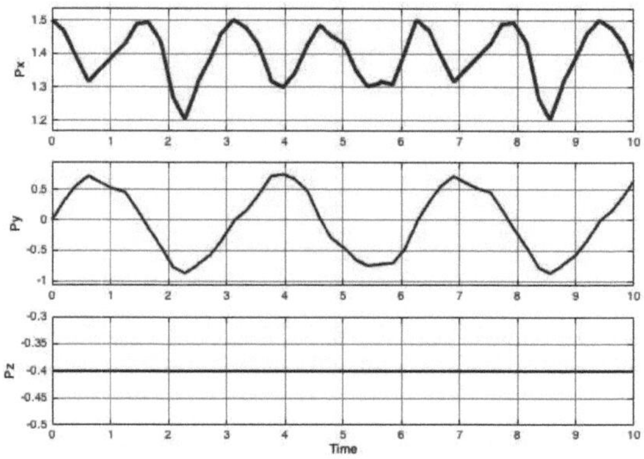

Fig. 5. Case 2: The robot output position (meter)

Case 3: To verify the orientation vector (see vector n, s, a in Figure 1). The s vector is aligned with Y4 axis (by virtue of the right hand rule) of the exoskeleton, θ1 and θ2 are set constant whilst θ4 becomes a variable to be manipulated.. This is due to the fact that θ4 is the variable that is coupled directly to the orientation of the ankle position of the robot. It is evident from Figure 6, that the position value only varies within the range of 0 to 1 for N unit vector. The value 1 as depicted through the Nx vector indicates that the X4 axis is aligned with the base axis, X0. It is also apparent from Figure 6, that the total magnitude vector at the specific time t will be equal to 1. This suggests that the orientation vector that has been developed is correct. The same result can be obtained for s and a vector.

IV. Conclusion and Future Works

In this paper, the kinematics model of the lower limb exoskeleton is established. The forward kinematics result suggests that the model is able to simulate the behaviour of the lower limb movement along the sagittal plane. This preliminary investigation is non-trivial as it paves the way towards the understanding the kinematics of the exoskeleton. The results may be further used to understand the dynamics

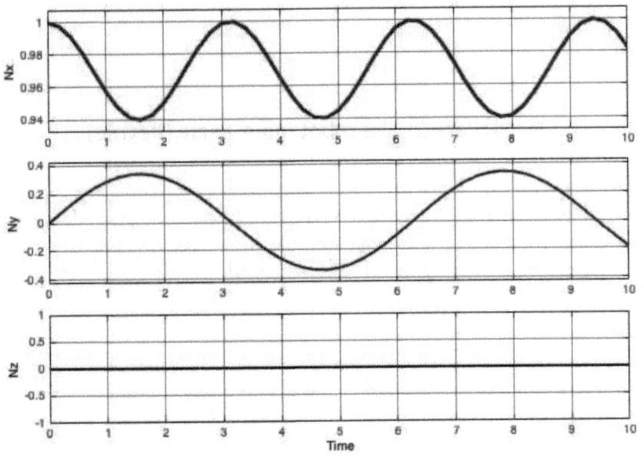

Fig. 6. Case 3: The unit vector N when sinusoidal wave input is given to θ_4.

of the lower limb exoskeleton which is crucial in the control study of the aforementioned rehabilitation device.

Acknowledgment

The research is supported by Faculty of Manufacturing Engineering, Universiti Malaysia Pahang, and research university research grant RDU160375.

References

[1] World Health Organization, World Health statistics 2014, 2014.

[2] Z. Taha, A. P. P. A. Majeed, M. Y. W. P. Tze, and A. G. A. Rahman, "Preliminary Investigation on the Development of a Lower Extremity Exoskeleton for Gait Rehabilitation: A Clinical Consideration," Journal of Medical and Bioengineering Vol, vol. 4, no. 1, 2015.

[3] M. N. A. Ab Patar, A. F. Said, J. Mahmud, A. P. P. A. Majeed, and M. A. Razman, "System integration and control of Dynamic Ankle Foot Orthosis for lower limb rehabilitation," in Technology Management and Emerging Technologies (ISTMET), 2014 International Symposium on. IEEE, 2014, pp. 82–85.

[4] L. Sudarsky and M. Ronthal, "Gait disorders among elderly patients: a survey study of 50 patients," Archives of neurology, vol. 40, no. 12, pp. 740–743, 1983.

[5] J. G. Nutt, C. D. Marsden, and P. D. Thompson, "Human walking and higherlevel gait disorders, particularly in the elderly," Neurology, vol. 43, no. 2, p. 268, 1993.

[6] O. Franch, L. Calandre, J. Alvarez-Linera, E. D. Louis, F. Bermejo-´ Pareja, and J. Benito-Leon, "Gait disorders of unknown cause in the´ elderly: Clinical and MRI findings," Journal of the neurological sciences, vol. 280, no. 1, pp. 84–86, 2009.

[7] S. Murray and M. Goldfarb, "Towards the use of a lower limb exoskeleton for locomotion assistance in individuals with neuromuscular locomotor deficits," in Engineering in Medicine and Biology Society (EMBC), 2012 Annual International Conference of the IEEE. IEEE, 2012, pp. 1912–1915.

[8] V. Dietz and S. J. Harkema, "Locomotor activity in spinal cord-injured persons," Journal of Applied Physiology, vol. 96, no. 5, pp. 1954–1960, 2004.

[9] I. D´ıaz, J. J. Gil, and E. Sanchez, "Lower-limb robotic rehabilitation:´ literature review and challenges," Journal of Robotics, vol. 2011, 2011.

[10] A. Wernig, S. Muller, A. Nanassy, and E. Cagol, "Laufband therapy¨ based on rules of spinal locomotion'is effective in spinal cord injured persons," European Journal of Neuroscience, vol. 7, no. 4, pp. 823–829, 1995.

[11] K. S. Fu, R. Gonzalez, and C. G. Lee, Robotics: Control Sensing. Vis.

[12] Tata McGraw-Hill Education, 1987.

Author: M. A. Zakaria
Institute: Universiti Malaysia Pahang
City: Pekan, Pahang
Country: Malaysia
Email: maizzat@ump.edu.my

Author Index

Keyword Index

MIX
Papier aus verantwortungsvollen Quellen
Paper from responsible sources
FSC® C105338

If you have any concerns about our products,
you can contact us on
ProductSafety@springernature.com

In case Publisher is established outside the EU,
the EU authorized representative is:
Springer Nature Customer Service Center GmbH
Europaplatz 3, 69115 Heidelberg, Germany

Printed by Libri Plureos GmbH
in Hamburg, Germany